THE AMERICAN ECONOMY

ISSN 1554-4400

THE AMERICAN ECONOMY

Kim Masters Evans

INFORMATION PLUS® REFERENCE SERIES
Formerly Published by Information Plus, Wylie, Texas

LIBRARY
FORSYTH TECHNICAL COMMUNITY COLLEGE
2100 SILAS CREEK PARKWAY
WINSTON SALEM, N.C. 27103

Farmington Hills, Mich • San Francisco • New York • Waterville, Maine
Meriden, Conn • Mason, Ohio • Chicago

GALE
CENGAGE Learning

The American Economy

Kim Masters Evans

Kepos Media, Inc.: Steven Long and Janice Jorgensen, Series Editors

Project Editors: Tracie Moy, Laura Avery

Rights Acquisition and Management: Ashley M. Maynard

Composition: Evi Abou-El-Seoud, Mary Beth Trimper

Manufacturing: Rita Wimberley

© 2015 Gale, Cengage Learning

ALL RIGHTS RESERVED. No part of this work covered by the copyright herein may be reproduced, transmitted, stored, or used in any form or by any means graphic, electronic, or mechanical, including but not limited to photocopying, recording, scanning, digitizing, taping, Web distribution, information networks, or information storage and retrieval systems, except as permitted under Section 107 or 108 of the 1976 United States Copyright Act, without the prior written permission of the publisher.

This publication is a creative work fully protected by all applicable copyright laws, as well as by misappropriation, trade secret, unfair competition, and other applicable laws. The authors and editors of this work have added value to the underlying factual material herein through one or more of the following: unique and original selection, coordination, expression, arrangement, and classification of the information.

For product information and technology assistance, contact us at
Gale Customer Support, 1-800-877-4253.
For permission to use material from this text or product,
submit all requests online at www.cengage.com/permissions.
Further permissions questions can be e-mailed to
permissionrequest@cengage.com

Cover photograph: © Stuart Monk/Shutterstock.com.

While every effort has been made to ensure the reliability of the information presented in this publication, Gale, a part of Cengage Learning, does not guarantee the accuracy of the data contained herein. Gale accepts no payment for listing; and inclusion in the publication of any organization, agency, institution, publication, service, or individual does not imply endorsement of the editors or publisher. Errors brought to the attention of the publisher and verified to the satisfaction of the publisher will be corrected in future editions.

Gale
27500 Drake Rd.
Farmington Hills, MI 48331-3535

ISBN-13: 978-0-7876-5103-9 (set)
ISBN-13: 978-1-57302-652-9

ISSN 1554-4400

This title is also available as an e-book.
ISBN-13: 978-1-57302-676-5 (set)
Contact your Gale sales representative for ordering information.

Printed in the United States of America
1 2 3 4 5   19 18 17 16 15

TABLE OF CONTENTS

PREFACE vii

CHAPTER 1
The U.S. Economy: Historical Overview 1
This chapter traces the development of the nation's capitalistic free-market economy and reviews the social and political events that have shaped the government's economic policy. It also introduces some basic terms of economics, such as supply and demand, inflation, gross national product, gross domestic product, recession, and depression.

CHAPTER 2
Economic Indicators and Public Perceptions 23
Economists use mathematical measures called economic indicators to gauge the overall health of the national economy. Various economic indicators are categorized and briefly described. One key measure—gross domestic product—is examined in detail. In addition, recent public opinion polls on economic issues are reported on.

CHAPTER 3
The American Consumer 39
American consumers fuel economic growth by spending large sums of money. This chapter surveys the rise of the consumer culture in the United States and delves into the role of personal consumption expenditures in the nation's gross domestic product. Consumer expenses that have risen dramatically in recent years, particularly those for medical care, are studied in detail.

CHAPTER 4
Personal Debt 55
The assumption of personal debt has both good and bad consequences for the economy. Historical viewpoints on debt, the role of interest rates, and the different kinds of debt, such as mortgages and consumer credit loans, are analyzed. This chapter includes a detailed discussion of the housing industry boom and bust during the first decade of the 21st century. Issues related to bankruptcy are also covered.

CHAPTER 5
The American Worker 77
This chapter presents the latest data on employment and compensation in the United States and investigates the economic performance and projections for various job sectors. Other topics include labor unions, government programs that protect worker rights, and the controversies over foreign workers in the United States and American jobs moving out of the country.

CHAPTER 6
U.S. Businesses 97
U.S. businesses show great diversity in size and legal structure, ranging from self-employed individuals to multinational corporations. Businesses are discussed in terms of their economic performance, federal regulation, market power, and perceived social responsibilities. Recent corporate scandals and concerns over lack of competition in some markets are also addressed.

CHAPTER 7
Saving and Investing 115
The U.S. economy offers many opportunities for saving and investing. This chapter outlines the major types of investments, with focus on homeownership and the stock and bond markets. Government regulation of these markets is also tracked.

CHAPTER 8
Wealth in the United States 127
Wealth is defined not only by income but also by ownership of assets—real estate, stocks, bonds, and other securities. Unequal wealth distribution among Americans is a major cause of controversy, viewed by some as a natural consequence of capitalism and by others as a symptom of deep sociopolitical problems.

CHAPTER 9
The Role of the Government 147
Government bodies at the local, state, and federal levels play a major role in the U.S. economy by redistributing money and providing employment. At the federal level, the government has assumed an extremely large national debt that has consequences for present and future generations of Americans. Finally, the federal government manipulates the economy through spending and monetary policies.

CHAPTER 10
International Trade and the United States' Place in the Global Economy 163
The U.S. economy is preeminent in the world when it comes to national production. Nevertheless, the United States buys far more from other countries than it sells to them. Economists disagree

about whether this trade imbalance is good or bad for the U.S. economy. Major trade agreements, economic sanctions, and the increasing trend toward global free trade, or globalization, are also highlighted in this chapter.

IMPORTANT NAMES AND ADDRESSES....... 173
RESOURCES........................ 175
INDEX 177

PREFACE

The American Economy is part of the *Information Plus Reference Series*. The purpose of each volume of the series is to present the latest facts on a topic of pressing concern in modern American life. These topics include the most controversial and studied social issues of the 21st century: abortion, capital punishment, crime, the environment, gambling, gun control, health care, national security, race and ethnicity, social welfare, women, youth, and many more. Although this series is written especially for high school and undergraduate students, it is an excellent resource for anyone in need of factual information on current affairs.

By presenting the facts, it is the intention of Gale, Cengage Learning, to provide its readers with everything they need to reach an informed opinion on current issues. To that end, there is a particular emphasis in this series on the presentation of scientific studies, surveys, and statistics. These data are generally presented in the form of tables, charts, and other graphics placed within the text of each book. Every graphic is directly referred to and carefully explained in the text. The source of each graphic is presented within the graphic itself. The data used in these graphics are drawn from the most reputable and reliable sources, such as from the various branches of the U.S. government and from private organizations and associations. Every effort has been made to secure the most recent information available. Readers should bear in mind that many major studies take years to conduct and that additional years often pass before the data from these studies are made available to the public. Therefore, in many cases the most recent information available in 2015 is dated from 2012 or 2013. Older statistics are sometimes presented as well, if they are landmark studies or of particular interest and no more-recent information exists.

Although statistics are a major focus of the *Information Plus Reference Series*, they are by no means its only content. Each book also presents the widely held positions and important ideas that shape how the book's subject is discussed in the United States. These positions are explained in detail and, where possible, in the words of their proponents. Some of the other material to be found in these books includes historical background, descriptions of major events related to the subject, relevant laws and court cases, and examples of how these issues play out in American life. Some books also feature primary documents or have pro and con debate sections that provide the words and opinions of prominent Americans on both sides of a controversial topic. All material is presented in an evenhanded and unbiased manner; readers will never be encouraged to accept one view of an issue over another.

HOW TO USE THIS BOOK

The U.S. economy in the 21st century is enormous and extremely complicated. Workers, employers large and small, consumers, the equities markets, the U.S. government, and the world economy are constantly interacting with each other to affect the U.S. economy and, through it, each other. The U.S. economy produces and consumes raw materials, services, manufactured goods, and intellectual property in vast amounts. This book describes the size and scope of the U.S. economy, explains how it functions, and examines some of the challenges it faces, such as unemployment, inflation, government debt, and corporate scandals.

The American Economy consists of 10 chapters and three appendixes. Each chapter is devoted to a particular aspect of the U.S economy. For a summary of the information that is covered in each chapter, please see the synopses that are provided in the Table of Contents. Chapters generally begin with an overview of the basic facts and background information on the chapter's topic, then proceed to look at subtopics of particular interest. For example, Chapter 9: The Role of the Government examines the finances of local and state governments and the federal government. These entities collect money, for example, by taxation, and spend money through investments, purchases of goods and services, intergovernmental transfers, and transfers to individuals. The chapter focuses in particular on federal budgets, taxes, and spending and the national debt. The future of Social Security and Medicare are discussed at length. The ways in which the federal government manipulates the nation's economy are also described. The chapter ends with predictions provided by the Congressional Budget Office for the performance of key economic indicators into the 2020s. Readers can find their way through a chapter by looking for the section and subsection headings, which are clearly set off from the text. They can also refer to the book's extensive Index, if they already know what they are looking for.

Statistical Information

The tables and figures featured throughout *The American Economy* will be of particular use to readers in learning about this topic. These tables and figures represent an extensive collection of the most recent and valuable statistics on the U.S. economy—for example, graphics cover the spending habits of the typical consumer, employment by industry type, the gross domestic product, the trade deficit, and consumer debt levels. Gale, Cengage Learning, believes that making this information available to readers is the most important way to fulfill the goal of this book: to help readers understand the issues and controversies surrounding the U.S. economy and reach their own conclusions.

Each table or figure has a unique identifier appearing above it, for ease of identification and reference. Titles for the tables and figures explain their purpose. At the end of each table or figure, the original source of the data is provided.

To help readers understand these often complicated statistics, all tables and figures are explained in the text. References in the text direct readers to the relevant statistics. Furthermore, the contents of all tables and figures are fully indexed. Please see the opening section of the Index at the back of this volume for a description of how to find tables and figures within it.

Appendixes

Besides the main body text and images, *The American Economy* has three appendixes. The first is the Important Names and Addresses directory. Here, readers will find contact information for a number of government and private organizations that can provide further information on aspects of the U.S. economy. The second appendix is the Resources section, which can also assist readers in conducting their own research. In this section, the author and editors of *The American Economy* describe some of the sources that were most useful during the compilation of this book. The final appendix is the Index. It has been greatly expanded from previous editions and should make it even easier to find specific topics in this book.

COMMENTS AND SUGGESTIONS

The editors of the *Information Plus Reference Series* welcome your feedback on *The American Economy*. Please direct all correspondence to:

Editors
Information Plus Reference Series
27500 Drake Rd.
Farmington Hills, MI 48331-3535

CHAPTER 1
THE U.S. ECONOMY: HISTORICAL OVERVIEW

It is not what we have that will make us a great nation; it is the way in which we use it.

—Theodore Roosevelt, 1886

The workings of the U.S. economy are complex and often mysterious, even to economists. At its simplest, the economy runs on three major sectors: consumers, businesses, and government. (See Figure 1.1.) Consumers earn money and exchange much of it for goods and services from businesses. These businesses use the money to produce more goods and services and to pay wages to their employees. Both consumers and businesses fund the government sector, which spends and transfers money back into the system. The banking system plays a crucial role in the economy by providing the means for all sectors to save and borrow money. Finally, there are the stock markets, which allow consumers to invest their money in the nation's businesses—an enterprise that further fuels economic growth for all sectors. Thus, the U.S. economy is a circular system based on interdependent relationships in which massive amounts of money change hands. The historical developments that produced this system are important to understand because they provide key information about what has made the U.S. economy such a powerful force in the world.

DEFINING THE U.S. ECONOMY

The term *market economy* describes an economy in which the forces of supply and demand dictate the way in which goods and resources are allocated and what prices will be set. The opposite of a market economy is a *planned economy*, in which the government determines what will be produced and what prices will be charged. In reality all nations have mixed economies somewhere between these two extremes; that is, there are no true market economies or true planned economies.

Another important term is *capitalism*. In a capitalistic economic system the means of production (i.e., the businesses) are owned and controlled by the private (nongovernmental) sector, rather than by the government. In addition, there is a labor market, meaning that workers compete against each other for jobs, and business owners compete against each other for workers. Wage and benefit amounts are driven by market factors, such as supply and demand. Although there can be various noncapitalistic alternatives, the most extreme is one in which the government owns and controls all means of production, assigns workers to jobs, and dictates wage and benefit amounts.

The United States favors capitalism and free markets in which private producers anticipate what products the market will be interested in and at what price. Producers make decisions about what products they will bring to market and how these products will be produced and priced. These factors foster competition among businesses, which typically leads to lower prices and is generally considered beneficial for both workers and consumers. This does not mean that the U.S. economy is free from government manipulation. As will be explained in this book the U.S. government takes many steps to influence economic factors. Nevertheless the U.S. economy is so market oriented that it is often called a market economy. The same holds true for the economies of other developed nations, such as Australia, Canada, and western European countries. All of these nations embrace (to various degrees) economic production and distribution by the private sector. There are stark differences between the market-leaning economies in terms of government intervention. This is particularly evident in regards to wealth redistribution (moving money from the most affluent members of society to the less affluent), a topic discussed in Chapter 8.

Countries whose governments exercise a great deal of control over national economic matters are said to have planned economies. These governments regulate

FIGURE 1.1

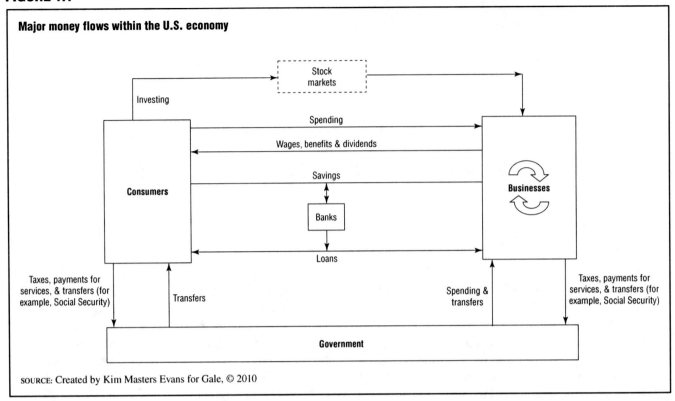

Major money flows within the U.S. economy

SOURCE: Created by Kim Masters Evans for Gale, © 2010

wages, production, and prices in accordance with centralized plans. They also typically own large companies in key industries, such as energy development. This is a hallmark of communist or socialist political systems in which adherents believe that economic production and distribution are best managed by the government for society as a whole. North Korea, China, and Cuba are said to have planned economies because their governments aggressively manipulate supply and demand factors. However, within this group the level of control is very rigid for some aspects of the economy and much looser for others. Every nation in the world blends elements of centralized control and free enterprise to achieve a unique economy.

The mixed economy of the United States combines aspects of a market economy with some government planning and control. This creates a system with both a high degree of market freedom and regulatory agencies and social programs that promote the public welfare. The U.S. economy did not develop overnight. Its origins date back to the nation's founding in the 1700s. The newly formed United States adopted economic principles that favored a competitive marketplace with little government interference. These principles would dominate the nation's economic policy for more than a century. During this time the U.S. government had a hands-off approach to business regulation, a tactic described by the French term *laissez-faire* (leave alone or "do as you please"). It was generally believed that the government should not interfere in economic affairs but should instead allow supply and demand and competition to operate unfettered, resulting in a free market.

Panics and Depressions in the 19th Century

In economic terms a panic is a widespread occurrence of public anxiety about financial affairs. People lose confidence in banks and investments and want to hold onto their money instead of spending it. This can lead to a severe downturn, or depression, in the economic condition of a nation. Economists argue about the exact definitions of panics and depressions, but in general it is agreed that they occurred in the United States in 1819, 1837, 1857, 1869, 1873, and 1893.

The crises were triggered by a variety of factors. Common problems included too much borrowing and speculation by investors and poor oversight of banks by the federal government. Speculation is the buying of assets on the hope that they will greatly increase in value in the future. During the 1800s many speculators borrowed money from banks to buy land. Huge demand caused land prices to increase dramatically, often above what the land was actually worth in the market. Poorly regulated banks extended too much credit to speculators and to each other. When a large bank failed, there was a domino effect through the industry, which caused other banks and businesses to fail.

A panic or depression results in a downward economic spiral in which individuals and businesses are afraid to make new investments. People rush to withdraw their money from banks. As panic spreads, banks demand that borrowers pay back money, but borrowers may lack the funds to do so. Consumers are reluctant to spend money, which negatively affects businesses. Demand for products goes down, and prices must be lowered to move merchandise off of shelves. This means less profit for business owners. To reduce their costs, businesses begin laying off employees and do not hire new employees. As more people become unemployed or fearful about their jobs, there is even less spending in the marketplace, which leads to more business cutbacks and so forth. The cycle continues until some compelling change takes place to nudge the economy back into a positive direction.

THE EARLY 20TH CENTURY

The early 20th century was a time of social and political change in the United States. Public disgust at the corruption and greed of previous decades encouraged the movement called progressivism. Progressives promoted civic responsibility, workers' rights, consumer protection, political and tax reform, "trust busting," and strong government action to achieve social improvements. The Progressive Era greatly affected the U.S. economy because of its focus on improving working conditions for average Americans. Successes for the progressives included child labor restrictions, improved working conditions in factories, compensation funds for injured workers, a growth surge in labor unions, federal regulation of food and drug industries, and the formation of the Federal Trade Commission to oversee business practices.

Despite its laissez-faire attitude, the federal government took two actions in 1913 that had long-lasting effects on the U.S. economy:

- Established the Federal Reserve System to serve as the nation's central bank, furnish currency, and supervise banking

- Ratified the 16th Amendment to the U.S. Constitution authorizing the collection of income taxes

World War I and Inflation

World War I erupted in Europe in August 1914. The United States entered the conflict in April 1917 and fought until the war ended in November 1918. Although the nation spent only 19 months at war, the U.S. economy underwent major changes during this period.

It is sometimes said that "war is good for the economy" because during a major war the federal government spends large amounts of money on weapons and machinery through contracts with private industries. These industries hire more employees, which reduces unemployment and puts more money into the hands of consumers to spend in the marketplace. This increase in production and hiring also benefits other businesses that are not directly involved in the war effort. On the surface, these economic effects appear positive. However, major wars almost always result in high inflation rates.

Inflation is an economic condition in which the purchasing power of money goes down because of price increases in goods and services. For example, if a nation experiences an inflation rate of 3% in a year, an item that cost $1.00 at the beginning of the year will cost $1.03 at the end of the year. Inflation causes the "value" of a dollar to go down over the course of the year. In general, small increases in inflation occur over time in a healthy growing economy because demand slightly outpaces supply. Economists consider an inflation rate of 3% or less per year to be tolerable. During a major war the supply and demand ratio becomes distorted. This occurs when the nation produces huge amounts of war goods and far fewer consumer goods, such as food, clothing, and cars. This lack of supply and anxiety about the future drive up the prices of consumer goods, making it difficult for people to afford things they need or want.

Although the government tried to impose some level of price control in the food and fuel industries, inflation still occurred. Figure 1.2 shows the average annual inflation rate between 1914 and 1924. The rate was unusually high between 1916 and 1920, peaking at 18% in 1918. Wartime inflation was particularly hard on nonworking citizens, such as the elderly and the sick, because they had no means of increasing their incomes, and, unlike later generations, they were not protected by government assistance programs.

A lasting legacy of World War I was the assumption of large amounts of debt by the federal government to fund the war effort. Figure 1.3 shows the enormous differences that occurred between government spending and revenues (receipts) during the war years. In 1919 government spending peaked at nearly $19 billion, whereas revenues for that year were just over $5 billion. The government made up the difference by borrowing money. One method used was the selling of Liberty bonds. Bonds are a type of financial asset—an IOU (an abbreviation for "I Owe yoU") that promises to pay back at some future date the original purchase price plus interest.

THE ROARING 20S AND THE GREAT DEPRESSION

The Roaring 20s featured several years of robust economic growth that were characterized by increases in mass production, the availability of electricity, consumer demand for goods, and consumer use of credit to fund purchases. However, the prosperity of the 1920s was not shared by all Americans. Financial problems rocked

FIGURE 1.2

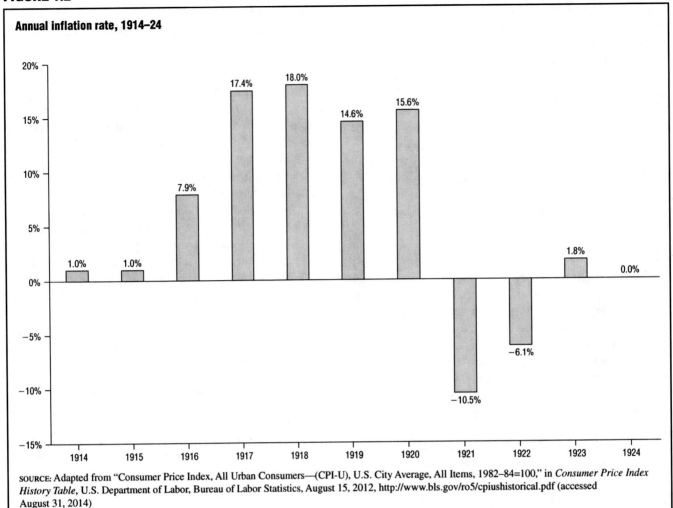

Annual inflation rate, 1914–24

SOURCE: Adapted from "Consumer Price Index, All Urban Consumers—(CPI-U), U.S. City Average, All Items, 1982–84=100," in *Consumer Price Index History Table*, U.S. Department of Labor, Bureau of Labor Statistics, August 15, 2012, http://www.bls.gov/ro5/cpiushistorical.pdf (accessed August 31, 2014)

the agricultural sector, and there were economic downturns in the coal mining and railroad industries.

The Stock Market Crash

During the late 1920s the stock market became a major factor in the U.S. economy as businesses sold stock (i.e., ownership shares) in their companies. Investors were richly rewarded when their stocks increased dramatically in value. Many people took out loans from banks to pay for stock or purchased stock by "buying on margin." In this arrangement an investor would make a small down payment (as little as 10%) on a stock purchase. The remainder of the balance would be paid (in theory) by the future increase in the stock value. Buying on margin was extremely risky but was widely practiced by investors of the time. Overoptimism caused stock prices to rise higher than the actual worth of companies. During the fall of 1929 investors began to get nervous and selling frenzies occurred as people tried to get rid of stocks they thought might be overvalued. On Tuesday, October 29, 1929 ("Black Tuesday"), panic selling took place all day. Stock values dropped dramatically. By the end of the day many margin buyers had lost their life savings and their stock. Those who managed to hold on to their stock found it was worth only a fraction of its former value.

According to Harold Bierman Jr. of Cornell University, in "The 1929 Stock Market Crash" (March 26, 2008, http://eh.net/encyclopedia/the-1929-stock-market-crash), the U.S. stock market lost 90% of its value between 1929 and 1932.

The Great Depression

The U.S. economy suffered a devastating downturn following the stock market crash. The depression was so deep and long (more than a decade) that it is called the Great Depression. It brought long-term unemployment and hardship to millions of people. The unemployment rate soared from 3.2% in 1929 to 24.9% in 1933. (See Figure 1.4.) It remained more than 10% throughout the 1930s. The public lost confidence in the stock market, the banking system, and big business and at the same time was saddled with large amounts of debt that had been taken on during the 1920s.

FIGURE 1.3

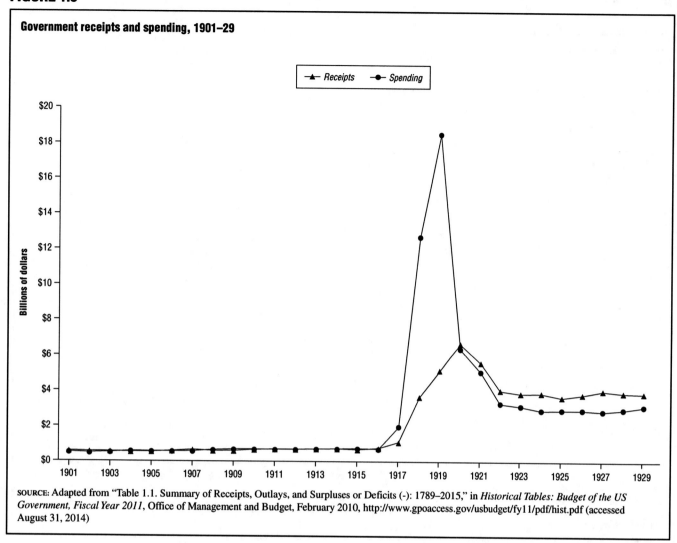

Government receipts and spending, 1901–29

SOURCE: Adapted from "Table 1.1. Summary of Receipts, Outlays, and Surpluses or Deficits (-): 1789–2015," in *Historical Tables: Budget of the US Government, Fiscal Year 2011*, Office of Management and Budget, February 2010, http://www.gpoaccess.gov/usbudget/fy11/pdf/hist.pdf (accessed August 31, 2014)

The Great Depression was aggravated by a crisis in the banking industry. Some banks had invested heavily in the stock market using their depositors' money or lent large amounts of money to stock market investors. These banks failed after the crash, and the depositors lost their savings. Fear of further failures caused so-called bank runs, in which large numbers of depositors rushed to withdraw their money at the same time. This caused more bank failures, which perpetuated the cycle. In addition, some economists believe that the banking market became oversaturated during the 1920s with underfunded and loosely regulated banks that lent money too easily. These institutions were already financially troubled before the crash and could not survive the stress.

When the Great Depression began, the laissez-faire attitude still dominated political opinion. However, in 1933 newly elected President Franklin D. Roosevelt (1882–1945) instituted what he called "a New Deal" for the nation. His administration acted aggressively in economic affairs by creating work programs, trying to revive farming and business, and spearheading laws that were designed to reform the stock market and banking industry. After nearly 80 years, economists still argue about whether the New Deal was actually "a good deal" for the nation. They all agree, however, that it was a turning point in U.S. economic history.

Perhaps the most significant legacy of Roosevelt's New Deal was the new role of the federal government as a manipulator of economic forces and a provider of benefits to the needy. In U.S. history the New Deal is considered the birth of big government.

By 1940 the unemployment rate was 14.6%. (See Figure 1.4.) Although the rate was down from a peak of 24.9% in 1933, it was still high by historical standards. The hardship suffered by many Americans had been softened by nearly a decade of New Deal programs, but the country was still gripped by the Great Depression.

WORLD WAR II AND THE COLD WAR

The United States was involved in World War II from 1941 to 1945. After the war began, businesses rushed to

FIGURE 1.4

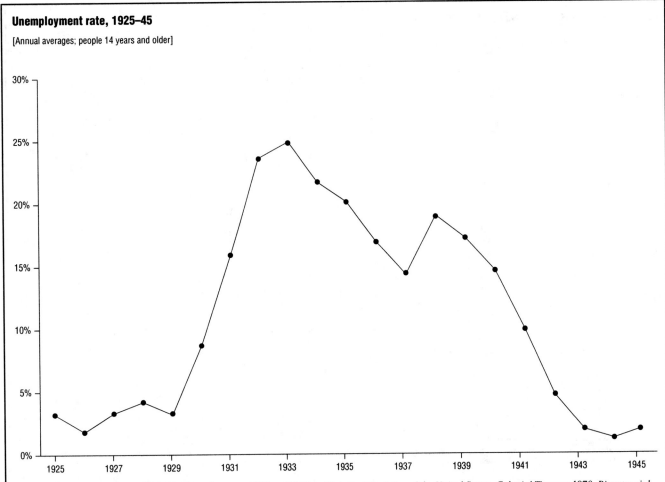

Unemployment rate, 1925–45

[Annual averages; people 14 years and older]

SOURCE: Adapted from "Series D85–86. Unemployment: 1890 to 1970," in *Historical Statistics of the United States, Colonial Times to 1970, Bicentennial Edition, Part 1*, U.S. Department of Commerce, U.S. Census Bureau, September 1975, http://www2.census.gov/prod2/statcomp/documents/CT1970p1-05.pdf (accessed August 31, 2014)

increase production and hire workers to produce the goods needed for the war effort. Unemployment dropped dramatically, and wages went up, particularly for workers in low-skilled factory jobs. Laborers found themselves in high demand and joined labor unions in record numbers to consolidate their power and seek better working conditions.

The federal government made efforts to avoid the huge inflation increase that had occurred during World War I. This required unprecedented interference in private markets and control of supply and demand factors. Rationing (tight controls over how much of an item a person can use or consume in a certain amount of time) was instituted on some goods to prevent dramatic price increases. Federal agencies oversaw wartime production, labor relations, wages, and prices. Overall, these efforts were successful. Figure 1.5 shows the annual rates of inflation that were experienced in the United States between 1940 and 1950. Inflation spiked during the early years of the war and immediately after but was not consistently high over the decade.

Keynesian Economics

World War II was an expensive endeavor for the United States. However, it was believed that the stakes were so high that the war had to be won at any cost. As shown in Figure 1.6, government spending during the war far outpaced revenues. By 1945 the federal government was spending about $90 billion per year and taking in revenues around half this amount. Once again, the difference was made up by borrowing.

Following World War II it appeared obvious that huge government spending had helped fuel recovery from the Great Depression. General belief in the laissez-faire approach to economics gave way to a new approach advocated by the economist John Maynard Keynes (1883–1946). Keynesian economics stresses strong government intervention in the economy and became the operating principle of the U.S. government during the post–World War II era. Although Keynes had his critics, and his methods have been revised over time, he is considered by many to be the father of the mixed economy system that is still being used in the United States.

FIGURE 1.5

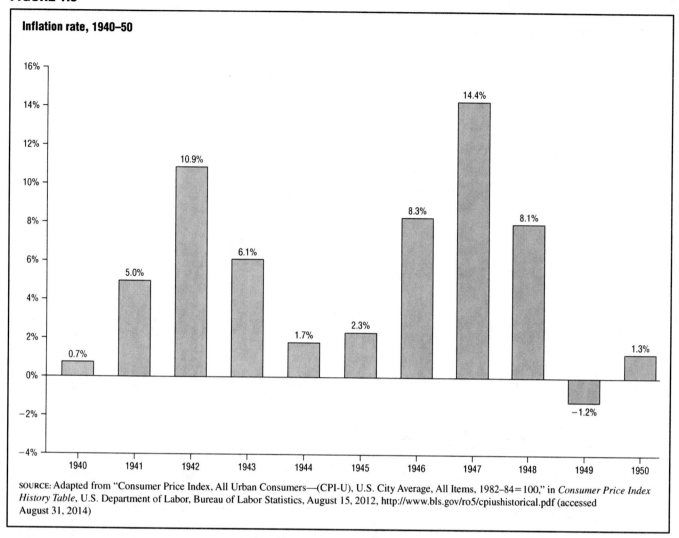

Inflation rate, 1940–50

SOURCE: Adapted from "Consumer Price Index, All Urban Consumers—(CPI-U), U.S. City Average, All Items, 1982–84=100," in *Consumer Price Index History Table*, U.S. Department of Labor, Bureau of Labor Statistics, August 15, 2012, http://www.bls.gov/ro5/cpiushistorical.pdf (accessed August 31, 2014)

Another innovation of this era was the compilation by the federal government of economic data on the nation's inputs and outputs, such as labor and production of goods and services. Researchers at the National Bureau of Economic Research (NBER) began estimating national income (e.g., wages, profits, and rent) as part of a program called the National Income and Product Accounts. Development was overseen by the U.S. Department of Commerce's Division of Economic Research, which evolved into the modern Bureau of Economic Analysis.

During World War II the federal government began compiling another economic measure called the gross national product (GNP). The GNP is the amount in dollars of the value of final goods and services that are produced by Americans over a particular period. It is calculated by summing consumer and government spending, business and residential investments, and the net value of U.S. exports (exports minus imports).

The GNP provides a valuable tool for tracking national productivity over time. The National Income and Product Accounts and GNP became important economic indicators of the state of the economy as a whole—that is, the macroeconomy.

Uneven Prosperity

The decades following World War II were generally prosperous times for the U.S. economy as the nation continued to grow industrially. Postwar euphoria drove a spending spree by consumers and a baby boom.

Dwight D. Eisenhower (1890–1969) was president from 1953 to 1961. His administration is associated with low inflation rates and general prosperity. However, the prosperity was not shared equally in American society. An oversupply of agricultural goods meant lower prices for consumers, but lower profits for farmers. Agriculture became increasingly an industry in which large factory farms run by corporations were able to survive, whereas many smaller farmers could not compete.

Minority populations (largely African American) also suffered financial hardship during this era. Figure 1.7 shows the dramatic difference between the unemployment

FIGURE 1.6

Government receipts and spending, 1930–50

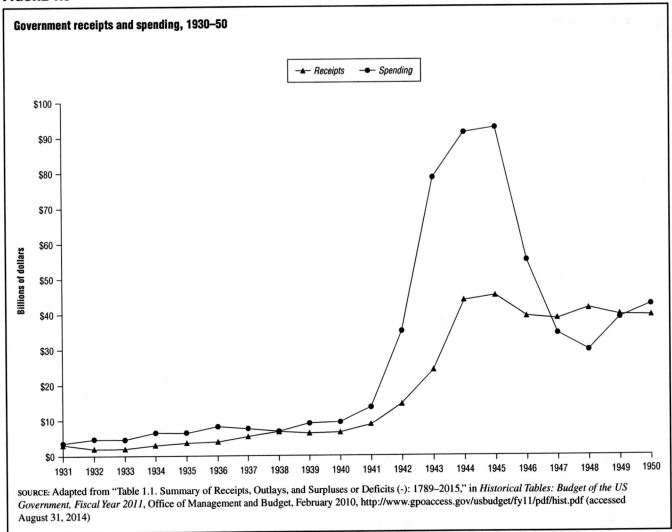

SOURCE: Adapted from "Table 1.1. Summary of Receipts, Outlays, and Surpluses or Deficits (-): 1789–2015," in *Historical Tables: Budget of the US Government, Fiscal Year 2011*, Office of Management and Budget, February 2010, http://www.gpoaccess.gov/usbudget/fy11/pdf/hist.pdf (accessed August 31, 2014)

rates for whites and minorities during the postwar decades. By the mid-1950s unemployment among minorities was twice as high as it was among white workers, a disparity that lingered well into the 1960s. It was in this atmosphere that the civil rights movement gained strength and urgency.

The Cold War, Korea, and Vietnam

The United States left World War II in sound economic shape. By contrast, all other industrialized nations had suffered great losses in their infrastructure, financial stability, and populations. As a result, the United States was able to invest heavily in the postwar economies of Western Europe and Japan, with the hope of instilling an atmosphere conducive to peace and the spread of capitalism. U.S. barriers to foreign trade were relaxed to build new markets for U.S. exports and to allow some war-ravaged nations to make money selling goods to American consumers.

The Union of Soviet Socialist Republics (or the Soviet Union) had been an ally of the United States during World War II, but relations became strained after the war. The Soviet Union had adopted communism during the early 1920s. During World War II it liberated a large part of eastern Europe from Nazi occupation and through various means assumed political control over these nations. The Soviet Union had been largely industrialized before World War II and quickly regained its industrial capabilities. It soon took a major role in international affairs, placing it in direct conflict with the only other superpower of the time: the United States. A cold war began between the two rich and powerful nations that had completely different political, economic, and social goals for the world. Unlike a "hot war" or direct and large-scale military conflict, the conflict between the United States and the Soviet Union was called a "cold war" because it was waged mostly by politicians and diplomats.

A direct and large-scale military conflict between U.S. and Soviet forces never occurred. Regardless, an expensive arms race began in which both sides produced and stockpiled large amounts of weaponry as a show of

FIGURE 1.7

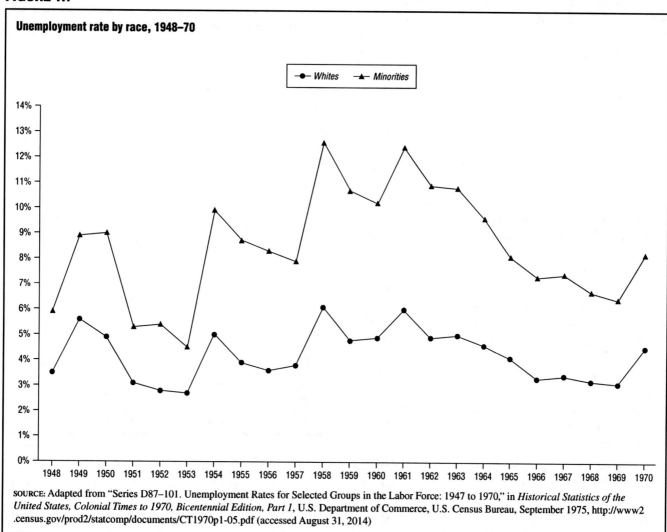

Unemployment rate by race, 1948–70

SOURCE: Adapted from "Series D87–101. Unemployment Rates for Selected Groups in the Labor Force: 1947 to 1970," in *Historical Statistics of the United States, Colonial Times to 1970, Bicentennial Edition, Part 1*, U.S. Department of Commerce, U.S. Census Bureau, September 1975, http://www2.census.gov/prod2/statcomp/documents/CT1970p1-05.pdf (accessed August 31, 2014)

force to deter a first strike by the enemy. In addition, both sides provided financial and military support to smaller countries throughout the world in an attempt to influence the political leanings of those populations. Communist China joined the Cold War during the 1950s and often partnered with the Soviet Union against U.S. interests.

By the 1950s the United States was embroiled in two Asian conflicts over communism: the Korean War (1950–1953) and the Vietnam War (1954–1975). In both wars the United States chose to fight in a limited manner without using its arsenal of nuclear weapons or engaging Chinese or Soviet troops directly for fear of sparking another world war. Unlike World War II, full-scale mobilization of U.S. industries was not required for these wars. Instead, a defense industry developed during the Cold War to supply the U.S. military on a continuous basis with the arms and matériel it needed.

Figure 1.8 shows the percentage of the national budget devoted to national defense between 1940 and 1970. Spending on national defense soared during World

FIGURE 1.8

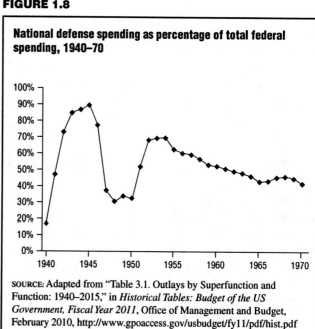

National defense spending as percentage of total federal spending, 1940–70

SOURCE: Adapted from "Table 3.1. Outlays by Superfunction and Function: 1940–2015," in *Historical Tables: Budget of the US Government, Fiscal Year 2011*, Office of Management and Budget, February 2010, http://www.gpoaccess.gov/usbudget/fy11/pdf/hist.pdf (accessed August 31, 2014)

War II and then declined dramatically afterward. However, military spending quickly climbed again as the Cold War intensified during the early 1950s and remained above 40% for nearly two decades.

THE 1960S: SOCIAL UPHEAVAL AND ECONOMIC GROWTH

The 1960s were a time of social and economic change for the United States. The decade began with the election of President John F. Kennedy (1917–1963), who promised to ensure economic growth and address growing social problems within the United States. In 1963 Kennedy's efforts were cut short by his assassination. Lyndon B. Johnson (1908–1973) took over as president and dramatically enlarged the federal government and its role in socioeconomic affairs. Johnson's administration initiated large-scale programs for the needy, including the health care programs Medicare (for the elderly and people with disabilities) and Medicaid (for the poor), jobs programs, federal aid to schools, and food stamps for low-income Americans. The so-called War on Poverty and the escalating war in Vietnam proved to be extremely expensive. At the same time, the United States was pursuing a costly (but ultimately successful) endeavor to land astronauts on the moon before the end of the decade.

Consumer and government spending drove the nation's GNP during the 1960s. However, inflation became a problem (as it often does in a fast-growing economy) during the late 1960s. At the macroeconomic level, there was too much money in the hands of consumers, which resulted in consumer demand that was higher than supply.

According to the NBER, in "US Business Cycle Expansions and Contractions" (November 2012, http://www.nber.org/cycles.html), the United States experienced the longest continuous stretch of positive GNP growth in history, from the first quarter of 1961 to the last quarter of 1969. However, high inflation was about to become a major problem.

THE 1970S: STAGFLATION AND ENERGY CRISES

The term *stagflation* was coined during the 1970s to describe an economy suffering stagnant growth, high inflation, and high unemployment all at the same time. This combination of economic problems was unprecedented in U.S. history. Previously, high inflation had occurred when the economy was growing quickly, such as during World War II, and high production had meant high employment levels. By contrast, economic downturns were associated with higher unemployment but lower inflation (and even deflation). These relationships had been considered natural and certain.

TABLE 1.1

Misery index, annual average, 1968–83

Year	Inflation rate	Unemployment rate	Misery index
1968	4.2%	3.6%	7.8%
1969	5.5%	3.5%	9.0%
1970	5.7%	4.9%	10.6%
1971	4.4%	5.9%	10.3%
1972	3.2%	5.6%	8.8%
1973	6.2%	4.9%	11.1%
1974	11.0%	5.6%	16.6%
1975	9.1%	8.5%	17.6%
1976	5.8%	7.7%	13.5%
1977	6.5%	7.1%	13.6%
1978	7.6%	6.1%	13.7%
1979	11.3%	5.8%	17.1%
1980	13.5%	7.1%	20.6%
1981	10.3%	7.6%	17.9%
1982	6.2%	9.7%	15.9%
1983	3.2%	9.6%	12.8%
1984	4.3%	7.5%	11.8%
1985	3.6%	7.2%	10.8%
1986	1.9%	7.0%	8.9%
1987	3.6%	6.2%	9.8%
1988	4.1%	5.5%	9.6%
1989	4.8%	5.3%	10.1%

SOURCE: Adapted from "Table 24. Historical Consumer Price Index for All Urban Consumers (CPI-U): U.S. City Average, All Items," in *CPI Detailed Report (Tables 1–29) June 2014*, U.S. Department of Labor, Bureau of Labor Statistics, July 22, 2014, http://www.bls.gov/cpi/cpid1406.pdf (accessed August 9, 2014), and "Employment Status of the Civilian Noninstitutional Population," in *Labor Force Statistics from the Current Population Survey*, U.S. Department of Labor, Bureau of Labor Statistics, February 26, 2014, http://data.bls.gov/cgi-bin/surveymost?bls (accessed August 7, 2014)

The 1970s were unique because both unemployment and inflation were high by historical standards. The economist Arthur Okun (1928–1980) created the term *discomfort factor* to describe this condition. His discomfort factor, which became popularly known as the Misery index, is computed by summing the unemployment rate and the inflation rate. Table 1.1 shows the annual average Misery index calculated between 1968 and 1989. It ranged from a low of 7.8% to a high of 20.6%.

There were three presidents during the 1970s: Richard M. Nixon (1913–1994), Gerald R. Ford (1913–2006), and Jimmy Carter (1924–). Although they tried a variety of measures to stem stagflation, their efforts were considered ineffective.

Foreign Oil and Competition

U.S. economic problems were aggravated by the country's dependence on foreign oil and competition from foreign industries. In 1973 the Middle Eastern members of the Organization of the Petroleum Exporting Countries halted oil exports to the United States in retaliation for U.S. support of Israel. The oil embargo lasted five months. When shipments resumed, the price of oil had dramatically increased. Americans faced high prices, long lines, and shortages at the gas pumps. During the late 1970s a political and cultural revolution in oil-rich Iran brought a second wave of shortages to U.S. energy supplies.

The energy crisis of the 1970s had a ripple effect throughout the U.S. economy, causing the prices of other goods and services to increase. Lower profits and uncertainty about the future caused businesses to slow down and reduce their workforces. At the same time, U.S. industries in steel, automobiles, and electronics endured stiff foreign competition, particularly from Japan. Small fuel-efficient Japanese cars became popular in the United States. U.S. carmakers struggled to compete, having always relied on consumer demand for large automobiles that were now considered "gas guzzlers."

Deregulation

One of the measures that President Carter instituted to combat stagflation was deregulation. For decades, certain U.S. industries had been given government immunity from market supply and demand factors. The railroad, trucking, and airline industries were prime examples. Companies in these industries were guaranteed rates and routes and were allowed to operate contrary to antitrust laws. In 1978 the airline industry was deregulated. The result was that airlines began competing with each other over fares and routes, and new companies entered the industry. Some of the large, well-established companies were unable to compete in the new environment and went out of business. However, demand increased as prices came down and air travel became available to many more Americans. By 1980 deregulation had been completed or was under way for the railroad, trucking, energy, financial services, and telecommunications industries.

THE 1980S: RECESSION AND REAGANOMICS

In November 1980 the American people elected Ronald Reagan (1911–2004) as the new president. Inflation was at 13.5% that year, which was incredibly high for a peacetime economy. (See Table 1.1.) Unemployment was at 7.1%, meaning that millions of people were unemployed and faced with rapidly increasing prices in the marketplace. The economic situation was dire, and drastic measures were required to turn the economy around.

Slaying the Inflationary Dragon

In late 1979 President Carter had appointed a new chair of the Federal Reserve board of governors, Paul A. Volcker (1927–), who promised to "slay the inflationary dragon." Volcker began by tightening the nation's money supply. This had the effect of making credit more difficult to obtain, which drove up interest rates. The government knew that rising interest rates would probably trigger a production slowdown (a recession) that would push unemployment even higher. It was a trade-off that policy makers during the previous decade had been unwilling to accept.

Volcker forged ahead with his policies, and by the early 1980s interest rates had reached historical highs. Figure 1.9 shows that the prime loan rate (the interest rate that banks charge their best customers) peaked at 21.5% in December 1980. According to the Federal Home Loan Mortgage Corporation, in "30-Year Conventional Mortgage Rate" (August 4, 2014, http://research.stlouisfed.org/fred2/data/MORTG.txt), in 1981 the average interest rate for a conventional 30-year mortgage soared to nearly 18.5%, the highest rate ever recorded.

The lack of credit caused a business slowdown—a reduction in GNP growth (or recession). As expected, the recession put more people out of work. Unemployment climbed at first, averaging 9.7% in 1982 and 9.6% in 1983, but then began to decline. (See Table 1.1.) By 1989 it was down to 5.3%. Likewise, the inflation rate dropped from a high of 13.5% in 1980 to 4.8% by 1989. Although the spike in unemployment had been painful for Americans, the inflationary dragon was finally dead.

Reaganomics

When Reagan took office in 1981, he brought a new approach to curing the nation's financial woes: supply-side economics. Traditionally, the government had focused on the demand side (the role of consumers in stimulating businesses to produce more). Reagan preferred economic policies that directly helped producers. In "Supply Side Economics" (2005, http://www.auburn.edu/~johnspm/gloss/supply_side), Paul M. Johnson of Auburn University describes the philosophy this way: "Supply-side policy analysts focus on barriers to higher productivity—identifying ways in which the government can promote faster economic growth over the long haul by removing impediments to the supply of, and efficient use of, the factors of production."

One of the cornerstones of supply-side economics is reducing taxes so that people and businesses have more money to invest in private enterprise. Reagan enacted tax cuts through two pieces of legislation: the Economic Recovery Tax Act of 1981 and the Tax Reform Act of 1986. The result was a much lower number of tax brackets (the various rates at which individuals are taxed based on their income), a broader tax base (wealth within a jurisdiction that is liable to taxation), and reduced tax rates on income and capital gains (the profit made from selling an investment, such as land).

At the same time, Reagan pushed for greater national defense spending as part of his "peace through strength" approach to the Soviet Union and for selective cuts in social services spending. However, no cuts were made to the largest and most expensive programs within the social services budget. The combination of all these factors resulted in high federal deficits during the 1980s. In other words, the federal government was spending more than it

FIGURE 1.9

Bank prime loan rate, January 1949–July 2014

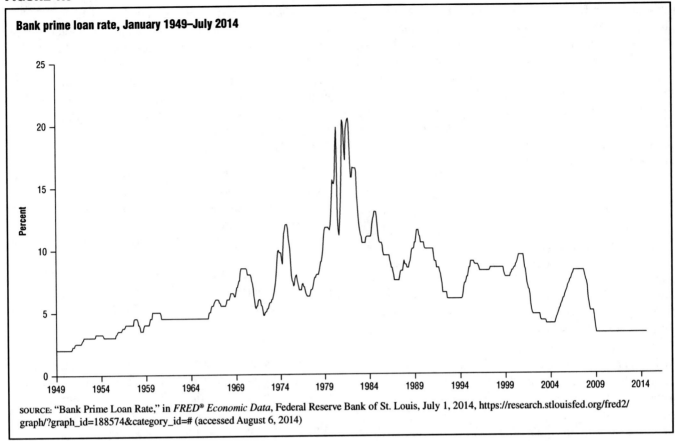

SOURCE: "Bank Prime Loan Rate," in *FRED® Economic Data*, Federal Reserve Bank of St. Louis, July 1, 2014, https://research.stlouisfed.org/fred2/graph/?graph_id=188574&category_id=# (accessed August 6, 2014)

was making each year. As shown in Figure 1.10, the federal deficits of the mid-1980s were more than three times what they had been during the mid-1970s. According to the article "U.S. Debt Past $1 Trillion" (NYTimes.com, October 23, 1981), the national debt (the sum of all accumulated federal deficits since the nation began) reached $1 trillion in 1981.

THE 1990S: SPARKLING ECONOMIC PERFORMANCE

The 1990s were a time of phenomenal economic growth for the United States. President George H. W. Bush (1924–) took office in 1989 and served until 1993. Bush had been elected in large part because of his promise not to raise taxes. During his presidential campaign he famously said, "Read my lips: No new taxes." The promise, however, was not one he could keep, given the economic realities of the time. During the late 1980s there had been a severe financial crisis in the savings and loan industry, which had been recently deregulated. A series of unwise loans and poor business decisions left most of the industry in shambles and necessitated a government bailout. At the same time, the government faced rapidly rising expenditures on health care programs for the elderly (Medicare) and the needy (Medicaid). Bush reluctantly agreed to a tax increase, a move that was politically damaging. In 1992 he lost his reelection bid to the Arkansas governor Bill Clinton (1946–), who was reelected in 1996.

Joseph Tracy, Henry Schneider, and Sewin Chan indicate in "Are Stocks Overtaking Real Estate in Household Portfolios?" (*Current Issues in Economics and Finance*, vol. 5, no. 5, April 1999) that, overall, the 1990s were a period of peace and prosperity for the United States: the Cold War ended when the Soviet Union disintegrated into individual republics; technological innovations, particularly in the computer industry, helped push the U.S. economy to new heights; and sterling business success led to robust investor confidence in the stock markets. From 1945 through 1998 real estate was the preferred investment in the United States. However, the 1990s saw tremendous increases in the holdings of corporate equity by the average American. During the mid-1980s the average household had only 10% of its assets in corporate equity. By 1998 this percentage had reached nearly 30%, roughly equal to the percentage held in real estate. The Dow Jones Industrial Average is a stock market index—a measure used by economists to gauge the value (and performance) of the stock of 30 large companies. Between the late 1970s and the late 1990s the index soared from about 1,000 points to 11,000 points, reflecting the tremendous value gained by these companies during this period.

FIGURE 1.10

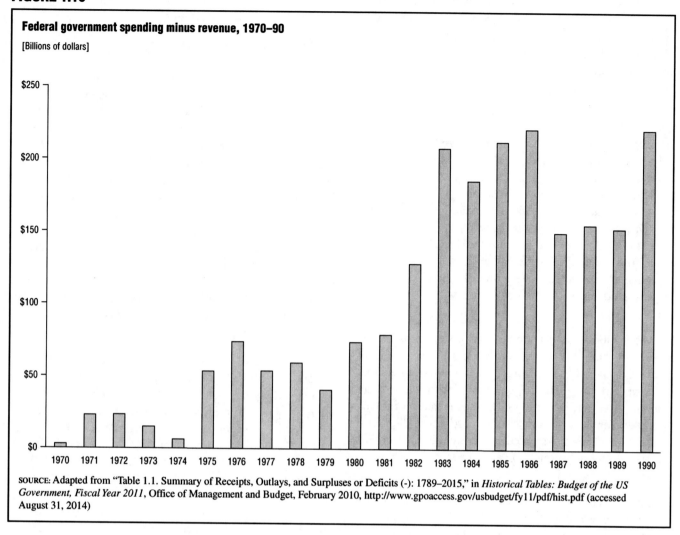

Federal government spending minus revenue, 1970–90
[Billions of dollars]

SOURCE: Adapted from "Table 1.1. Summary of Receipts, Outlays, and Surpluses or Deficits (-): 1789–2015," in *Historical Tables: Budget of the US Government, Fiscal Year 2011*, Office of Management and Budget, February 2010, http://www.gpoaccess.gov/usbudget/fy11/pdf/hist.pdf (accessed August 31, 2014)

The combination of low interest rates, low unemployment, and high investment rates and business growth combined to greatly expand the U.S. economy. According to the article "Excerpts from Federal Reserve Chairman's Testimony" (NYTimes.com, January 21, 1999), Alan Greenspan (1926–), the chair of the Federal Reserve board of governors, described this expansion as "America's sparkling economic performance."

THE EARLY 21ST CENTURY

During the early years of the first decade of the 21st century the United States endured the September 11, 2001, terrorist attacks, the outbreak of wars in Afghanistan and Iraq, and devastating hurricanes. Nevertheless, the overall economy was robust at first. As the decade progressed, the nation's economic soundness began to unravel, culminating in problems of historic proportion.

The Internet Bubble Bursts

During the late 1990s the stock market witnessed tremendous growth, driven in large part by investor enthusiasm for Internet-related businesses. Access to the Internet became widespread in the United States and in much of the developed world, which created many new market opportunities for entrepreneurs. Investors enthusiastically poured money into the stock of these new businesses. The National Association of Securities Dealers Automated Quotation System (NASDAQ) is a U.S.-based stock market on which the stock of many technology companies is traded. The NASDAQ composite index is a measure of the performance of many of the stocks on the NASDAQ. In 1990 the index was less than 500. In early 2000 it peaked above 4,000 during the height of the Internet stock craze. Many of the stocks had become overvalued, and their high prices could not be sustained based on the actual financial results that the companies were producing. What followed was a sharp market correction, as investors sold off many Internet-based stocks and prices plummeted. By late 2002 the NASDAQ composite index was about 1,200, from which it slowly began to climb again.

In economics a bubble is a phenomenon in which investors overzealously invest (speculate) in a particular commodity or market sector that becomes overvalued.

Excitement about possible gains overrules frank analysis of the underlying financial factors. What frustrates investors and analysts alike is that the very existence of a bubble is not evident until after the fact, when the bubble has burst and much value has been lost in the investments and the businesses involved.

The Great Recession

Overall, the U.S. economy prospered through the early part of the first decade of the 21st century. This soundness was evidenced by relatively low unemployment rates, moderate rates of inflation, and growth in the nation's production. National production is tracked through a numerical measure called the gross domestic product (GDP). The GDP is similar to the GNP described earlier, but the GNP includes production of U.S. companies outside the United States, whereas the GDP considers only production of U.S. companies within the United States.

The nation's GDP is considered a key measure (or metric) of how the economy is doing. The GDP is expected to increase over time, for example, from quarter to quarter. If the GDP stagnates or declines, then the economy is ailing. Declining GDP means that businesses are producing less. Because they need fewer workers, unemployment rises. As noted earlier, the term *recession* refers to a national production slowdown.

In "Determination of the December 2007 Peak in Economic Activity" (December 11, 2008, http://www.nber.org/cycles/dec2008.html), the NBER officially defines a recession as "a significant decline in economic activity spread across the economy, lasting more than a few months, normally visible in production, employment, real income, and other indicators. A recession begins when the economy reaches a peak of activity and ends when the economy reaches its trough [lowest point]." According to the NBER, the U.S. economy reached a peak in December 2007 and then went into a recession.

The NBER (September 20, 2010, http://www.nber.org/cycles/sept2010.pdf) notes that its analysis of national economic data indicates the recession ended in June 2009. In total, the recession spanned 18 months, making it the longest recession since World War II.

There have been many recessions in U.S. history; most were short and rather unremarkable. The exception is, of course, the Great Depression. The economic downturn that occurred from December 2007 to June 2009 was so deep and so damaging that many economists dubbed it "the Great Recession." One contributing factor was the bursting of another bubble, this time in the housing industry, which is described in detail in Chapter 4. The financial industry consequently suffered deep financial losses. Some banks and investment corporations failed, whereas others were rescued by enormous inflows of cash from the U.S. government.

GOVERNMENT INTERVENTION. The Great Recession officially began during the second administration of President George W. Bush (1946–) and ended during the first administration of President Barack Obama (1961–). Bush was a Republican conservative and thus generally favored a laissez-faire attitude toward the economy. However, the extreme depth of the recession prompted Bush and Congress to take action. Congress passed the Emergency Economic Stabilization Act of 2008, which Bush signed in October 2008. The law created the Troubled Asset Relief Program and authorized the U.S. Department of the Treasury to spend up to $700 billion to purchase or insure "troubled assets." These included mortgages and related investments (i.e., securities and financial instruments). The federal government also provided troubled companies in the auto industry and other sectors with loans and purchased ownership shares in them to keep them from failing.

The bailout of big businesses proved extremely unpopular politically, but the policy was continued by Obama when he assumed the presidency in 2009. In addition, Obama spearheaded passage of the American Recovery and Reinvestment Act (ARRA) of 2009, a multibillion-dollar stimulus package that was designed to revive the economy. In "The Recovery Act" (2014, http://www.recovery.gov/Pages/default.aspx), the Recovery Accountability and Transparency Board notes that ARRA funding totaled $840 billion. It was directed toward tax cuts and benefits for families and businesses; funding for entitlement programs, such as extending unemployment benefits for unemployed workers; and funding federal contracts, grants, and loans for the private sector.

The Federal Reserve, the nation's central bank, also took aggressive actions to boost the economy. It lowered interest rates in an effort to encourage consumers and businesses to borrow money for spending or investing. As is explained in Chapter 9, the Federal Reserve also created billions of dollars of "new" money which it transferred to banks by buying investment assets, such as bonds, from them.

The Great Recession officially ended in June 2009 as the economy began to recover. Analysts disagree about whether the government's actions alleviated or aggravated the economic downturn. Republicans generally criticized Obama's Keynesian actions, whereas some Democrats complained that the federal government should have injected even more money into the economy. Despite these disagreements, Obama secured another presidential term during the November 2012 general election.

Health Care System Reform

During his first term President Obama spearheaded a comprehensive reform of the nation's health care system. Two primary objectives drove this effort: reducing the number of uninsured Americans and reducing health care spending. In 2010, after significant debate, the Congress passed the Patient Protection and Affordable Care Act. It is commonly called the Affordable Care Act (ACA) or Obamacare. A follow-up law, the Health Care and Education Reconciliation Act of 2010, amended the ACA. Overall, the ACA represents a significant change to the nation's health care system. To better understand the law's economic impacts it is first necessary to examine the nation's health care system and the roles of the private and public sectors in it.

PRIVATELY FUNDED HEALTH PLANS. Privately funded health plans are financed by businesses and individuals. Insurers have long preferred to provide health plans on a group basis, for example, to a large group of employees. Each group includes members with varying medical needs. The healthiest members (i.e., those who use insurance the least) help pay for the higher medical costs of the more unhealthy members. Thus, the overall costs are spread across the group. Private insurers negotiate payment rates with medical providers (i.e., hospitals, doctors, etc.). Each insurer has a specific network of providers that have agreed to the payment rates and other terms and conditions.

Many Americans have access to group health insurance plans through their employers. Some employers subsidize (pay part of the cost of) health insurance plans as a benefit for their employees. Subsidized and non-subsidized group plans may be offered by employers, religious groups, labor unions, and trade and professional organizations. People who do not have access to group plans can buy so-called individual plans directly from insurance companies; however, individual plans are typically more expensive and impose more restrictions than do group plans.

In 1996 Congress passed the Health Insurance Portability and Accountability Act (HIPAA). The law prohibits employer-sponsored group plans from discriminating against plan members based on their health. HIPAA essentially ended the practice of excluding new members from group plans because they had preexisting health problems (i.e., preexisting conditions). HIPAA did not affect preexisting condition coverage in individual health plans. As a result people with preexisting conditions who did not have access to group health plans found it very difficult to obtain individual plans.

PUBLICLY FUNDED HEALTH PLANS. Publicly funded health plans are financed by taxpayer dollars and cover very specific population segments. As noted earlier in this chapter, the Medicare and Medicaid programs were created during the 1960s to assist with health care expenses of elderly and low-income Americans, respectively. Medicare is wholly administered by the Centers for Medicare and Medicaid Services (CMS), an agency within the U.S. Department of Health and Human Services. Originally designed only for people aged 65 years and older, Medicare has been expanded over the decades to include younger people with certain disabilities and diseases. Since its inception Medicare has been partially funded by taxes paid by employers and employees (including self-employed persons).

Medicaid is a joint federal-state program. Each state sets its own eligibility criteria and benefits (within broad federal guidelines) and provides part of the funding. Additional funding is provided by the federal government. The Children's Health Insurance Program (CHIP) is funded similarly to Medicaid. It provides health coverage for children in families whose incomes are too high to qualify for Medicaid, but too low to afford private health plans. Government agencies set payment rates for covered services in the Medicare, Medicaid, and CHIP programs. In other words the prices paid to medical providers for particular services are determined by the government, not by the market.

The U.S. Department of Defense and U.S. Department of Veterans Affairs operate their own medical facilities that provide care to certain existing and former members of the military and their families. In addition, the Defense Department's TRICARE program includes health plans operated by private insurance companies.

INSURANCE COVERAGE BEFORE THE ACA. Data regarding health insurance coverage in the United States are available from a variety of government and private sources. Most sources calculate national estimates based on polling and survey data. For example, the U.S. Census Bureau collects data as part of the Current Population Survey Annual Social and Economic Supplement, a sample survey of approximately 100,000 U.S. households.

Because nearly all of the nation's elderly population (i.e., those aged 65 years and older) is covered by Medicare, discussions of the uninsured typically center around nonelderly people. The Kaiser Family Foundation (KFF) is a nonprofit organization that focuses on national health issues. In *Overview of the Uninsured* (March 2012, http://kff.org/interactive/uninsured-tutorial), Rachel Garfield of the KFF indicates that in 2000 an estimated 36.3 million nonelderly people in the United States were uninsured. By 2010 the number had increased to 49.1 million or 18.5% of the total nonelderly population. According to Garfield, these values are based on Current Population Survey Annual Social and Economic Supplement data.

The Gallup Organization has conducted polling in which it asks respondents of all ages about their health insurance coverage. In *In U.S., Uninsured Rate Sinks to 13.4% in Second Quarter* (July 10, 2014, http://www.gallup.com/poll/172403/uninsured-rate-sinks-second-quarter.aspx), Jenna Levy of the Gallup Organization notes that the uninsured rate from 2008 through 2013 varied from 14.4% to 18%.

In July 2014 the Board of Governors of the Federal Reserve System (the Fed) published *Report on the Economic Well-Being of U.S. Households in 2013* (http://www.federalreserve.gov/econresdata/2013-report-economic-well-being-us-households-201407.pdf). The report contains data gleaned from the Fed's Survey of Household Economics and Decisionmaking, which was conducted in September and October 2013 using a sample population of 4,134 people. According to the Fed, "The sample is designed to be representative of the U.S. population." As shown in Table 1.2, nearly 84% of the respondents indicated they had health insurance coverage. The highest coverage rate (99.6%) was for those aged 65 years and older. The lowest coverage rate (75.8%) was for those aged 18 to 29 years. A breakdown by income for nonelderly respondents is provided in Table 1.3 for the 3,102 respondents who reported their incomes. Overall, 81.5% of them were insured. The coverage rate was lowest (69.2%) for respondents making less than $25,000 annually. The rate increased with income, with 95.8% of those earning $100,000 or more per year having health insurance coverage.

PREMIUMS AND OUT-OF-POCKET COSTS. There are four major cost components paid by persons with health insurance. These costs are associated with all privately funded health plans and some publicly funded health plans:

- Premium—an amount paid (usually monthly) by the insured to be covered by an insurance plan. A premium for a privately funded plan can total many hundreds of dollars per month. Employers that provide subsidized coverage typically pay at least half of the premium cost for each covered employee.
- Co-pay—an amount paid upon each visit to a doctor or other medical provider.
- Deductible—an amount that must be paid by the insured toward covered medical expenses each year before the insurer will begin paying.
- Coinsurance—the percentage of covered medical expenses that the insured must pay each year after paying co-pays and the deductible. For example, the insurer may pay 80% of the total leaving the insured with 20% to pay.

Together, co-pays, deductibles, and coinsurance amounts are called out-of-pocket expenses for the insured.

NATIONAL HEALTH SPENDING. Every year the CMS calculates the nation's total health spending. In January 2014 it released data indicating that $2.8 trillion was paid in 2012. A breakdown by source is shown in Figure 1.11. The sources included health insurance plans; consumer out-of-pocket payments; private and government investments in medical facilities, equipment, and research (excluding commercial research, for example, to develop new drugs); government public health activities; and various other payers and programs. Health insurance (both private and public) comprised the largest portion (72%) of the spending.

In "Sponsor Highlights" (January 2014, http://www.cms.gov/Research-Statistics-Data-and-Systems/Statistics-Trends-and-Reports/NationalHealthExpendData/NationalHealthAccountsHistorical.html), the CMS provides a breakdown by specific spending source in 2012:

- Households—28% of total
- Federal government—26% of total

TABLE 1.2

Health insurance coverage, by age, September 2013

[Percent, except as noted]

Age categories	Insured	Uninsured
18–29	75.8	24.2
30–44	79.4	20.6
45–64	84.0	16.0
65+	99.6	0.4
Overall	83.8	16.2
Total number of respondents		**4,134**

SOURCE: "Table 21. Health Insurance Coverage (by Age)," in *Report on the Economic Well-Being of U.S. Households in 2013*, Board of Governors of the Federal Reserve System, July 2014, http://www.federalreserve.gov/econresdata/2013-report-economic-well-being-us-households-201407.pdf (accessed August 9, 2014)

TABLE 1.3

Health insurance coverage among those under age 65, by income, September 2013

[Percent, except as noted]

Income categories	Insured	Uninsured
Less than $25,000	69.2	30.8
$25,000–$49,999	76.2	23.8
$50,000–$74,999	87.7	12.3
$75,000–$99,999	94.6	5.4
$100,000 and greater	95.8	4.2
Overall	81.5	18.6
Total number of respondents		**3,102**

Note: Among those who reported their income.

SOURCE: "Table 22. Health Insurance Coverage (by Income among Those under Age 65)," in *Report on the Economic Well-Being of U.S. Households in 2013*, Board of Governors of the Federal Reserve System, July 2014, http://www.federalreserve.gov/econresdata/2013-report-economic-well-being-us-households-201407.pdf (accessed August 9, 2014)

FIGURE 1.11

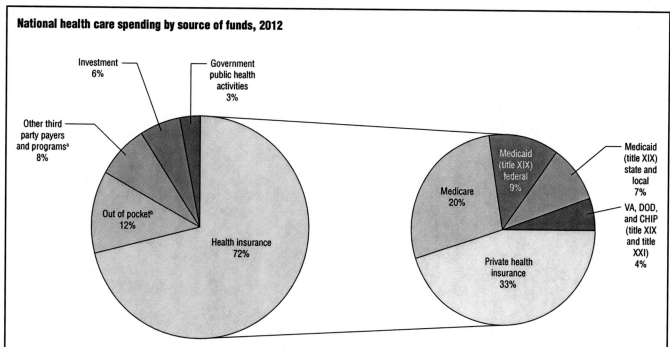

National health care spending by source of funds, 2012

aIncludes worksite health care, other private revenues, Indian Health Service, workers' compensation, general assistance, maternal and child health, vocational rehabilitation, Substance Abuse and Mental Health Services Administration, school health, and other federal and state local programs.
bIncludes co-payments, deductibles, and any amounts not covered by health insurance.
VA = U.S. Department of Veterans Affairs.
DOD = U.S. Department of Defense.
CHIP = Children's Health Insurance Program.
Note: Sum of pieces may not equal 100% due to rounding.

SOURCE: "The Nation's Health Dollar ($2.8 Trillion), Calendar Year 2012: Where It Came From," in *Nation's Health Dollar—Where It Came from, Where It Went*, U.S. Department of Health and Human Services, Centers for Medicare and Medicaid Services, Office of the Actuary, National Health Statistics Group, January 7, 2014, http://cms.hhs.gov/Research-Statistics-Data-and-Systems/Statistics-Trends-and-Reports/NationalHealthExpendData/Downloads/PieChartSourcesExpenditures2012.pdf (accessed August 11, 2014)

- Private businesses—21% of total
- State and local governments—18% of total
- Other private revenues (e.g., charitable donations)—7% of total

Figure 1.12 shows the major end uses for the $2.8 trillion spent on health care during 2012. Nearly a third (32%) was spent on hospital care, 20% went to physicians and clinics, and 9% was spent on prescription drugs. According to the CMS (December 2013, http://cms.hhs.gov/Research-Statistics-Data-and-Systems/Statistics-Trends-and-Reports/NationalHealthExpendData/Downloads/tables.pdf), national health spending totaled $26 billion in 1960. Spending has grown dramatically since that time. Of course, the U.S. population also grew over this period. To compensate for this effect analysts examine spending on a per capita (per person) basis. Figure 1.13 shows the annual growth rate in per capita health spending for 1961 through 2013. As noted earlier inflation causes the costs of goods and services to increase over time. The values shown in Figure 1.13 are said to be "real," meaning that inflationary effects have been removed. Real per capita spending grew at rates above 3.5% per year through the early 1990s. After falling briefly the rate soared again in the first decade of the 21st century before beginning to decline. According to the Executive Office of the President of the United States, in *Economic Report of the President: Together with the Annual Report of the Council of Economic Advisers* (March 2014, http://www.whitehouse.gov/sites/default/files/docs/full_2014_economic_report_of_the_president.pdf), the real per capita spending rate grew on average by 4.6% annually from 1960 through 2010.

Figure 1.14 shows national health spending as a percent of GDP. The value has grown significantly from 5% in 1960 to 17.2% in 2012. A variety of factors are behind the increased spending:

- Technological advances in medical diagnosis and treatment options.

- An aging U.S. population coupled with longer life spans. Elderly people tend to have more medical problems than younger people.

- Lower mortality rates for people suffering from certain serious illnesses and conditions. Improved survival rates for these patients can mean long-term medical care costs for them.

FIGURE 1.12

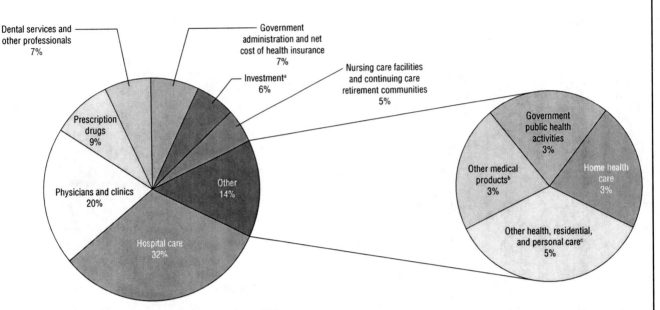

National health care spending by purpose, 2012

[a]Includes non-commercial research (2%) and structures and equipment (4%).
[b]Includes durable (1%) and non-durable (2%) goods.
[c]Includes expenditures for residential care facilities, ambulance providers, medical care delivered in non-traditional settings (such as community centers, senior citizens centers, schools, and military field stations), and expenditures for home and community waiver programs under Medicaid.
Note: Sum of pieces may not equal 100% due to rounding.

SOURCE: "The Nation's Health Dollar ($2.8 Trillion), Calendar Year 2012: Where It Went," in *Nation's Health Dollar—Where It Came from, Where It Went*, U.S. Department of Health and Human Services, Centers for Medicare and Medicaid Services, Office of the Actuary, National Health Statistics Group, January 7, 2014, http://cms.hhs.gov/Research-Statistics-Data-and-Systems/Statistics-Trends-and-Reports/NationalHealthExpendData/Downloads/PieChartSourcesExpenditures2012.pdf (accessed August 11, 2014)

FIGURE 1.13

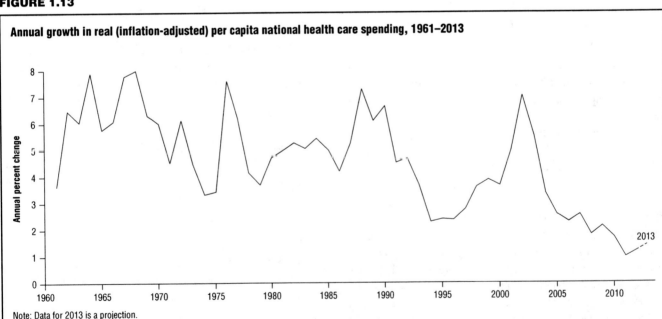

Annual growth in real (inflation-adjusted) per capita national health care spending, 1961–2013

Note: Data for 2013 is a projection.

SOURCE: "Figure 1-8. Growth in Real per Capita National Health Spending, 1961–2013," in *Economic Report of the President: Together with the Annual Report of the Council of Economic Advisers*, Executive Office of the President of the United States, March 2014, http://www.whitehouse.gov/sites/default/files/docs/full_2014_economic_report_of_the_president.pdf (accessed August 11, 2014)

FIGURE 1.14

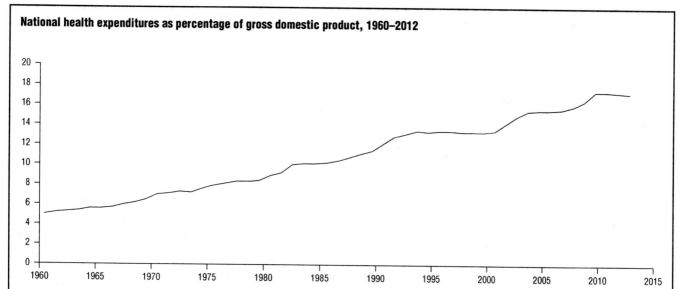

National health expenditures as percentage of gross domestic product, 1960–2012

SOURCE: Adapted from "Table 1. National Health Expenditures; Aggregate and per Capita Amounts, Annual Percent Change and Percent Distribution: Calendar Years 1960–2012," in *National Health Expenditures by Type of Service and Source of Funds, CY 1960–2012*, U.S. Department of Health and Human Services, Centers for Medicare and Medicaid Services, Office of the Actuary, National Health Statistics Group, January 7, 2014, http://cms.hhs.gov/Research-Statistics-Data-and-Systems/Statistics-Trends-and-Reports/NationalHealthExpendData/NationalHealthAccountsHistorical.html (accessed August 31, 2014)

- Greater prevalence of obesity and other health problems linked to unhealthy lifestyles.
- A growing population of uninsured people, which has forced medical care providers and taxpayers to cover more unpaid medical bills.
- Greater practice of "defensive" medicine by health care providers who conduct tests or procedures to protect themselves from being sued for malpractice.
- Rising prescription drug costs, particularly for so-called branded drugs for which there are no generic substitutes.

In *The Cost Disease: Why Computers Get Cheaper and Health Care Doesn't* (2012), the economist William J. Baumol points out that medical services are highly labor-intensive, and this increases their costs compared to other sectors of the economy. Medical professionals are required to be well-educated. They are highly skilled and hence highly paid compared with service providers in some other industries. In addition, some analysts believe there are cost drivers inherent to the U.S. health insurance system, which is based on a fee-for-service model. In other words, insurers pay medical providers for each service they provide (assuming the services are considered reasonable). There is little focus on the relative value, quality, or efficiency of the services that are provided.

ACA PROVISIONS. Table 1.4 lists and briefly describes the major provisions of the ACA. Two of the most controversial provisions are the individual mandate and Medicaid expansion. The individual mandate requires most legal residents of the United States to either obtain health insurance or pay a penalty to the government. Originally the ACA required all states to expand their Medicaid programs to cover more low-income people. These and other provisions of the law were the subject of much litigation. Their constitutionality was ultimately decided by the U.S. Supreme Court in 2012. In *National Federation of Independent Business et al. v. Sebelius, Secretary of Health and Human Services, et al.* (567 U.S. ___ [2012]) the court upheld numerous ACA provisions, including the individual mandate. However, the court ruled that Congress could not force the states to expand their Medicaid programs.

As the deadline for the individual mandate approached in 2014 the federal government (and some state governments) established online insurance exchanges at which consumers could purchase individual health plans from various insurance companies. The federal exchange, in particular, suffered massive technical glitches after its rollout in late 2013. These problems elicited sharp criticism from ACA critics, particularly Republican governors and legislators.

As shown in Table 1.4, a key requirement of the ACA is that eligible applicants must be accepted regardless of preexisting conditions. In addition, certain low-income applicants are eligible for government-subsidized premiums and out-of-pocket expenses. Employers and insurance companies must meet specific requirements under the ACA, which also implements major reforms

TABLE 1.4

Major provisions of the Patient Protection and Affordable Care Act

Applicant acceptance and premium requirements	Most insurers offering policies either for purchase through the exchanges or directly to consumers outside of the exchanges must meet several requirements: For example, they must accept all applicants regardless of health status; they may vary premiums only by age, smoking status, and geographic location; and they may not limit coverage for preexisting medical conditions.
Cost-sharing subsidy	Lower-income applicants that purchase certain insurance plans on the exchanges will pay lower out-of-pocket costs (i.e., deductibles, copayments, and coinsurance) than higher-income applicants.
Employer mandate	Certain employers that decline to offer their employees health insurance plans meeting affordability and coverage standards will be assessed penalties.
Excise tax on high-premium plans	A federal excise tax will be imposed on certain employer-sponsored health insurance plans with high premiums. Also known as the "Cadillac tax."
Excise tax/fee on certain business sectors	Federal excise taxes or fees will be imposed on certain business sectors, including health insurers, medical device manufacturers, indoor tanning services, and manufacturers and importers of brand name pharmaceutical drugs.
Individual mandate	Most legal residents of the United States must either obtain health insurance or pay a penalty for not doing so.
Medicaid expansion	States are permitted but not required to expand eligibility for Medicaid to include nonelderly adults with incomes up to 138% of the federal poverty level.
Medicare reform	Various programs targeting waste and fraud; incentives to health care providers to reduce costs; reduced reimbursements to certain health care providers (including hospitals with high readmissions rates) and to private Medicare Advantage programs; increased Medicare taxes for certain higher-income persons.
Premium tax credit	Lower-income applicants that purchase insurance plans on the exchanges and meet certain requirements are eligible to receive a tax credit. Some or all of the tax credit can be paid in advance directly to the insurer to lower monthly premiums or the insured can pay full price for the monthly premiums and receive the tax credit when filing their federal income tax return form.
Small-employer tax credit	Certain small employers that provide health insurance to their employees are eligible to receive a tax credit of up to 50% of the cost of that insurance.

SOURCE: Adapted from *Updated Estimates of the Effects of the Insurance Coverage Provisions of the Affordable Care Act, April 2014*, Congressional Budget Office, April 2014, https://www.cbo.gov/sites/default/files/cbofiles/attachments/45231-ACA_Estimates.pdf (accessed August 14, 2014), and *Affordable Care Act Tax Provisions*, Internal Revenue Service, August 12, 2014, http://www.irs.gov/uac/Affordable-Care-Act-Tax-Provisions (accessed August 14, 2014)

FIGURE 1.15

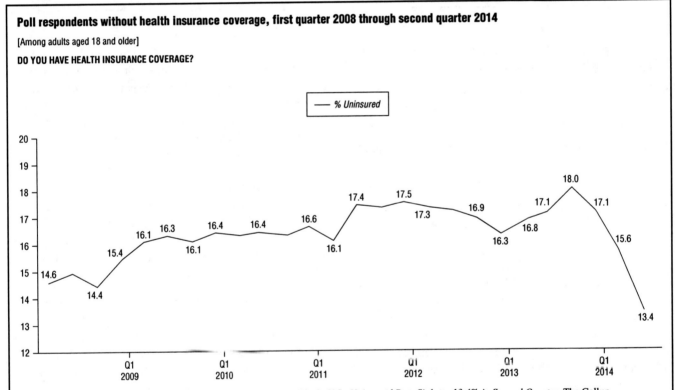

Poll respondents without health insurance coverage, first quarter 2008 through second quarter 2014

SOURCE: Jenna Levy, "Percentage Uninsured in the U.S., by Quarter," in *In U.S., Uninsured Rate Sinks to 13.4% in Second Quarter*, The Gallup Organization, July 10, 2014, http://www.gallup.com/poll/172403/uninsured-rate-sinks-second-quarter.aspx (accessed August 12, 2014). Copyright © 2014 Gallup, Inc. All rights reserved. The content is used with permission; however, Gallup retains all rights of republication.

of the Medicaid and Medicare programs. Details about specific provisions of the law and their effects on consumers, businesses, and the government are provided in subsequent chapters.

INSURANCE COVERAGE SINCE ACA IMPLEMENTATION. As noted earlier, polls and surveys conducted from 2008 through 2013 found that roughly 14% to 18% of Americans were uninsured. As shown in Figure 1.15, Gallup data indicate the uninsured rate declined to 13.4% in the first quarter of 2014. This was down from an average of 17.1% in the fourth quarter of 2013 (i.e., immediately before the individual mandate went into effect). Table 1.5 provides a coverage breakdown by state and indicates which states

TABLE 1.5

Percentage changes in uninsured populations between 2013 and midyear 2014 based on polling results, by state

"DO YOU HAVE HEALTH INSURANCE?" (% NO)

State	% Uninsured, 2013	% Uninsured, midyear 2014	Change in uninsured (pct. pts.)	Medicaid expansion and state/partnership exchange in 2014
Alabama	17.7	15.1	−2.6	No
Alaska	18.9	16.2	−2.7	No
Arizona	20.4	17.2	−3.2	No
Arkansas	22.5	12.4	−10.1	Yes
California	21.6	16.3	−5.3	Yes
Colorado	17.0	11.0	−6.0	Yes
Connecticut	12.3	7.4	−4.9	Yes
Delaware	10.5	3.3	−7.2	Yes
Florida	22.1	18.9	−3.2	No
Georgia	21.4	20.2	−1.2	No
Hawaii	7.1	6.9	−0.2	Yes
Idaho	19.9	16.6	−3.3	No
Illinois	15.5	12.3	−3.2	Yes
Indiana	15.3	15.0	−0.3	No
Iowa	9.7	10.3	0.6	Yes
Kansas	12.5	17.6	5.1	No
Kentucky	20.4	11.9	−8.5	Yes
Louisiana	21.7	18.4	−3.3	No
Maine	16.1	13.3	−2.8	No
Maryland	12.9	8.4	−4.5	Yes
Massachusetts	4.9	4.9	0.0	Yes
Michigan	12.5	11.9	−0.6	Yes
Minnesota	9.5	8.8	−0.7	Yes
Mississippi	22.4	20.6	−1.8	No
Missouri	15.2	15.1	−0.1	No
Montana	20.7	17.9	−2.8	No
Nebraska	14.5	13.4	−1.1	No
Nevada	20.0	16.0	−4.0	Yes
New Hampshire	13.8	12.4	−1.4	No
New Jersey	14.9	11.8	−3.1	No
New Mexico	20.2	15.2	−5.0	Yes
New York	12.6	10.3	−2.3	Yes
North Carolina	20.4	16.7	−3.7	No
North Dakota	15.0	13.0	−2.0	No
Ohio	13.9	11.5	−2.4	No
Oklahoma	21.4	17.5	−3.9	No
Oregon	19.4	14.0	−5.4	Yes
Pennsylvania	11.0	10.1	−0.9	No
Rhode Island	13.3	9.3	−4.0	Yes
South Carolina	18.7	16.8	−1.9	No
South Dakota	14.0	11.3	−2.7	No
Tennessee	16.8	14.4	−2.4	No
Texas	27.0	24.0	−3.0	No
Utah	15.6	15.6	0.0	No
Vermont	8.9	8.5	−0.4	Yes
Virginia	13.3	13.4	0.1	No
Washington	16.8	10.7	−6.1	Yes
West Virginia	17.6	11.9	−5.7	Yes
Wisconsin	11.7	9.6	−2.1	No
Wyoming	16.6	12.8	−3.8	No

SOURCE: Dan Witters, "Change in Percentage of Uninsured by State, 2013 vs. Midyear 2014," in *Arkansas, Kentucky Report Sharpest Drops in Uninsured Rate*, The Gallup Organization, August 5, 2014, http://www.gallup.com/poll/174290/arkansas-kentucky-report-sharpest-drops-uninsured-rate.aspx (accessed August 12, 2014). Copyright © 2014 Gallup, Inc. All rights reserved. The content is used with permission; however, Gallup retains all rights of republication.

expanded their Medicaid programs and operated either a state exchange or a state-federal exchange for ACA enrollees. Texas had the highest uninsured rate (27%) in 2013. Its Republican governor, Rick Perry (1950–) declined to expand the state's Medicaid program or operate an ACA exchange. As of mid-2014, Texas had the highest uninsured rate (24%) in the nation. It was followed by Mississippi (20.6%), Georgia (20.2%), Florida (18.9%), and Louisiana (18.4%). These states also had Republican governors in 2014 and did not expand their Medicaid programs or operate ACA exchanges.

In "Obamacare Enrollment Falls Slightly to 7.3 Million in August" (LATimes.com, September 18, 2014), Noam N. Levey reports that the CMS announced that 7.3 million people were covered under plans purchased through the ACA exchanges as of August 2014. This value was higher than the 6 million enrollees that the Congressional Budget Office (CBO) predicted would be covered in 2014. (See Table 1.6.) That prediction and projected coverage data for future years are included in the CBO report *Updated Estimates of the Effects of the Insurance Coverage Provisions of the Affordable Care Act* (April 2014, https://www.cbo.gov/sites/default/files/cbofiles/attachments/45231-ACA_Estimates.pdf). As shown in Table 1.6, the CBO predicts that by the year 2024 the insured rate for the nonelderly population will be 89% for all U.S. residents and 92% for all U.S. residents excluding unauthorized immigrants.

TABLE 1.6

Predicted effects of the Affordable Care Act on health insurance coverage, 2014–24

[Millions of nonelderly people, by calendar year]

	2014	2015	2016	2017	2018	2019	2020	2021	2022	2023	2024
Insurance coverage without the ACA[a]											
Medicaid and CHIP	35	35	34	33	33	34	34	34	35	35	35
Employment-based coverage	156	158	160	163	164	165	165	165	166	166	166
Nongroup and other coverage[b]	24	24	25	25	26	26	26	26	27	27	27
Uninsured[c]	54	55	55	55	55	56	56	56	57	57	57
Total	270	272	274	277	278	280	281	282	283	284	285
Change in insurance coverage under the ACA											
Insurance exchanges	6	13	24	25	25	25	25	25	25	25	25
Medicaid and CHIP	7	11	12	12	13	13	13	13	13	13	13
Employment-based coverage[d]	*	−2	−7	−7	−8	−8	−8	−8	−8	−7	−7
Nongroup and other coverage[b]	−1	−3	−4	−4	−4	−4	−4	−4	−4	−5	−5
Uninsured[c]	−12	−19	−25	−26	−26	−26	−26	−26	−26	−26	−26
Uninsured under the ACA											
Number of uninsured nonelderly people[c]	42	36	30	30	29	30	30	31	31	31	31
Insured as a percentage of the nonelderly population including all U.S. residents	84	87	89	89	89	89	89	89	89	89	89
Excluding unauthorized immigrants	86	89	91	92	92	92	92	92	92	92	92
Memorandum:											
Exchange enrollees and subsidies											
Number with unaffordable offer from employer[e]	**	**	**	**	**	**	**	**	**	**	**
Number of unsubsidized exchange enrollees (millions of people)[f]	1	3	5	6	6	6	6	6	6	6	6
Average exchange subsidy per subsidized enrollee (dollars)	4,410	4,250	4,830	4,930	5,300	5,570	5,880	6,220	6,580	6,890	7,170

* = between −500,000 and zero.
** = between zero and 500,000.
ACA = Affordable Care Act.
CHIP = Children's Health Insurance Program.
[a]Figures reflect average enrollment over the course of a year and include spouses and dependents covered under family policies; people reporting multiple sources of coverage are assigned a primary source.
[b]"Other" includes Medicare; the changes under the ACA are almost entirely for nongroup coverage.
[c]The uninsured population includes people who will be unauthorized immigrants and thus ineligible either for exchange subsidies or for most Medicaid benefits; people who will be ineligible for Medicaid because they live in a state that has chosen not to expand coverage; people who will be eligible for Medicaid but will choose not to enroll; and people who will not purchase insurance to which they have access through an employer, an exchange, or directly from an insurer.
[d]The change in employment-based coverage is the net result of projected increases and decreases in offers of health insurance from employers and changes in enrollment by workers and their families.
[e]Workers who would have to pay more than a specified share of their income (9.5 percent in 2014) for employment-based coverage could receive subsidies through an exchange.
[f]Excludes coverage purchased directly from insurers outside of an exchange.
Notes: Figures for the nonelderly population include residents of the 50 states and the District of Columbia who are younger than 65.

SOURCE: "Table 2. Effects of the Affordable Care Act on Health Insurance Coverage," in *Updated Estimates of the Effects of the Insurance Coverage Provisions of the Affordable Care Act*, Congressional Budget Office, April 2014, https://www.cbo.gov/sites/default/files/cbofiles/attachments/45231-ACA_Estimates.pdf (accessed August 14, 2014)

CHAPTER 2
ECONOMIC INDICATORS AND PUBLIC PERCEPTIONS

The U.S. economy is extremely large and complex, with many different interacting factors at play. Macroeconomics is the study of the overall condition of an economy; that is, its big picture. By contrast, microeconomics is the study of one or more small pieces of an economy, for example, how a particular industry or region is performing. Government agencies and private organizations collect voluminous amounts of data on many different aspects of the U.S. economy. Some of these data are aggregated (combined for consideration as a whole) so that the macroeconomic condition of the country can be analyzed and tracked over time. The most common economic indicators are widely broadcast in the media and are the subject of much analysis by leaders in government, industry, and the financial markets. The American public consists of hundreds of millions of individuals experiencing and interpreting various microeconomies and the national macroeconomy. Public opinion polls aggregate data reflecting these individual economic experiences and expectations to provide a picture of the national economic mood of the populace.

DEFINING ECONOMIC INDICATORS

An economic indicator is a data set that provides information about an economy's condition. Indicators compiled in a consistent fashion over time are particularly useful because they show economic trends. Many thousands of economic indicators are measured in the United States. For example, in "FRED Redesign Focuses on Improved Functionality, Enhanced Graphs" (March 17, 2014, http://www.stlouisfed.org/newsroom/displayNews.cfm?article=2114), the Federal Reserve Bank of St. Louis notes that as of March 2014, its online database called FRED (Federal Reserve Economic Data) included 213,000 data series from 64 sources.

Federal, state and local government agencies, educational institutions, private businesses, nonprofit organizations, and other types of organizations compile economic indicators. In general, economic data collected about individual businesses and people are aggregated (combined to represent the whole). Most government-compiled indicators are published openly. Nongovernmental entities typically publish only summaries or select portions of the data and charge membership or subscription fees for full access to all of the data.

Indicators by Economic Sector

Economic indicators can be broadly categorized by the economic sectors that they represent: business, personal, or government.

BUSINESS SECTOR. The U.S. business sector is extremely diverse with businesses ranging in size from large corporations to small home-based companies. Businesses may be publicly held (meaning that at least some of their ownership shares have been sold to the public) or privately held (meaning that none of their ownership shares have been sold to the public). This distinction is important because publicly held companies are required by federal law to report certain operating data, such as earnings, to the government and to stockholders. This is a key source of information for entities that compile economic data. In addition, businesses of all sizes and types are required to file tax returns with the Internal Revenue Service and meet other government reporting requirements, particularly if the businesses have employees. The data contained in these documents may or may not be available to the public but are accessible to government agencies, such as the Bureau of Economic Analysis (BEA), a part of the U.S. Department of Commerce. Industry organizations and trade associations are also key sources of economic data about business activities.

Economic indicators for the business sector track expenses, profits, sales, prices, taxes, assets, numbers of establishments and employees, wages and salaries, product inventories, capacity, productivity, and other financial factors.

PERSONAL SECTOR. The population of the United States exceeds 300 million people. For almost their entire lives people earn, spend, save, and invest money. These personal financial activities have an enormous impact on the U.S. economy.

Polls and surveys are widely used to obtain economic data from individual households; the data are then aggregated to represent the entire nation. It should be noted that federal government agencies often lump together individuals and nonprofit organizations in their economic indicators. For example, the BEA considers nonprofit institutions serving households as people, rather than as businesses.

Indicators of particular interest for the personal sector include income, assets, spending, saving, investing, debts, bankruptcy filings, and attitudes and expectations about the economy.

GOVERNMENT SECTOR. The local, state, and federal governments engage in economic activities with far-reaching effects on the nation's business and personal sectors. The indicators of interest for the government sector include revenues (particularly taxes), expenditures, budget surpluses and deficits, and the national debt. All of these indicators are described in detail in Chapter 9.

PRINCIPAL FEDERAL ECONOMIC INDICATORS

The Office of Management and Budget (OMB) within the White House designates certain data sets as Principal Federal Economic Indicators (PFEIs). In "Statistical Policy Directive on Compilation, Release, and Evaluation of Principal Federal Economic Indicators" (September 25, 1985, http://www.whitehouse.gov/sites/default/files/omb/assets/omb/inforeg/statpolicy/dir_3_fr_09251985.pdf), the OMB defines PFEIs as "statistical series that are widely watched and heavily relied upon by government and the private sector as indicators of the current condition and direction of the economy." The OMB further notes, "These series can have substantial effects upon market decisions and government policy." Table 2.1 lists the 38 data sets considered PFEIs as of 2014. The table briefly describes each data set and lists the federal agency that compiles it and the frequency with which it is published.

Some of the key agencies and bureaus involved in this effort are the BEA, the U.S. Department of Labor's Bureau of Labor Statistics (BLS), the U.S. Department of Agriculture, the U.S. Census Bureau, and the Federal Reserve Board. The agencies collect raw data from many different sources, both public and private, to prepare the PFEIs.

Nearly all of the data sets listed in Table 2.1 contain multiple economic indicators. For example, the Employment Situation compiled by the BLS includes the number of employees on company payrolls, the number of jobs added by employers during the previous month, employee earnings, the number of people that are unemployed, unemployment rates, and other labor statistics.

The PFEIs are published in agency press releases that are released on a regular basis, usually monthly or quarterly. Most of the press releases have unique titles that include the economic indicator name and date (e.g., "The Employment Situation—August 2014" or "U.S. International Transactions: First Quarter 2014 and Annual Revisions"). The press releases for some economic indicators continuously use the same title. For example, the PFEI called Hogs and Pigs is reported in a press release titled "Quarterly Hogs and Pigs." Thus, the reader must be careful to note the exact date range of the data reported.

Media sources frequently use informal names for economic indicators and the press releases in which they appear. For example, the Employment Situation press release is often referred to as the "jobs report" or the "monthly jobs report." Likewise, a press release covering Manufacturers' Shipments, Inventories, and Orders may be called the "factory orders report" or simply "factory orders." Some of the most common informal names used to reference PFEIs and their component data parts are listed in Table 2.1.

Many of the PFEIs shown in Table 2.1 are discussed in subsequent chapters. The gross domestic product (GDP) is examined in depth in this chapter because it is so widely reported and closely watched and provides a good big-picture view of how the U.S. economy is doing.

UNDERSTANDING ECONOMIC DATA SETS
Levels and Rates

Economists are interested in two main types of economic data: levels and rates. Levels provide a snapshot of conditions at a particular time. For example, the number of employed people during a particular month is a level. Levels change over time as they rise and fall. Economists are keenly interested in these changes, which are expressed as rates. Imagine that employment in January was at a level of 2 million and in February was at a level of 3 million; thus, employment increased by a rate of 50% between January and February.

Data Revisions

As shown in Table 2.1, the PFEIs are published on a systematic schedule that includes quarterly, monthly, or weekly releases to the public. In some cases the data released are considered preliminary as analysts continue to collect and assess information and make their calculations. Thus, the PFEIs and other economic indicators

TABLE 2.1

Principal federal economic indicators

Economic indicator data set	Source	Frequency	Description
Advance monthly retail sales	Census Bureau	Monthly	Provides the dollar value of retail sales of a broad range of goods, from cars and gasoline to furniture, food services, and clothing.
Advance report on durable goods	Census Bureau	Monthly	A measure of the number of orders for a broad range of products—from computers and furniture to autos and defense aircraft—with an expected life of at least three years.
Agricultural prices	Department of Agriculture	Monthly	Prices received by farmers for principal crops, livestock and livestock products; indexes of prices received by farmers; feed price ratios; indexes of prices paid by farmers; and parity prices.
Cattle on feed	Department of Agriculture	Monthly	Total number of cattle and calves on feed, placements, marketings, and other disappearances; by class and feedlot capacity for selected states; number of feedlots and fed cattle marketings by size groups for selected states.
Consumer credit	Federal Reserve Board	Monthly	Also known as "consumer installment credit." Outstanding credit extended to individuals for household, family, and other personal expenditures, excluding loans secured by real estate. Also includes selected terms of credit, including interest rates on new car loans, personal loans, and credit card plans at commercial banks.
Consumer price index	Bureau of Labor Statistics	Monthly	Data on changes in the prices paid by urban consumers for a representative basket of goods and services.
Construction spending	Census Bureau	Monthly	Also known as "value of construction put in place." A measure of the dollar value of new construction activity. Includes data on residential projects (such as homes and apartment buildings), nonresidential projects (such as privately funded office buildings), and public projects (such as schools and highways funded by the local, state, or federal government).
Corporate profits	Bureau of Economic Analysis	Quarterly	Tracks profits from current production (corporate profits with inventory valuation adjustment and capital consumption adjustment), taxes on corporate income, and dividends.
Crop production	Department of Agriculture	Monthly	Crop production data for the United States, including acreage, area harvested, yield, etc.
Current account balance	Bureau of Economic Analysis	Quarterly	Also known as the "current account deficit," it is a measure of net U.S. trade in merchandise, services, and certain financial transactions.
Employment cost index	Bureau of Labor Statistics	Monthly	Compensation costs (including wages, salaries, and benefits) for civilian workers in private industry and local and state governments.
Employment situation	Bureau of Labor Statistics	Monthly	Provides detailed data on non-farm payroll employment, jobs added, average workweek and hourly earnings, unemployment rates, reasons for unemployment, duration of unemployment, and other labor statistics.
Factors affecting reserve balances of depository institutions and condition statement of federal reserve banks	Federal Reserve Board	Weekly	Presents a balance sheet for each Federal Reserve Bank, a consolidated balance sheet for all 12 Reserve Banks, an associated statement that lists the factors affecting reserve balances of depository institutions, and several other tables presenting information on the assets, liabilities, and commitments of the Federal Reserve Banks.
Grain stocks	Department of Agriculture	Quarterly	Stocks of specific grains by location and by position (on-farm or off-farm storage); includes number and capacity of off-farm storage facilities and capacity of on-farm storage facilities.
Gross domestic product (GDP)	Bureau of Economic Analysis	Quarterly	GDP is a comprehensive measure of the economic health of the nation. It represents the total value of the country's production and consists of purchases of domestically produced goods and services by individuals, businesses, foreigners, and the government.
Hogs and pigs	Department of Agriculture	Quarterly	Data on the U.S. pig crop including inventory number by class, weight group, and value of hogs and pigs, farrowings, farrowing intentions. Also includes the number of operations keeping hogs, the number of hog operations and percent of inventory by size groups.
Housing vacancies and homeownership	Census Bureau	Quarterly	Provides current information on the rental and homeowner vacancy rates, and characteristics of units available for occupancy.
Industrial production and capacity utilization	Federal Reserve Board	Monthly	A monthly index of industrial production and associated capacity indexes and capacity utilization rates for manufacturing and mining companies and electric and gas utilities
Manufacturers' shipments, inventories, and orders	Census Bureau	Monthly	Also known as "factory orders," this data set comprises the dollar level of new orders for manufactured durable goods (products with an expected life of at least three years) and nondurable goods. The report gives more complete information than the Advance Report on Durable Goods, which is released earlier in the month.
Manufacturing and trade inventories and sales	Census Bureau	Monthly	Also known as "business inventories," this is the dollar value of product inventories held by manufacturers, wholesalers, and retailers. Included in the data set is the inventories/sales ratio, a gauge of the number of months it would take to deplete existing inventories at the current rate of sales.
Money stock measures	Federal Reserve Board	Weekly	Measures of the monetary aggregates (M1 and M2) and their components. M1 and M2 are progressively more inclusive measures of money: M1 is included in M2. M1, the more narrowly defined measure, consists of the most liquid forms of money, namely currency and checkable deposits. The non-M1 components of M2 are primarily household holdings of savings deposits, small time deposits, and retail money market mutual funds.
New residential construction	Census Bureau and the Department of Housing and Urban Development	Monthly	Also known as "housing starts and building permits," this includes data on the construction of private residential structures, such as single-family homes and apartment buildings.

are often revised over time. For example, the BEA releases three GDP estimates for a particular quarter; they are called the "advance" estimate, the "second" estimate, and the "third" estimate. In the blog post "Revising Economic Indicators: Here's Why the Numbers Can Change" (July 23, 2012, http://blog.bea.gov/2012/07/23/

TABLE 2.1

Principal federal economic indicators [CONTINUED]

Economic indicator data set	Source	Frequency	Description
New residential sales	Census Bureau and the Department of Housing and Urban Development	Monthly	National and regional data on the number of new single-family houses sold and for sale. It also provides national data on median and average prices for new homes, the number of new houses sold and for sale by stage of construction, and other statistics.
Personal income and spending	Bureau of Economic Analysis	Monthly	The income that households receive from all sources, such as wages and salaries, employer contributions to pension plans, rental properties, and dividends and interest. It also includes data on personal spending for durable goods (products with an expected life of more than one year) and nondurable goods and services, as well as information on the percentage of their income that households are saving.
Producer price indexes	Bureau of Labor Statistics	Monthly	A group of indexes that measure the average change over time in the prices received by U.S. producers for their goods and services
Productivity and costs	Bureau of Labor Statistics	Quarterly	Non-farm business and manufacturing sector labor productivity (i.e., output per hour), unit labor costs, and hourly compensation.
Prospective plantings	Department of Agriculture	March and June	Also known as "plantings." Provides expected acreages of plantings and harvests of various crops.
Quarterly estimates for selected service industries	Census Bureau	Quarterly	Also known as "quarterly services." Provides revenue data for selected service industries.
Quarterly financial report: large U.S. retail trade corporations	Census Bureau	Quarterly	Includes financial data, such as after-tax profits and sales, for large U.S. companies engaged in retail trade.
Quarterly financial report: U.S. manufacturing, mining, wholesale trade, and selected service industries	Census Bureau	Quarterly	Includes financial data, such as after-tax profits and sales, for U.S. companies in specific industries.
Real earnings	Bureau of Labor Statistics	Monthly	Current and real average hourly and weekly earnings for all employees and for production and non-supervisory employees on private non-farm payrolls.
U.S. import and export price indexes	Bureau of Labor Statistics	Monthly	Indexes of prices for U.S. imports and exports categorized by commodity. Import prices are categorized by country or area of origin. Also includes price indexes for selected international transportation services, such as air freight.
U.S. international trade in goods and services	Census Bureau and Bureau of Economic Analysis	Monthly	Also known as "U.S. trade balance." It provides the difference between the dollar value of exports and imports. Foreign trade is an important component of aggregate economic activity, representing a significant portion of gross domestic product.
U.S. international transactions	Bureau of Economic Analysis	Quarterly	A measure of transactions between the United States and the rest of the world in goods, services, primary income (investment income and compensation), and secondary income (current transfers).
Weekly natural gas storage report	Department of Energy	Weekly	Data on amount of natural gas in storage around the United States.
Wholesale trade	Census Bureau	Monthly	A tally of sales and inventories of U.S. merchant wholesalers.
World agricultural production	Department of Agriculture	Monthly	Data on crop acreage, yield and production in major countries worldwide.
World agricultural supply and demand estimates	Department of Agriculture	Monthly	World and U.S. supply, production, and use data for various agricultural commodities.

SOURCE: Adapted from *Schedule of Release Dates for Principal Federal Economic Indicators for 2014*, White House, 2014, http://www.whitehouse.gov/sites/default/files/omb/inforeg/statpolicy/pfei-schedule-of-release-dates-2014.pdf (accessed August 8, 2014), and *About Economic Indicators*, U.S. Department of Commerce, Economics and Statistics Administration, October 19, 2010, http://www.esa.doc.gov/about-economic-indicators (accessed August 8, 2014)

revising-economic-indicators), the BEA explains, "The public wants accurate data and wants it as soon as possible. To meet that need, BEA publishes early estimates that are based on partial data." As more incoming data are analyzed, the agency revises its early GDP estimates. The BEA also notes that it conducts regular reassessments of previous GDP data it has released dating all the way back to 1929 and revises the values as necessary.

Mathematical and Statistical Adjustments

Economists sometimes use mathematical and statistical adjustments to make raw economic data more useful for analysis. Three common adjustments are seasonal adjustment, annualization, and inflation adjustment.

In "Seasonally Adjusting Data" (2014, http://www.dallasfed.org/research/basics/seasonally.cfm), the Federal Reserve Bank of Dallas explains that seasonal adjustment is applied to some monthly and quarterly economic data. There are certain predictable seasonal fluctuations in the U.S. economy each year. For example, retail sales and employment increase every December due to holiday shopping. Housing construction increases every spring as the weather improves. Educator employment declines every summer. Economists use statistical tools to adjust seasonal-dependent data so they can see underlying trends. Then they can determine if changes from month to month or from quarter to quarter are actually significant.

Annualization is a mathematical method used to adjust monthly or quarterly economic data to predict annual values. For example, sales data reported in January 2014 are annualized to predict sales data for the entire year. Annualization can be used on levels or rates. In "Annualizing Data" (2014, http://www.dallasfed.org/research/basics/annualizing.cfm), the Federal Reserve Bank of Dallas explains how a monthly increase in employment is adjusted to provide an annual rate noting, "This rate represents the amount employment would have increased for the year had it expanded at that monthly rate all 12 months."

Inflation adjustment is used on economic data expressed in monetary units, for example, prices, profits, incomes, and the value of sales, investments, and inventories. The adjustment is made to account for the fact that inflation lowers the purchasing power of money over time. As a result, prices, incomes, and other money-dependent indicators typically rise over time. This makes it difficult for economists to tell whether changes are due to inflation alone or to other supply and demand factors. Unadjusted values are called nominal values; adjusted values are called real values. Figure 2.1 shows nominal and real prices for a gallon of regular-grade gasoline for 1976 through 2013. The nominal price was $0.57 per gallon in 1976. However, if a dollar had the same purchasing power in 1976 that it had in 2013 the price in 1976 would have been $2.48 per gallon. This is the real (inflation-adjusted) price. The graph of real gasoline prices over time shows changes that were due to supply and demand factors other than inflation. There are various mathematical methods for converting nominal data to real data. One of the most common methods utilizes the consumer price index which is explained in detail in Chapter 3.

GROSS DOMESTIC PRODUCT

The gross domestic product (GDP) is one of the PFEIs listed in Table 2.1 and arguably the most important measure of the health of the economy. The GDP measures in dollars the total value of U.S. goods and services that are newly produced or newly provided during a given period. It is calculated and published by the BEA. The GDP can be calculated in different ways. One method is the expenditure approach. It sums all the spending that takes place during a specified period on final goods and services that are newly produced or newly provided. The components of this calculation are:

- Personal consumption expenditures (PCE)—the amount spent by consumers on final goods and services. This includes food, clothing, household appliances, and so on, and payments for medical care, haircuts, dry cleaning, and other types of services. One major item not included in this category is the purchase of residential housing, which is considered an investment, rather than a consumption expense.

- Gross private domestic investment—this category has three components. One is the amount spent by businesses on assets they will use to provide goods and services (e.g., new machines and equipment, warehouses, software, and company vehicles). Also included are changes in the value of business inventories. This amount can be positive or negative.

FIGURE 2.1

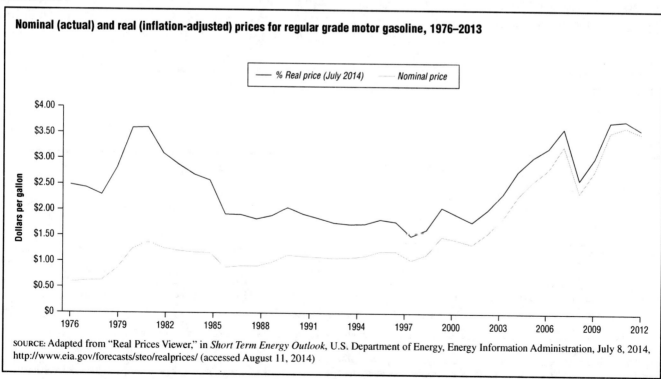

Nominal (actual) and real (inflation-adjusted) prices for regular grade motor gasoline, 1976–2013

SOURCE: Adapted from "Real Prices Viewer," in *Short Term Energy Outlook*, U.S. Department of Energy, Energy Information Administration, July 8, 2014, http://www.eia.gov/forecasts/steo/realprices/ (accessed August 11, 2014)

Residential housing is the third component of the investment category.

- Net exports of goods and services—the difference between the value of U.S. exports and imports. In other words, net exports equal the amount that foreigners paid for American goods and services minus the amount that Americans spent on foreign goods and services. If the United States exports more than it imports, this value will be positive. If the country imports more than it exports, this value will be negative.
- Government consumption expenditures and gross investment—the amount spent by the government (local, state, and federal) on final goods and services. This category does not include transfer payments to the public (such as Social Security and unemployment compensation) because these expenditures do not represent goods or services purchased.

It should be noted that the GDP counts only the final value paid for goods and services, not the value of intermediate transactions. For example, the value of steel sold by a steel company to an automaker is not counted. The value of the car made from the steel is counted when the car is sold in the marketplace. As mentioned earlier, to be counted in the GDP, a product must be new; thus, the sales of used items are not included in the GDP.

Real GDP

Economists often convert nominal GDP values to real values in order to remove the effects of inflation on the data. Consider a simple example in which a nation's only production is 1,000 identical new cars produced and sold each year. Assume this nation suffers from inflation, meaning that the price charged and paid for each car increases each year. A graph of this nation's nominal GDP over time would go upward, indicating that production increases each year, when actually it is just the price of the cars that is increasing. A graph of the real GDP over time would be horizontal showing accurately that the nation's production is constant year after year.

The statistical methods used by the BEA to convert nominal GDP values to real GDP values are quite complex and are described in *Concepts and Methods of the U.S. National Income and Product Accounts (Chapters 1–11 and 13)* (November 2014, http://www.bea.gov/national/pdf/chapters1-4.pdf). Each of the hundreds of components that contribute to the GDP is adjusted separately. As a result the real component values do not sum to the real total. This is illustrated in Table 2.2, which lists real GDP values for 2010 through 2013 and the major component categories contributing to them. Table 2.2 also provides seasonally adjusted and annualized GDP data for the first quarter of 2014. These data predict values for 2014 as a whole.

Figure 2.2 shows real GDP values for the U.S. economy from 1929 through 2013. These values were calculated assuming that a dollar had the exact same value over time (the value it had in 2009). Figure 2.3 shows the annual percentage change in the real GDP between 1998 and 2013. In general, GDP growth of 2% to 4% per year is considered optimal for the U.S. economy. The real GDP experienced negative growth between 2007 and 2008 (−0.3%) and between 2008 and 2009 (−2.8%), reflecting the poor state of the economy. Since that time the GDP has grown each year; between 2012 and 2013 it increased by 2.2%.

Figure 2.4 shows the quarterly percentage change in the real GDP for the fourth quarter of 2010 through the second quarter of 2014. The values were all positive with two exceptions: the first quarter of 2011 (down 1.5%) and the first quarter of 2014 (down 2.1%). Each of the quarters that followed these down quarters showed healthy economic growth. For example, the GDP grew by 4% in the second quarter of 2014.

TABLE 2.2

Major components of the real gross domestic product (GDP), 2010–13 and first and second quarters 2014

[Billions of chained (2009) dollars]

					2014	
	2010	2011	2012	2013	I	II
Gross domestic product	14,783.8	15,020.6	15,369.2	15,710.3	15,831.7	15,985.7
Personal consumption expenditures	10,036.3	10,263.5	10,449.7	10,699.7	10,844.3	10,910.5
Gross private domestic investment	2,120.4	2,230.4	2,435.9	2,556.2	2,588.2	2,691.8
Net exports of goods and services	−458.8	−459.4	−452.5	−420.4	−447.2	−470.3
Government consumption expenditures and gross investment	3,091.4	2,997.4	2,953.9	2,894.5	2,868.5	2,880.0

Note: Values are in billions of chained (2009) dollars. Quarterly data are seasonally adjusted at annual rates.

SOURCE: Adapted from "Table 3B. Real Gross Domestic Product and Related Measures," in *National Income and Product Accounts—Gross Domestic Product: Second Quarter 2014 (Advance Estimate) Annual Revision: 1999 through First Quarter 2014*, U.S. Department of Commerce, Bureau of Economic Analysis, July 30, 2014, http://www.bea.gov/newsreleases/national/gdp/2014/pdf/gdp2q14_adv.pdf (accessed August 7, 2014)

FIGURE 2.2

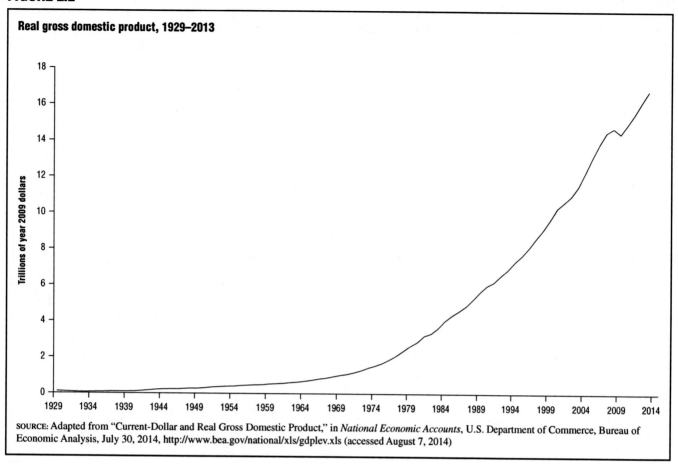

Real gross domestic product, 1929–2013

SOURCE: Adapted from "Current-Dollar and Real Gross Domestic Product," in *National Economic Accounts*, U.S. Department of Commerce, Bureau of Economic Analysis, July 30, 2014, http://www.bea.gov/national/xls/gdplev.xls (accessed August 7, 2014)

FIGURE 2.3

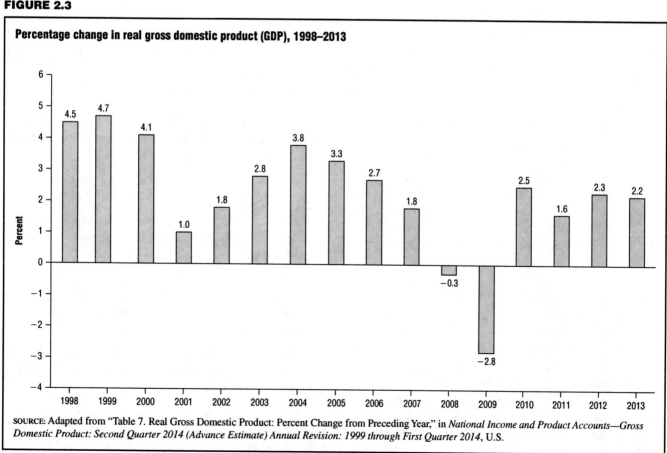

Percentage change in real gross domestic product (GDP), 1998–2013

SOURCE: Adapted from "Table 7. Real Gross Domestic Product: Percent Change from Preceding Year," in *National Income and Product Accounts—Gross Domestic Product: Second Quarter 2014 (Advance Estimate) Annual Revision: 1999 through First Quarter 2014*, U.S.

FIGURE 2.4

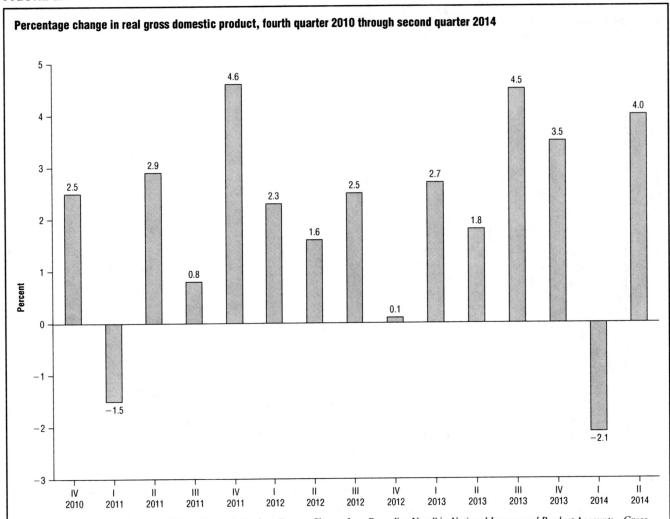

Percentage change in real gross domestic product, fourth quarter 2010 through second quarter 2014

SOURCE: Adapted from "Table 7. Real Gross Domestic Product: Percent Change from Preceding Year," in *National Income and Product Accounts—Gross Domestic Product: Second Quarter 2014 (Advance Estimate) Annual Revision: 1999 through First Quarter 2014*, U.S. Department of Commerce, Bureau of Economic Analysis, July 30, 2014, http://www.bea.gov/newsreleases/national/gdp/2014/pdf/gdp2q14_adv.pdf (accessed August 7, 2014)

Comparison with the Gross National Product

Before 1991 the U.S. government relied on an economic indicator called the gross national product (GNP) to measure U.S. economic output. The GNP is calculated the same way that the GDP is calculated, except that the GNP includes the contribution of U.S. production in foreign countries (e.g., an American-owned factory in Mexico). The GDP includes only production that occurs within the boundaries of the United States.

Comparison with the Gross Domestic Income

As described in Chapter 1, the National Income and Product Accounts (NIPAs) were initiated following World War II. In "The National Income and Product Accounts" (May 2–4, 2004, http://www.bea.gov/papers/pdf/NABEspeakernotes.pdf), Brooks Robinson and Shelly Smith of the BEA explain that the two key components of the NIPAs are the GDP and the gross domestic income (GDI). Although calculated separately, the two indicators provide very similar values. This is because the GDI measures national output based on the incomes earned and the costs incurred during production. GDI is calculated by summing the following:

- Compensation of employees including private and public (government) wages and salaries and supplements

- Taxes on production and imports (such as sales taxes, property taxes, and custom duties) minus subsidies paid to businesses by the government

- Profits and surpluses of private and public enterprises including corporate profits, the net earnings of unincorporated businesses, and the surpluses of government enterprises

- Consumption of fixed capital, which is the value or cost of using certain assets (property, buildings, and equipment) in production

In *A Guide to the National Income and Product Accounts of the United States* (September 2006, http://www.bea.gov/national/pdf/nipaguid.pdf), the BEA notes,

"In theory, GDP should equal GDI, but in practice, they differ because their components are estimated using largely independent and less than perfect source data. This difference is termed the 'statistical discrepancy.'" Major GDP and GDI components are discussed in detail in subsequent chapters.

THE BUSINESS CYCLE

In a "perfect" economy all economic indicators would be at optimum levels; for example, the employment and productivity rates would constantly be at 100%, and the GDP would always grow at a healthy rate. In reality, the economy rises and falls over time. It expands for a while and then contracts, and then recovers and begins to expand again. This is known as the business cycle. Note that the word *cycle* is misleading because the expansions and contractions do not happen on a set time table; they occur irregularly.

As explained in Chapter 1, the National Bureau of Economic Research (NBER) at the University of New York designates the starting and ending dates of recessions. In "Determination of the December 2007 Peak in Economic Activity" (December 11, 2008, http://www.nber.org/cycles/dec2008.html), the NBER officially defines a recession as "a significant decline in economic activity spread across the economy, lasting more than a few months, normally visible in production, employment, real income, and other indicators. A recession begins when the economy reaches a peak of activity and ends when the economy reaches its trough [lowest point]."

Table 2.3 lists business cycle peak and trough dates dating back to 1854. As of 2014 the most recent recession occurred from December 2007 (when the economy peaked) through June 2009 (when the economy reached a low point). Thus, following June 2009 the economy began to grow again.

TABLE 2.3

Business cycles as designated by the National Bureau of Economic Research

Business cycle reference dates		Duration in months			
		Contraction	Expansion	Cycle	
Peak	Trough	Peak to trough	Previous trough to this peak	Trough from previous trough	Peak from previous peak
Quarterly dates are in parentheses	Quarterly dates are in parentheses				
	December 1854 (IV)	—	—	—	—
June 1857 (II)	December 1858 (IV)	18	30	48	—
October 1860 (III)	June 1861 (III)	8	22	30	40
April 1865 (I)	December 1867 (I)	32	46	78	54
June 1869 (II)	December 1870 (IV)	18	18	36	50
October 1873 (III)	March 1879 (I)	65	34	99	52
March 1882 (I)	May 1885 (II)	38	36	74	101
March 1887 (I)	April 1888 (I)	13	22	35	60
July 1890 (III)	May 1891 (II)	10	27	37	40
January 1893 (I)	June 1894 (II)	17	20	37	30
December 1895 (IV)	June 1897 (II)	18	18	36	35
June 1899 (III)	December 1900 (IV)	18	24	42	42
September 1902 (IV)	August 1904 (III)	23	21	44	39
May 1907 (II)	June 1908 (II)	13	33	46	56
January 1910 (I)	January 1912 (IV)	24	19	43	32
January 1913 (I)	December 1914 (IV)	23	12	35	36
August 1918 (III)	March 1919 (I)	7	44	51	67
January 1920 (I)	July 1921 (III)	18	10	28	17
May 1923 (II)	July 1924 (III)	14	22	36	40
October 1926 (III)	November 1927 (IV)	13	27	40	41
August 1929 (III)	March 1933 (I)	43	21	64	34
May 1937 (II)	June 1938 (II)	13	50	63	93
February 1945 (I)	October 1945 (IV)	8	80	88	93
November 1948 (IV)	October 1949 (IV)	11	37	48	45
July 1953 (II)	May 1954 (II)	10	45	55	56
August 1957 (III)	April 1958 (II)	8	39	47	49
April 1960 (II)	February 1961 (I)	10	24	34	32
December 1969 (IV)	November 1970 (IV)	11	106	117	116
November 1973 (IV)	March 1975 (I)	16	36	52	47
January 1980 (I)	July 1980 (III)	6	58	64	74
July 1981 (III)	November 1982 (IV)	16	12	28	18
July 1990 (III)	March 1991 (I)	8	92	100	108
March 2001 (I)	November 2001 (IV)	8	120	128	128
December 2007 (IV)	June 2009 (II)	18	73	91	81

SOURCE: Adapted from *US Business Cycle Expansions and Contractions*, National Bureau of Economic Research, April 23, 2012, http://www.nber.org/cycles.html (accessed August 8, 2014)

Figure 2.5 shows real GDP values for 1980 through 2013 in relation to recession timing. The GDP slumped during each recession. In fact, many economic indicators decline during recessions. The Conference Board is a nonprofit company that compiles economic data collected from numerous sources, both government and private. In *Business Cycle Indicators Handbook* (2000, http://www.conference-board.org/pdf_free/economics/bci/BCI-Handbook.pdf), the Conference Board notes that economists divide economic indicators into three categories, depending on the timing relationship between the indicators and the business cycle:

- Leading indicators—tend to show peaks and troughs soon before the peaks and troughs of the business cycle
- Coincident indicators—tend to show peaks and troughs that coincide with the peaks and troughs of the business cycle
- Lagging indicators—tend to show peaks and troughs soon after the peaks and troughs of the business cycle

It should be noted that the word *soon* in this context is not time measured in days or weeks, but in months or quarters. For example, a leading indicator might peak three to six months before the business cycle does; likewise, a lagging indicator might peak up to a year or two after the business cycle does.

The GDP and many other economic indicators are coincident indicators. Other indicators either lead or lag the business cycle. For example, one leading indicator is the average weekly hours of employees in the manufacturing sector, a data set compiled by the BLS. The Conference Board notes "this component tends to lead the business cycle because employers usually adjust work hours before increasing or decreasing their workforce."

Leading indicators are closely watched because of their predictive nature; that is, they provide clues about what lies ahead for the economy at large. Investors, in particular, monitor leading indicators to help them make decisions about how best to invest their money for maximum profit. However, the clues can be difficult to decipher. Sometimes different leading indicators move in opposite directions. Also, it is impossible to know until after the fact that a certain indicator has definitely

FIGURE 2.5

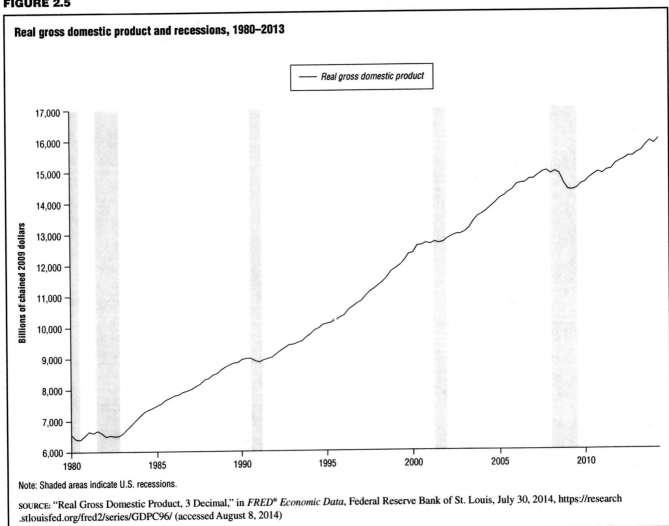

Real gross domestic product and recessions, 1980–2013

Note: Shaded areas indicate U.S. recessions.

SOURCE: "Real Gross Domestic Product, 3 Decimal," in *FRED® Economic Data*, Federal Reserve Bank of St. Louis, July 30, 2014, https://research.stlouisfed.org/fred2/series/GDPC96/ (accessed August 8, 2014)

peaked or troughed because indicators fluctuate over the short term.

Lagging indicators (like coincident indicators) have no predictive nature. An example lagging indicator is the ratio of consumer debt to personal income. As the Conference Board explains, "This ratio usually shows a trough many months after the recession ends, because consumers tend to initially hold off personal borrowing until personal income has risen substantially."

REGIONAL AND LOCAL ECONOMIES

National economic indicators provide a picture of the overall condition of the U.S. economy. The state of regional and local economies can be quite different from the national condition.

Part of microeconomics study (an analysis of the behavior of economic units such as companies, industries, or households) involves the economic health of various geographical regions and communities. A regional economy can be an area as small as a neighborhood or as large as a group of states with certain factors in common, such as climate, geography, industry, or culture.

The economy of the American South was in shambles after the Civil War (1861–1865). As a result northern cities attracted millions of workers to their rapidly growing industrial centers, such as the steel mills of Pittsburgh, Pennsylvania, and the car factories of Detroit, Michigan. The economy of the upper Midwest soared through the early 20th century. Later the auto and steel manufacturers starting losing business to foreign competitors. The region became known as the Rust Belt as its economy declined and many workers migrated elsewhere. During the late 20th century, western states experienced economic growth with the rise of the computer industry. This was due in large part to Microsoft, which is headquartered in Seattle, Washington, and the dot-com companies centered in California's Silicon Valley. Other major economic regions of the United States include the Farm Belt of the Great Plains and the Sun Belt states of the South and Southwest. The latter have warm climates that make them popular tourist destinations and strong agricultural regions.

Microeconomic data are compiled by various sources including the government, academia, businesses, and organizations. The BEA (November 20, 2014, http://www.bea.gov/newsreleases/news_release_sort_regional.htm) publishes GDP and other economic indicators for regions, states, and municipalities. In "Widespread but Slower Growth in 2013: Advance 2013 and Revised 1997–2012 Statistics of GDP by State" (June 11, 2014, http://www.bea.gov/newsreleases/regional/gdp_state/2014/pdf/gsp0614.pdf), the agency provides an overview of state-by-state GDP growth rates. At that time the nation's overall GDP was believed to have grown by 1.8% in 2013. (This value was later revised to 2.2% as shown in Figure 2.3.) According to the BEA, GDP growth by state in 2013 varied greatly from 9.7% for Wyoming to −2.5% for Alaska.

Eight times per year the Federal Reserve publishes qualitative information about regional economies in a report titled "Summary of Commentary on Current Economic Conditions by Federal Reserve District." Each report is known informally as "the beige book" and summarizes information collected about each of the 12 Federal Reserve districts throughout the country. According to the Federal Reserve (December 3, 2014, http://www.federalreserve.gov/monetarypolicy/beigebook), "Each Federal Reserve Bank gathers anecdotal information on current economic conditions in its District through reports from Bank and Branch directors and interviews with key business contacts, economists, market experts, and other sources."

THE ROLE OF THE INDIVIDUAL IN THE ECONOMY

Almost every aspect of American life is influenced by, and further influences, the economy. Whether a person drives or flies during his or her next vacation, how he or she will pay for college and save for retirement, what advertisements he or she sees, what movies he or she watches, and what magazines he or she reads all involve making economic decisions, which then affect the way the economy functions. If a person works or plans to work, that person is a small but important part of the economy. Likewise, every time an individual buys goods or saves money, he or she is participating in economic activity.

In a healthy economy people tend to have more job security, earn more money, and are able to increase opportunities for themselves and their family, thus improving their overall quality of life. By contrast, in an unstable, bad economy (such as during the Great Depression and during recessions) people are less certain of the future, face increasing pressures at work and may lose their job, and have less flexibility in being able to pay for goods and services, which in turn affects trends in employment, interest rates, the cost of living, the money supply, and all other aspects of the economy, on both the macro and micro levels.

PUBLIC SURVEYS AND POLLS

Surveys and polls provide valuable information about the economic conditions and opinions of persons living in the United States.

Problems Facing the Country

The Gallup Organization is a U.S.-based firm that conducts frequent public opinion polls to gauge the mood and attitudes of the American people on a variety of subjects. In July 2014 Gallup asked Americans to name "the most important problem facing this country today." The responses receiving the most mentions during the poll are listed in Table 2.4. Overall, eight of the 17 responses involved the economy including "economy in general" which was cited by 15% of respondents. It was the third most mentioned of all the problems facing the country. Other economic problems rating high on the list included unemployment/jobs (cited by 14%), poor healthcare/hospitals and high cost of healthcare (cited by 8%), and federal budget deficit/federal debt (cited by 6%). Figure 2.6 shows the percentage of respondents that named at least one economic problem as the nation's most important problem in Gallup polls dating back to 2001. The percentage soared to 86% in 2009 during the Great Recession. Since that time it has fallen and was at 41% in 2014.

The National Economy

Gallup conducts daily polls in which it asks respondents to qualitatively rate the overall economic condition of the country from "excellent," to "poor." In addition, the poll participants are asked if the economy is "getting better" or "getting worse." The responses to these two questions are used to calculate the Gallup Economic Confidence Index (GECI). As shown in Figure 2.7, the GECI for July 2014 was −17. The GECI reached its lowest level (−60) in late 2008 during the Great Recession.

Another respected indicator of the nation's economic mood is the Consumer Confidence Index (CCI), which is maintained by the Conference Board. The CCI is calculated from the responses to monthly surveys of 5,000 U.S. households. The respondents are asked to rate the state of the existing economy and its likely direction over the next six months. In "The Conference Board Consumer Confidence Index Improves Again" (November 25, 2014, http://www.conference-board.org/data/consumerconfidence.cfm), the Conference Board reports that the CCI was 88.7 in November 2014, down from 94.1 the previous month. An arbitrary baseline value of 100 is assigned to the year 1985. According to Jason Bram and Sydney Ludvigson of the Federal Reserve Bank of New York, in "Does Consumer Confidence Forecast Household Expenditure? A Sentiment Index Horse Race" (*Economic Policy Review*, vol. 4, no. 2, June 1998), the CCI ranged between 95 and 130 during the late 1990s, a time of strong economic performance. In "Consumer Confidence Plummets" (CNN.com, February 24, 2009), Ben Rooney notes that the CCI fell to 25 in early 2009, the lowest point ever recorded.

Personal Economic Well-Being

Since 2001 Gallup has asked poll participants to rate their financial situations as excellent, good, only fair, or poor. The results are shown in Figure 2.8. In April 2014 more than half (52%) of the respondents said their financial situation was only fair or poor, compared with 48% who said it was excellent or good. The ratings were decidedly bad from early 2009 through early 2012, reflecting public pessimism over the Great Recession and its aftermath. Gallup has also quizzed the public about their monetary status. As shown in Figure 2.9, nearly three-quarters (71%) of those asked in April 2014 said they had "enough money to live comfortably." This value has risen since 2012, when only 60% of respondents expressed the same level of confidence.

In April 2014 Gallup asked poll participants to express their level of concern about various financial matters based on their current financial situation. Table 2.5 shows the responses ranked from the highest to the lowest level of concern. Overall, 59% of respondents were very worried or moderately worried about having enough money for retirement. Slightly more than half (53%) felt that way about being able to pay their medical costs in the event of a serious illness or accident. Nearly half (48%) said they were very worried or moderately worried about being able to maintain the standard of living they currently enjoy.

TABLE 2.4

Public opinion about the most important problem facing the United States, January, June, and July 2014

WHAT DO YOU THINK IS THE MOST IMPORTANT PROBLEM FACING THIS COUNTRY TODAY?*

	July 2014 %	June 2014 %	January 2014 %
Immigration/illegal aliens	17	5	3
Dissatisfaction with government/Congress/politicians; poor leadership/corruption/abuse of power	16	19	21
Economy in general	15	20	18
Unemployment/jobs	14	16	16
Poor healthcare/hospitals; high cost of healthcare	8	10	16
Federal budget deficit/federal debt	6	5	8
Education/poor education/access to education	5	4	4
Ethics/moral/religious/family decline; dishonesty	4	7	5
Poverty/hunger/homelessness	3	2	4
Foreign aid/focus overseas	3	3	3
Judicial system/Courts/laws	3	2	1
Lack of money	2	2	4
Taxes	2	1	1
Wage issues	2	1	1
Crime/violence	2	3	1
Race relations/racism	2	1	1
Lack of respect for each other	2	2	2

*Problems mentioned by at least 2% of Americans.

SOURCE: Lydia Saad, "U.S. Most Important Problem—Recent Selected Trend," in *One in Six Say Immigration Most Important U.S. Problem*, The Gallup Organization, July 16, 2014, http://www.gallup.com/poll/173306/one-six-say-immigration-important-problem.aspx (accessed August 7, 2014). Copyright © 2014 Gallup, Inc. All rights reserved. The content is used with permission; however, Gallup retains all rights of republication.

FIGURE 2.6

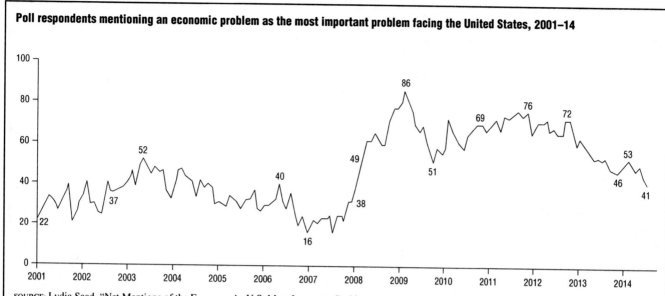

Poll respondents mentioning an economic problem as the most important problem facing the United States, 2001–14

SOURCE: Lydia Saad, "Net Mentions of the Economy As U.S. Most Important Problem," in *One in Six Say Immigration Most Important U.S. Problem*, The Gallup Organization, July 16, 2014, http://www.gallup.com/poll/173306/one-six-say-immigration-important-problem.aspx (accessed August 7, 2014). Copyright © 2014 Gallup, Inc. All rights reserved. The content is used with permission; however, Gallup retains all rights of republication.

FIGURE 2.7

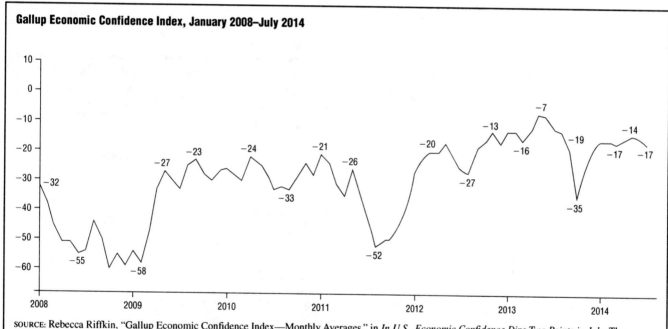

Gallup Economic Confidence Index, January 2008–July 2014

SOURCE: Rebecca Riffkin, "Gallup Economic Confidence Index—Monthly Averages," in *In U.S., Economic Confidence Dips Two Points in July*, The Gallup Organization, August 5, 2014, http://www.gallup.com/poll/174308/economic-confidence-dips-two-points-july.aspx (accessed August 7, 2014). Copyright © 2014 Gallup, Inc. All rights reserved. The content is used with permission; however, Gallup retains all rights of republication.

As noted in Chapter 1, in September and October 2013 the Board of Governors of the Federal Reserve System (the Fed) conducted its Survey of Household Economics and Decisionmaking (SHED). The results were published in *Report on the Economic Well-Being of U.S. Households in 2013* (July 2014, http://www.federalreserve.gov/econresdata/2013-report-economic-well-being-us-households-201407.pdf). The total sample population included 4,134 people whom the Fed says are representative of the U.S. population. The survey participants were asked to rate how well they were "managing financially these days." As shown in Table 2.6, nearly a quarter (23%) said they were "living comfortably," whereas 37.3% said they were "doing okay." Another 25.3% said they were "just getting by," and 13.3% said they were "finding it very difficult to get by." The remaining 1.1% refused to answer the question. The Fed also asked participants to compare their current

FIGURE 2.8

Poll respondents rate their personal financial situations, 2001–14

HOW WOULD YOU RATE YOUR FINANCIAL SITUATION TODAY—AS EXCELLENT, GOOD, ONLY FAIR, OR POOR?

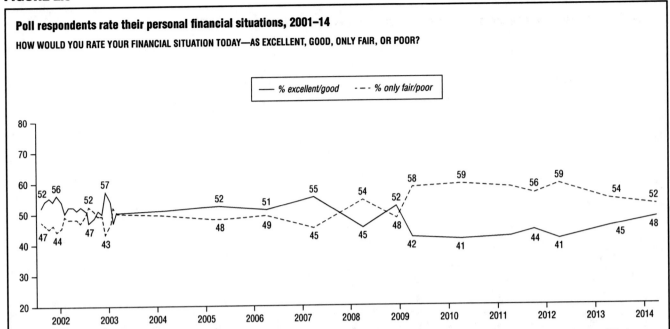

SOURCE: Frank Newport, "How would you rate your financial situation today—as excellent, good, only fair, or poor?" in *Americans' Ratings of Their Personal Finances Inch Up*, The Gallup Organization, April 23, 2014, http://www.gallup.com/poll/168638/americans-ratings-personal-finances-inch.aspx (accessed August 8, 2014). Copyright © 2014 Gallup, Inc. All rights reserved. The content is used with permission; however, Gallup retains all rights of republication.

FIGURE 2.9

Poll respondents rate their financial comfort, 2002–14

RIGHT NOW, DO YOU HAVE ENOUGH MONEY TO LIVE COMFORTABLY, OR NOT?

SOURCE: Frank Newport, "Right now, do you have enough money to live comfortably, or not?" in *Americans' Ratings of Their Personal Finances Inch Up*, The Gallup Organization, April 23, 2014, http://www.gallup.com/poll/168638/americans-ratings-personal-finances-inch.aspx (accessed August 8, 2014). Copyright © 2014 Gallup, Inc. All rights reserved. The content is used with permission; however, Gallup retains all rights of republication.

TABLE 2.5

Poll respondents rate their top financial concerns, April 2014

NEXT, PLEASE TELL ME HOW CONCERNED YOU ARE RIGHT NOW ABOUT EACH OF THE FOLLOWING FINANCIAL MATTERS, BASED ON YOUR CURRENT FINANCIAL SITUATION—ARE YOU VERY WORRIED, MODERATELY WORRIED, NOT TOO WORRIED, OR NOT WORRIED AT ALL?

	Very worried/ moderately worried %	Not too worried/not at all worried %
Not having enough money for retirement	59	35
Not being able to pay medical costs in the event of a serious illness or accident	53	45
Not being able to maintain the standard of living you enjoy	48	52
Not having enough money to pay off your debt	40	48
Not being able to pay medical costs for normal healthcare	39	57
Not having enough money to pay your normal monthly bills	36	62
Not having enough money to pay for your children's college	35	31
Not being able to pay your rent, mortgage, or other housing costs	31	64
Not being able to make the minimum payments on your credit cards	16	65

SOURCE: Andrew Dugan, "Americans' Top Financial Concerns," in *Retirement Remains Americans' Top Financial Worry*, The Gallup Organization, April 22, 2014, http://www.gallup.com/poll/168626/retirement-remains-americans-top-financial-worry.aspx (accessed August 8, 2014). Copyright © 2014 Gallup, Inc. All rights reserved. The content is used with permission; however, Gallup retains all rights of republication.

financial situation to what it was five years ago. (See Figure 2.10.) The responses were roughly evenly split with about a third each indicating they were better off, about the same, or worse off.

Another topic of interest was home values. As explained in Chapter 1, the housing market experienced a sharp downturn during the first decade of the 21st century. This downturn was a major contributing factor to the Great Recession, which lasted from December

TABLE 2.6

Survey respondents rate how well they are managing financially, September 2013

[Percent, except as noted]

Response	Rate
Refused	1.1
Finding it very difficult to get by	13.3
Just getting by	25.3
Doing okay	37.3
Living comfortably	23.0
Number of respondents	4,134

SOURCE: "Table C.53. Which one of the following best describes how well you are managing financially these days?" in *Report on the Economic Well-Being of U.S. Households in 2013*, Board of Governors of the Federal Reserve System, July 2014, http://www.federalreserve.gov/econresdata/2013-report-economic-well-being-us-households-201407.pdf (accessed August 9, 2014)

TABLE 2.7

Survey respondents rate changes in their home values between 2008 and 2013

[Percent, except as noted]

Response	Rate
Refused	0.8
Lower value	45.2
Value has stayed the same	19.5
Higher value	26.9
Don't know	7.5
Number of respondents	2,222

SOURCE: "Table C.22. Compared to Five Years Ago (since 2008), Do You Think the Value of Your Home Today Is Higher, Lower, or Stayed the Same?" in *Report on the Economic Well-Being of U.S. Households in 2013*, Board of Governors of the Federal Reserve System, July 2014, http://www.federalreserve.gov/econresdata/2013-report-econom-ic-well-being-us-households-201407.pdf (accessed August 9, 2014)

FIGURE 2.10

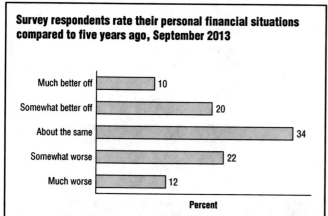

Survey respondents rate their personal financial situations compared to five years ago, September 2013

- Much better off: 10
- Somewhat better off: 20
- About the same: 34
- Somewhat worse: 22
- Much worse: 12

Note: Totals do not add to 100 percent due to rounding and question non-response.

SOURCE: "Figure 1. Compared to five years ago, are you better off, worse off, or the same financially?" in *Report on the Economic Well-Being of U.S. Households in 2013*, Board of Governors of the Federal Reserve System, July 2014, http://www.federalreserve.gov/econresdata/2013-report-economic-well-being-us-households-201407.pdf (accessed August 9, 2014)

2007 through June 2009. Home values were slow to recover even after the recession ended. As shown in Table 2.7, nearly half (45.2%) of the respondents in the survey said their home's value was lower in 2013 than it had been in 2008. Only 26.9% thought their home had increased in value, while 19.5% said the value had stayed the same. Another 7.5% were not sure, and 0.8% refused to answer the question.

The SHED participants were also asked to name their current financial challenges. The results are shown in Figure 2.11 for those respondents naming retirement, education, and jobs. Overall, retirement received the most mentions from Americans aged 55 years and older. Younger respondents were more worried about jobs. This was particularly true for those aged 18 to 24 years or 25 to 34 years. Twenty-one percent of the respondents in each age group named jobs as a current financial challenge.

FIGURE 2.11

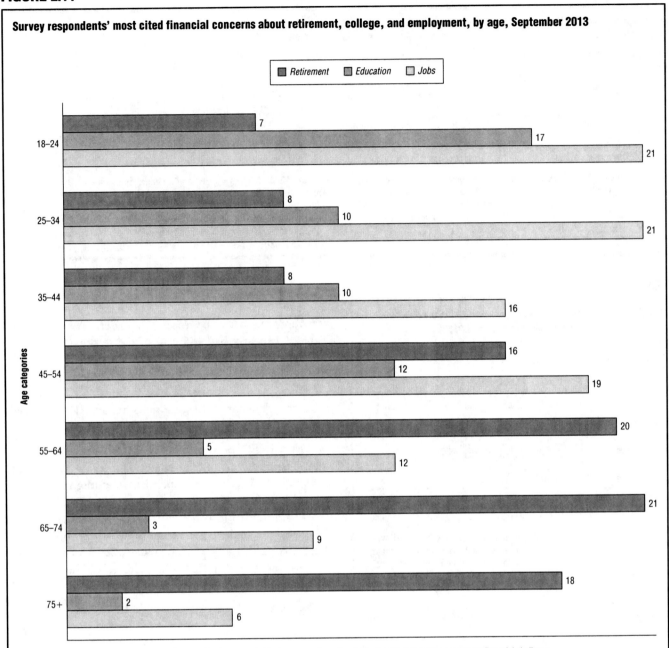

Survey respondents' most cited financial concerns about retirement, college, and employment, by age, September 2013

Note: Unweighted percent of respondents who completed open-ended text response in each cohort who mention topic as a current financial challenge.

SOURCE: "Figure 3. Most Frequently Cited Concerns about Retirement, College, and Employment in Open-Ended Text Responses (by Age)," in *Report on the Economic Well-Being of U.S. Households in 2013*, Board of Governors of the Federal Reserve System, July 2014, http://www.federalreserve.gov/econ-resdata/2013-report-economic-well-being-us-households-201407.pdf (accessed August 9, 2014)

CHAPTER 3
THE AMERICAN CONSUMER

The use of money is all the advantage there is in having money.

—Benjamin Franklin, *Poor Richard's Almanack* (1737)

Americans love to spend money, and their aggressive spending helps fuel both the U.S. and global economies, as imported goods are widely available and popular in the U.S. market. Consumer spending is a major driver of the nation's gross domestic product (GDP), which is the total value of national production.

When consumers spend more money, businesses can earn more profits, which they can use to hire more workers and/or raise worker wages. This, in turn, gives consumers more money to spend. It is a circular process that in good economic times pushes the nation's GDP ever higher. However, the reverse is also true. When consumers spend less money, businesses earn less profit and may lay off workers, freeze hiring, and/or cut wages. These actions give consumers less money to spend, leading to a downward spiral that is reflected in poor GDP performance.

Consumers spend in proportion to their incomes; thus, higher incomes promote greater spending and vice versa. Another factor that affects spending is consumer confidence. People who are optimistic about their future economic conditions tend to spend more freely. Uncertainty and fear about the future have the opposite effect. They spur people to hang on to their money, to save it rather than spend it.

PERSONAL INCOME AND OUTLAYS

The Bureau of Economic Analysis (BEA) is an agency within the U.S. Department of Commerce. The BEA tracks personal income and outlays for the National Income and Product Accounts, which are national economic accounts. In *Concepts and Methods of the U.S. National Income and Product Accounts (Chapters 1–11 and 13)* (November 2014, http://www.bea.gov/national/pdf/chapters1-4.pdf), the agency defines "persons" as including households, nonprofit institutions serving households (NPISHs), private noninsured welfare funds, and private trust funds.

Table 3.1 shows personal income and outlays for 2012, 2013, and the first two quarters of 2014. Note that the quarterly data are seasonally adjusted and annualized. Personal income in 2013 was $14.2 trillion. Employee compensation accounted for 62% of the total at $8.8 trillion. Persons also collected income from noncorporate businesses (including farms) that they operated. This is known as proprietor's income and was $1.3 trillion in 2013. Other sources of income were rental income ($595.8 billion) and receipts on assets ($2.1 trillion). The latter includes interest income (for example, from savings accounts or bonds) and dividend income (e.g., payments received by the stockholders of corporations). Persons also obtained income from transfer receipts. In *State Personal Income 2005 Methodology* (October 2006, http://www.bea.gov/regional/pdf/spi2005/Complete_Methodology.pdf), the BEA explains, "Personal current transfer receipts are benefits received by persons for which no current services are performed." Government social programs, such as Social Security, Medicare, and Medicaid, are the primary sources of transfer receipts.

As shown in Table 3.1 personal contributions (payments) to government social programs and personal tax payments are subtracted from personal income to calculate disposable personal income (DPI). This is the money that persons have available for spending, saving, and investing. In 2013 the national DPI was $12.5 trillion. Nearly all ($11.9 trillion or 95%) of this money went to personal outlays which included personal consumption expenditures (PCE; $11.5 trillion), nonmortgage interest payments ($247.1 billion), and transfer payments ($165.6 billion).

Figure 3.1 shows real (inflation-adjusted) DPI for June 2004 through June 2014. Overall, real DPI grew during this period from about $10 trillion to almost

TABLE 3.1

Personal income and personal outlays, 2012–13 and first and second quarters of 2014

	2012	2013	Seasonally adjusted at annual rates 2014 I	II
Personal income	13,887.7	14,166.9	14,488.3	14,696.3
Compensation of employees	8,606.5	8,844.8	9,100.2	9,235.0
Proprietors' income with inventory valuation and capital consumption adjustments	1,260.2	1,336.6	1,351.0	1,364.1
Rental income of persons with capital consumption adjustment	533.0	595.8	622.9	635.9
Personal income receipts on assets	2,088.6	2,079.7	2,090.4	2,120.7
Personal current transfer receipts	2,350.7	2,414.5	2,470.9	2,504.4
Less: Contributions for government social insurance, domestic	951.2	1,104.5	1,147.0	1,163.7
Less: Personal current taxes	1,503.7	1,661.8	1,712.5	1,727.7
Equals: Disposable personal income	12,384.0	12,505.1	12,775.0	12,968.5
Less: Personal outlays	11,487.9	11,897.1	12,146.9	12,285.7
Personal consumption expenditures	11,083.1	11,484.3	11,728.5	11,867.9
Personal interest payments*	241.6	247.1	249.8	249.7
Personal current transfer payments	163.1	165.6	168.6	168.0

*Consists of nonmortgage interest paid by households.

SOURCE: Adapted from "Table 2. Personal Income and Its Disposition (Years and Quarters)," in *Personal Income and Outlays: June 2014, Revised Estimates: 1999 through May 2014*, U.S. Department of Commerce, Bureau of Economic Analysis, August 1, 2014, http://www.bea.gov/newsreleases/national/pi/2014/pdf/pi0614.pdf (accessed August 19, 2014)

FIGURE 3.1

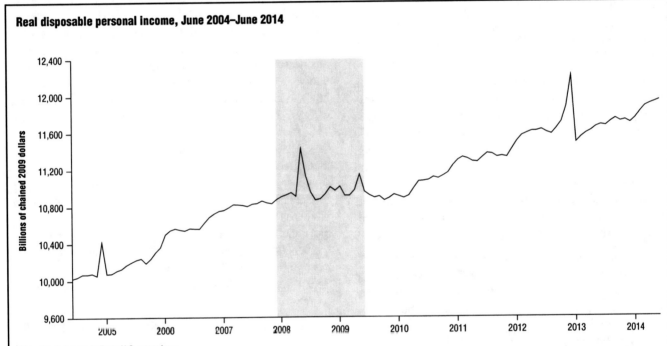

Real disposable personal income, June 2004–June 2014

Note: Shaded areas indicate U.S. recessions.

SOURCE: "Real Disposable Personal Income," in *FRED® Economic Data*, Federal Reserve Bank of St. Louis, August 1, 2014, http://research.stlouisfed.org/fredgraph.pdf?hires=1&type=application/pdf&chart_type=line&recession_bars=on&log_scales=&bgcolor=%23e1e9f0&graph_bgcolor=%23ffffff&fo=verdana&ts=12&tts=12&txtcolor=%23444444&show_legend=yes&show_axis_titles=yes&drp=0&cosd=2004-06-01&coed=2014-06-01&width=670&height=445&stacking=&range=10yrs&mode=fred&id=DSPIC96&transformation=lin&nd=&ost=-99999&oet=99999&scale=left&line_color=%234572a7&line_style=solid&lw=2&mark_type=none&mw=1&mma=0&fml=a&fgst=lin&fq=Monthly&fam=avg&vintage_date=&revision_date= (accessed August 22, 2014)

$12 trillion. There are some noticeable spikes in the DPI graph that reflect unique income events. A spike in mid-2008 coincided with tax rebates provided by the federal government as part of the Economic Stimulus Act of 2008. Another spike in December 2012 was precipitated by fears that the federal government was going to significantly increase the tax on dividends in January 2013. Corporations made early payments of dividends scheduled to be paid in 2013 so that recipients could enjoy the lower 2012 tax rate.

CONSUMER SPENDING

One measure of national consumer spending is provided by the BEA in the Personal Consumption Expenditures (PCE), a major component of the GDP. In *Concepts and Methods of the U.S. National Income and Product Accounts* (November 2014, http://www.bea.gov/national/pdf/chapter5.pdf), the BEA notes that data used to calculate the PCE are collected from numerous sources, including government agencies and trade associations. Primary sources include the U.S. Census Bureau and the U.S. Bureau of Labor Statistics (BLS). The PCE is based on aggregate data (data summed to represent the entire population).

In addition, the BLS publishes the annual Consumer Expenditure Survey (CEX) that estimates the consumer spending of an average U.S. household during a given year. The PCE and the CEX are examined in this chapter.

Personal Consumption Expenditures

The PCE is the largest contributing factor, by far, to the nation's GDP. As shown in Table 2.2 in Chapter 2, the real (inflation-adjusted) GDP was $15.7 trillion in 2013. The PCE totaled $10.7 trillion, or 68% of the GDP.

The PCE primarily covers purchases by households of new goods and services provided by private businesses. However, the following types of spending are also included:

- Purchases of new goods and of services by households from government and government enterprises
- Costs incurred by NPISHs in providing services on behalf of households
- Net purchases of used goods by households
- Purchases abroad of goods and services by U.S. residents traveling, working, or attending school in foreign countries for up to one year
- Expenditures financed by third-party payers on behalf of households, such as employer-paid health insurance and medical care financed through government programs
- Expenses associated with life insurance and with private and government employee pension plans
- Imputed purchases that keep PCE invariant to changes in the way that certain activities are carried out—for example, whether housing is rented or owned or whether employees are paid in cash or in kind

The last item requires some explanation. To "impute" means to attribute. In this context the BEA assigns market values to activities that are not actually market transactions. A simple example given by the BEA concerns a free meal that an employer provides to an employee. The BEA would estimate the market value of the meal to the employee and include it in the PCE as an imputed purchase. In a much more complicated calculation the agency also imputes the "space rent of nonfarm owner-occupied housing." This basically means that the BEA estimates how much in rent would have been paid by a homeowner-occupier to rent space of similar type and size to the owned home. The values of numerous other nonmarket transactions are also imputed by the BEA.

As shown in Table 3.2, the nation's real PCE grew from $10 trillion in 2010 to $10.7 trillion in 2013. The annualized values for the first and second quarter of 2014 are also shown in Table 3.2. These computed values predict the total PCE for 2014 by assuming that spending continues at the quarterly rate for the rest of the year.

PCE COMPONENTS. Major PCE categories include durable goods, nondurable goods, and services. (See Table 3.2.) In *Concepts and Methods of the U.S. National Income and Product Accounts*, the BEA defines goods as "tangible commodities that can be stored or inventoried." Durable goods are goods that have an "average useful life" of at least three years. Thus, nondurable goods are goods with an "average usual life" of less than three years. Services are defined as "commodities that cannot be stored or inventoried and that are usually consumed at the place and time of purchase." Note that housing is listed under services. This category includes rent paid by renters and the imputed cost for homeowners-occupiers.

Services accounted for the largest major category of the real PCE in 2013, totaling nearly $7.1 trillion. (See Table 3.2.) Housing and utilities ($2 trillion) and health care ($1.8 trillion) were the two largest components of services spending. Americans spent $2.3 trillion on nondurable goods in 2013. Nearly one-third ($809 billion) of this amount was devoted to food and beverages purchased for consumption elsewhere. Another $1.3 trillion was spent on durable goods, such as motor vehicles and furniture. Note that many PCE values are based on industry information. In other words, the BEA estimates consumer spending by calculating the final value of goods and services sold by businesses.

Table 3.3 shows the percent change from the preceding year for major PCE categories for 1994 through 2013. Overall, the durable goods PCE shows the largest year-to-year changes over the period. For example, from 2012 to 2013 the durable goods PCE increased by 6.7%. This compares to 1.9% for nondurable goods and 1.9% for services.

As shown in Table 3.2, durable goods, nondurable goods, housing and utilities, and health care services have historically been the four largest single components of the PCE. It is interesting to see how expenditures for these components have changed over several decades. Figure 3.2 shows spending on each component as a percentage of the total PCE for 1973, 1983, 1993, 2003, and 2013. The percentage of the PCE dedicated to housing and utilities remained relatively steady between

TABLE 3.2

Personal consumption expenditures, by major type of good and service, 2010–13 and first and second quarters 2014

	2010	2011	2012	2013	2014 I	2014 II
Personal consumption expenditures	10,036.3	10,263.5	10,449.7	10,699.7	10,844.3	10,910.5
Goods	3,308.7	3,411.8	3,506.5	3,626.0	3,678.3	3,733.7
Durable goods	1,085.7	1,151.5	1,235.7	1,319.0	1,355.0	1,400.2
Motor vehicles and parts	323.4	333.8	357.9	376.0	385.7	401.6
Furnishings and durable household equipment	261.5	276.6	288.4	305.1	312.7	322.5
Recreational goods and vehicles	336.8	370.2	410.8	452.0	468.3	483.2
Other durable goods	164.9	173.9	183.5	194.4	198.1	202.4
Nondurable goods	2,223.5	2,263.2	2,280.1	2,322.6	2,341.9	2,356.7
Food and beverages purchased for off-premises consumption	786.5	795.1	801.6	809.4	811.9	808.9
Clothing and footwear	322.7	335.3	337.7	341.2	338.3	344.0
Gasoline and other energy goods	282.2	274.3	269.2	271.7	274.4	271.8
Other nondurable goods	833.0	863.2	879.7	909.9	926.8	943.4
Services	6,727.6	6,851.4	6,942.4	7,073.1	7,165.4	7,177.3
Household consumption expenditures (for services)	6,449.3	6,575.9	6,653.4	6,772.5	6,857.1	6,863.0
Housing and utilities	1,904.3	1,928.0	1,940.4	1,965.7	1,996.0	1,979.7
Health care	1,649.2	1,690.3	1,745.0	1,781.1	1,798.4	1,801.4
Transportation services	287.1	294.1	299.6	307.6	311.4	313.2
Recreation services	381.0	389.6	396.1	405.7	406.2	409.0
Food services and accommodations	609.6	625.3	641.1	655.4	661.5	668.2
Financial services and insurance	733.9	747.2	713.7	728.5	746.2	748.2
Other services	884.3	901.2	918.1	928.9	937.3	943.0
Final consumption expenditures of nonprofit institutions serving households	278.3	275.2	289.3	301.2	309.1	315.5
Gross output of nonprofit institutions	1,086.0	1,097.1	1,131.2	1,151.4	1,159.4	1,161.5
Less: receipts from sales of goods and services by nonprofit institutions	807.7	821.6	842.0	850.8	851.3	847.5

Note: Users are cautioned that particularly for components that exhibit rapid change in prices relative to other prices in the economy, the chained-dollar estimates should not be used to measure the component's relative importance or its contribution to the growth rate of more aggregate series. Values are in billions of chained (2009) dollars. Quarterly data are seasonally adjusted at annual rates.

SOURCE: Adapted from "Table 3B. Real Gross Domestic Product and Related Measures," in *National Income and Product Accounts—Gross Domestic Product: Second Quarter 2014 (Advance Estimate) Annual Revision: 1999 through First Quarter 2014*, U.S. Department of Commerce, Bureau of Economic Analysis, July 30, 2014, http://www.bea.gov/newsreleases/national/gdp/2014/pdf/gdp2q14_adv.pdf (accessed August 7, 2014)

1973 and 2013. The percentage spent on durable goods slightly decreased over time. A much more dramatic decline is seen in the percentage of the PCE devoted to nondurable goods—from 33.6% in 1973 to 22.7% in 2013. However, the percentage of the PCE devoted to health care services more than doubled from 7.9% in 1973 to 16.9% in 2013.

Consumer Expenditure Survey

Consumer Expenditure Survey (CEX) data are compiled by the BLS based on consumer-supplied information. Purchase diaries are sent to sample households throughout the country. Participants record their everyday purchases and expenses in the diaries. Periodic interviews are conducted to collect diary information and quiz participants about their finances and spending habits. CEX data are collected from households (called "consumer units") that are representative of the civilian noninstitutional population of the United States (i.e., people not in the military and not in institutions, such as prisons or long-term care facilities).

According to the BLS, the average U.S. household spent $51,442 in 2012. (See Table 3.4.) The largest components were housing ($16,887), transportation ($8,998), and food ($6,599). It should be noted that the housing component does not include mortgage principal payments because they are considered to be repayment of a loan, rather than a consumer expense. Thus, CEX data underreport the true cost of housing for Americans with mortgages.

The BLS notes that the average consumer unit in 2013 included 2.5 people and had a pretax annual income of $65,596. (See Table 3.4.) Most (64%) of the consumer units owned their home.

The CEX results provide detailed data on American spending habits. For example, participants break down food purchases as to meal location (at home or away from home). The survey indicates that in 2012 the average household spent $3,921 on food at home and $2,678 eating out. (See Table 3.4.) Each household averaged $109 per year for reading materials and $1,207 per year for educational expenses. Another $451 per year was spent on alcohol and $332 per year on tobacco products and smoking supplies. The average U.S. household reported spending $3,556 on health care and donating $1,913 in cash to charitable causes.

TABLE 3.3

Percentage change in personal consumption expenditures, 1994–2013

	1994	1995	1996	1997	1998	1999	2000	2001	2002	2003	2004	2005	2006	2007	2008	2009	2010	2011	2012	2013
Personal consumption expenditures	3.9	3.0	3.5	3.8	5.3	5.3	5.1	2.6	2.6	3.1	3.8	3.5	3.0	2.2	−0.3	−1.6	1.9	2.3	1.8	2.4
Goods	5.3	3.0	4.5	4.8	6.7	7.9	5.2	3.0	3.9	4.8	5.1	4.1	3.6	2.7	−2.5	−3.0	3.4	3.1	2.8	3.4
Durable goods	8.0	3.9	7.5	8.2	12.1	12.8	8.6	5.2	7.3	7.1	8.2	5.4	4.3	4.6	−5.1	−5.5	6.1	6.1	7.3	6.7
Nondurable goods	3.9	2.5	2.9	2.9	3.7	5.0	3.2	1.7	1.9	3.5	3.3	3.3	3.3	1.7	−1.1	−1.8	2.2	1.8	0.7	1.9
Services	3.1	3.0	2.9	3.2	4.6	3.9	5.0	2.4	1.9	2.2	3.2	3.2	2.7	2.0	0.8	−0.9	1.2	1.8	1.3	1.9

Note: Food excludes personal consumption expenditures for purchased meals and beverages, which are classified in food services.

SOURCE: Adapted from "Table 7. Real Gross Domestic Product: Percent Change from Preceding Year," in *National Income and Product Accounts—Gross Domestic Product: Second Quarter 2014 (Advance Estimate) Annual Revision: 1999 through First Quarter 2014*, U.S. Department of Commerce, Bureau of Economic Analysis, July 30, 2014, http://www.bea.gov/newsreleases/national/gdp/2014/pdf/gdp2q14_adv.pdf (accessed August 7, 2014)

FIGURE 3.2

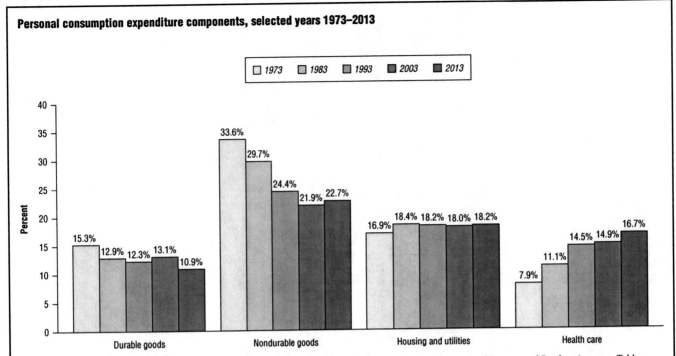

SOURCE: Adapted from "Table 2.3.5. Personal Consumption Expenditures by Major Type of Product," in *National Income and Product Accounts Tables*, U.S. Department of Commerce, Bureau of Economic Analysis, July 30, 2014, http://www.bea.gov/iTable/iTable.cfm?ReqID=9&step=1#reqid=9&step= 3&isuri=1&904=1970&903=65&906=a&905=2013&910=x&911=0 (accessed August 11, 2014)

PRICE INFLATION

As noted in previous chapters, inflation occurs when demand outpaces supply. Price inflation is a well-known phenomenon to consumers. Over time the prices of most goods and services increase. There are various methods used to calculate inflation rates. Two methods will be discussed in this chapter: the consumer price index (CPI) and the PCE index. Both methods determine price changes in a "basket" of goods and services. Each basket contains items such as apparel, beverages, communication, education, food, haircuts, housing, medical care, recreation, and transportation.

An index is a useful tool for comparing changes over time. The index for the basket price for a selected period is arbitrarily set to a value, such as 100. This is the reference index. All other basket prices are compared with the reference index using this equation: index = (basket price/reference basket price) × 100. Therefore, a graph of index data over time does not show the actual prices paid for the baskets but a series of index numbers that are useful for determining price changes over time.

Consumer Price Index

The CPI provides a measure of price changes in consumer goods and services over a specific period. Details about how the CPI is calculated are provided by the BLS in "Consumer Price Index: Frequently Asked Questions" (September 17, 2014, http://stats.bls.gov/cpi/cpifaq.htm). The agency notes, "The CPI market basket is developed from detailed expenditure information provided by families and individuals on what they actually bought. For the current CPI, this information was collected from the Consumer Expenditure Surveys for 2011 and 2012." The "basket" considered by the BLS contains more than 200 item categories. The CES data are used to determine the relative weight (importance) of each item category in the basket. For example, households probably buy more bread than limes. Then, the BLS collects price data from thousands of businesses selling the basket items and uses the weight values to obtain a total basket price. This price includes sales and excise taxes paid on purchased items. Payments for income taxes and investments, such as stocks and bonds, are not included.

The BLS calculates national CPI values for two reference populations: all urban consumers (CPI-U) and urban wage earners and clerical workers (CPI-W). The latter population is a subset of all urban consumers. In "Consumer Price Index: Frequently Asked Questions," the BLS notes that the CPI-U represents the buying habits of approximately 87% of the U.S. population. The CPI-W represents approximately 32% of the U.S. population. In addition, the agency publishes CPI data for specific regions of the United States and for dozens of metropolitan areas.

CPI values are published monthly. The percent changes in prices between two dates can be determined based on the

TABLE 3.4

Consumer expenditure survey results, 2012

Item	2012
Number of consumer units (in thousands)	124,416
Consumer unit characteristics:	
Income before taxes[a]	$65,596
Income after taxes[a]	63,370
Age of reference person	50.0
Average number in consumer unit:	
Persons	2.5
Children under 18	0.6
Persons 65 and older	0.3
Earners	1.3
Vehicles	1.9
Percent distribution:	
Sex of reference person:	
Male	47
Female	53
Housing tenure:	
Home owner	64
With mortgage	39
Without mortgage	26
Renter	36
Race of reference person:	
Black or African-American	13
White, Asian, and all other races	87
Hispanic or Latino origin of reference person:	
Hispanic or Latino	13
Not Hispanic or Latino	87
Education of reference person:	
Elementary (1–8)	4
High school (9–12)	33
College	62
Never attended and other	0
At least one vehicle owned or leased	88
Average annual expenditures	$51,442
Food	6,599
Food at home	3,921
Cereals and bakery products	538
Cereals and cereal products	182
Bakery products	356
Meats, poultry, fish, and eggs	852
Beef	226
Pork	166
Other meats	122
Poultry	159
Fish and seafood	126
Eggs	53
Dairy products	419
Fresh milk and cream	152
Other dairy products	267
Fruits and vegetables	731
Fresh fruits	261
Fresh vegetables	226
Processed fruits	114
Processed vegetables	130
Other food at home	1,380
Sugar and other sweets	147
Fats and oils	114
Miscellaneous foods	699
Nonalcoholic beverages	370
Food prepared by consumer unit on out-of-town trips	50
Food away from home	2,678
Alcoholic beverages	451

TABLE 3.4

Consumer expenditure survey results, 2012 [CONTINUED]

Item	2012
Housing	16,887
Shelter	9,891
Owned dwellings	6,056
Mortgage interest and charges	3,067
Property taxes	1,836
Maintenance, repairs, insurance, other expenses	1,153
Rented dwellings	3,186
Other lodging	649
Utilities, fuels, and public services	3,648
Natural gas	359
Electricity	1,388
Fuel oil and other fuels	137
Telephone services	1,239
Water and other public services	525
Household operations	1,159
Personal services	368
Other household expenses	791
Housekeeping supplies	610
Laundry and cleaning supplies	155
Other household products	319
Postage and stationery	136
Household furnishings and equipment	1,580
Household textiles	123
Furniture	391
Floor coverings	16
Major appliances	197
Small appliances, miscellaneous housewares	98
Miscellaneous household equipment	754
Apparel and services	1,736
Men and boys	408
Men, 16 and over	320
Boys, 2 to 15	88
Women and girls	688
Women, 16 and over	573
Girls, 2 to 15	116
Children under 2	63
Footwear	347
Other apparel products and services	230
Transportation	8,998
Vehicle purchases (net outlay)	3,210
Cars and trucks, new	1,639
Cars and trucks, used	1,516
Other vehicles	56
Gasoline and motor oil	2,756
Other vehicle expenses	2,490
Vehicle finance charges	223
Maintenance and repairs	814
Vehicle insurance	1,018
Vehicle rental, leases, licenses, and other charges	434
Public and other transportation	542
Health care	3,556
Health insurance	2,061
Medical services	839
Drugs	515
Medical supplies	142
Entertainment	2,605
Fees and admissions	614
Audio and visual equipment and services[b]	979
Pets, toys, hobbies, and playground equipment	648
Other entertainment supplies, equipment, and services	363
Personal care products and services	628
Reading	109
Education	1,207
Tobacco products and smoking supplies	332

difference in CPI values. For example, according to the BLS, in "CPI Detailed Report: Data for July 2014" (August 19, 2014, http://www.bls.gov/cpi/cpid1407.pdf), the CPI-U was 148.4 in July 1994. By July 2014 it had risen to 238.3, a difference of 89.9 index points. Dividing 89.9 by 148.4 and multiplying by 100 provides a percentage difference of 60.6%; thus, on average, prices increased by 60.6% between July 1994 and July 2014.

TABLE 3.4

Consumer expenditure survey results, 2012 [CONTINUED]

Item	2012
Miscellaneous	829
Cash contributions	1,913
Personal insurance and pensions	**5,591**
Life and other personal insurance	353
Pensions and Social Security	5,238
Sources of income and personal taxes:	
Money income before taxes[a]	$65,596
Wages and salaries	51,730
Self-employment income	2,917
Social Security, private and government retirement	8,021
Interest, dividends, rental income, other property income	1,358
Unemployment and workers' compensation, veterans' benefits	428
Public assistance, supplemental security income, food stamps	534
Regular contributions for support	380
Other income	229
Personal taxes (missing values not imputed)[a]	2,226
Federal income taxes	1,568
2008 Tax stimulus	n.a.
State and local income taxes	526
Other taxes	132
Income after taxes[a]	63,370
Addenda:	
Net change in total assets and liabilities	−$5,092
Net change in total assets	5,073
Net change in total liabilities	10,165
Other financial information:	
Other money receipts	712
Mortgage principal paid on owned property	−1,935
Estimated market value of owned home	149,574
Estimated monthly rental value of owned home	869
Gifts of goods and services, total	**1,116**
Food	104
Alcoholic beverages[c]	12
Housing	191
Housekeeping supplies	29
Household textiles	10
Appliances and miscellaneous housewares	18
Major appliances	6
Small appliances and miscellaneous housewares	12
Miscellaneous household equipment	48
Other housing	86
Apparel and services	215
Males, 2 and over	53
Females, 2 and over	86
Children under 2	20
Other apparel products and services	56
Jewelry and watches	24
All other apparel products and services	32
Transportation	111
Health care	53
Entertainment	**82**
Toys, games, arts and crafts, and tricycles	28
Other entertainment	54
Personal care products and services[c]	**16**
Reading[c]	**2**
Education	**260**
All other gifts[c]	**71**

n.a. = Not applicable.
[a] Components of income and taxes are derived from "complete income reporters" only through 2003. Beginning in 2004 income imputation was implemented. As a result, all consumer units are considered to be complete income reporters.
[b] Prior to 2005, the title of "Audio and visual equipment and services" was "Televisions, radio, sound equipment."
[c] Prior to 2000, gifts of "Alcoholic beverages," "Personal care products and services," and "Rading materials" were included in "All other gifts."
Note: All values have been rounded, and as a result some cell values have been rounded to zero. This is particularly evident in the characteristic section. When data are not reported or are not applicable (i.e., missing values), tabulated cell values have been set to zero.

SOURCE: Adapted from "Average Annual Expenditures and Characteristics of All Consumer Units, Consumer Expenditure Survey, 2006–2012," in *Average Annual Expenditures and Characteristics of All Consumer Units, Consumer Expenditure Survey, 2006–2012*, U.S. Department of Labor, Bureau of LaborStatistics, May 23, 2014, http://www.bls.gov/cex/2012/standard/multiyr.pdf (accessed August 9, 2014)

INFLATION RATES BASED ON THE CPI. The CPI provides economists with a tool for quantifying inflation rates. As noted in Chapter 2, small increases in inflation (up to 3% per year) are considered a sign of a healthy growing economy in which demand slightly outpaces supply. Larger inflation rates can be problematic because consumers find it difficult to afford the goods and services that they want to buy.

Figure 3.3 shows the average annual CPI-U between 1913 and 2013. It uses the period of 1982 to 1984 as the reference period for which the CPI value is arbitrarily set to 100. The inflation rate between two years can be calculated based on the annual average CPI for each year or on a December-to-December basis. The annual average CPI for a particular year is simply the average of the 12 monthly CPI values. Figure 3.4 shows the average annual inflation rate for the U.S. economy for each year between 1913 and 2013 based on annual average CPI-U data. The inflation rate for 2013 was 1.5%. This means that, on average, the price of consumer goods and services increased by 1.5% between 2012 and 2013. It also means that the purchasing power of a dollar decreased by 1.5% during this period. For example, an item that cost $100 in 2012 cost $101.50 in 2013.

Table 3.5 lists the annual percent change in the CPI-U between 2006 and 2013 on a December-to-December basis for "all items" and for specific item categories in the CPI basket. The "all items" values represent overall inflation rates. The inflation rate for 2013 on a December-to-December basis was 1.5%, the same value calculated using the annual average method. The "all items" inflation rate was exceeded in 2013 by the inflation rates for housing (2.2%), medical care (2%), other goods and services (1.8%), and education and communication (1.6%). In other words, prices grew for these items at faster rates during 2013 than did the prices for the overall basket. The inflation rates for these items are said to have outpaced the overall inflation rate for that year.

As shown in Table 3.5, the inflation rates for many items were negative during 2008 or 2009. The United

FIGURE 3.3

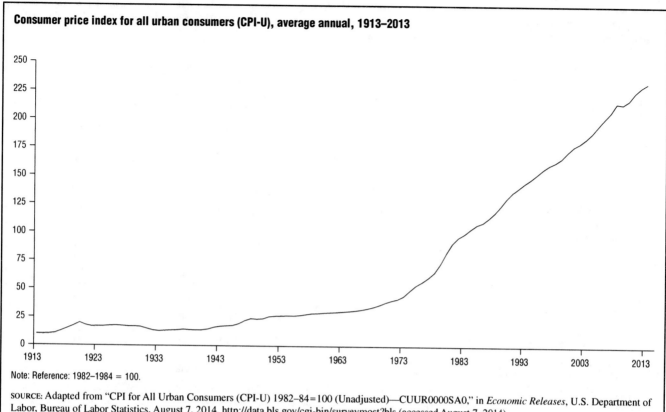

Consumer price index for all urban consumers (CPI-U), average annual, 1913–2013

Note: Reference: 1982–1984 = 100.

SOURCE: Adapted from "CPI for All Urban Consumers (CPI-U) 1982–84=100 (Unadjusted)—CUUR0000SA0," in *Economic Releases*, U.S. Department of Labor, Bureau of Labor Statistics, August 7, 2014, http://data.bls.gov/cgi-bin/surveymost?bls (accessed August 7, 2014)

FIGURE 3.4

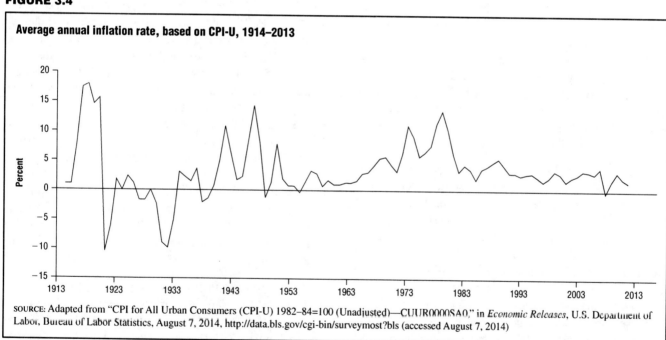

Average annual inflation rate, based on CPI-U, 1914–2013

SOURCE: Adapted from "CPI for All Urban Consumers (CPI-U) 1982–84=100 (Unadjusted)—CUUR0000SA0," in *Economic Releases*, U.S. Department of Labor, Bureau of Labor Statistics, August 7, 2014, http://data.bls.gov/cgi-bin/surveymost?bls (accessed August 7, 2014)

States endured an economic downturn called the Great Recession from December 2007 through June 2009. Consumers decreased their spending during this period, and lower demand pushed prices downward for many goods and services. Negative inflation is called deflation (or disinflation) and is relatively rare in U.S. history.

Over the seven-year period depicted in Table 3.5, some items in the CPI basket experienced very volatile inflation rates. This is particularly true for energy. Its annual inflation rate varied from a high of 18.2% to a low of −21.3%. Historically food and energy prices have shown considerable volatility. Food prices are dependent

TABLE 3.5

Percentage change in the urban Consumer Price Index (CPI-U) for selected expenditure categories, December 2006–December 2013, December 2013–June 2014

	Percent change from previous December								
	December								
Item and group	2006	2007	2008	2009	2010	2011	2012	2013	June 2014
Expenditure category									
All items	2.5	4.1	0.1	2.7	1.5	3.0	1.7	1.5	2.3
Food and beverages	2.2	4.8	5.8	−0.4	1.5	4.5	1.8	1.1	1.8
Housing	3.3	3.0	2.4	−0.3	0.3	1.9	1.7	2.2	2.2
Apparel	0.9	−0.3	−1.0	1.9	−1.1	4.6	1.8	0.6	0.7
Transportation	1.6	8.3	−13.3	14.4	5.3	5.2	1.6	0.5	5.0
Medical care	3.6	5.2	2.6	3.4	3.3	3.5	3.2	2.0	1.9
Recreation*	1.0	0.8	1.8	−0.4	−0.8	1.0	0.8	0.4	1.0
Education and communication*	2.3	3.0	3.6	2.4	1.3	1.7	1.5	1.6	0.3
Other goods and services	3.0	3.3	3.4	8.0	1.9	1.7	1.5	1.8	1.0
Special aggregate indexes									
Commodities	1.3	5.2	−4.1	5.5	2.0	4.2	1.0	0.2	2.5
Services	3.4	3.3	3.0	0.9	1.2	2.2	2.2	2.3	2.1
Energy	2.9	17.4	−21.3	18.2	7.7	6.6	0.5	0.5	10.8
All items less energy	2.5	2.8	2.4	1.4	0.9	2.6	1.9	1.6	1.4
All items less food and energy	2.6	2.4	1.8	1.8	0.8	2.2	1.9	1.7	1.3
Commodities less food and energy commodities	−0.1	0.1	−0.6	3.0	−0.4	2.2	0.3	−0.1	0.6
Energy commodities	6.1	29.4	−40.5	46.5	13.9	10.6	1.5	−0.8	11.6
Services less energy services	3.7	3.3	2.7	1.4	1.3	2.3	2.5	2.3	1.6
Domestically produced farm food	1.2	6.0	6.5	−2.7	2.0	6.0	1.5	0.5	2.6

*Indexes on a December 1997 = 100 base.
Notes: Index applies to a month as a whole, not to any specific date. All other item stratum index series were calculated using a geometric means estimator.

SOURCE: Adapted from "Table 26. Historical Consumer Price Index for All Urban Consumers (CPI-U): U.S. City Average, by Commodity and Service Group and Detailed Expenditure Categories," in *CPI Detailed Report (Tables 1–29) June 2014*, U.S. Department of Labor, Bureau of Labor Statistics, July 22, 2014, http://www.bls.gov/cpi/cpid1406.pdf (accessed August 9, 2014)

on a variety of factors, including weather conditions (which affect growing costs), transportation and processing costs, and subsidies paid to farmers by the government that influence supply and demand ratios. Energy prices, particularly for oil, are affected by political and economic factors in the Middle East. Economists calculate a "core" inflation rate that excludes food and energy because of the volatile nature of their prices. As shown in Table 3.5 the core inflation rate for 2013 was 1.7%.

Inflation rates based on the CPI are commonly reported in the media. The rates are also important benchmarks used by the government and some businesses to raise certain payments or wages to keep pace with inflation. For example, in "Latest Cost-of-Living Adjustment" (2014, http://www.ssa.gov/oact/cola/latestCOLA.html), the Social Security Administration notes that Social Security payments are subject to cost-of-living adjustments that are based on changes in the CPI-W. Likewise, labor contracts negotiated by unions typically feature an escalation (or escalator) clause that requires wages to rise in the future in accordance with CPI values.

Inflation Rates Based on the PCE Index

Another method for measuring price inflation is based on a PCE index calculated by the BEA. Some economists argue that the PCE index provides a more accurate depiction of inflation than does the CPI. In "I Say CPI, You Say PCE" (*Region*, vol. 25, no. 4, December 2011), Phil Davies of the Federal Reserve Bank of Minneapolis describes the differences between the two indexes. The PCE index, unlike the CPI, captures the spending of rural consumers and NPISHs. It also includes medical care expenses paid by businesses (e.g., employer-subsidized health insurance) and the government (e.g., Medicare payments). The PCE index also accounts for consumer substitutions in reaction to rising prices. For example, if the price of beef increases, then consumers may purchase chicken instead. Davies notes that the Federal Reserve (the nation's central banking system) and "many economists" prefer the PCE index over the CPI. In fact, in 2000 the Federal Reserve switched from the CPI to the PCE index for its inflation measuring tool.

In "PCE and CPI Inflation: What's the Difference?" (April 17, 2014, http://www.clevelandfed.org/research/trends/2014/0514/01infpri.cfm), Joseph G. Haubrich and Sara Millington note that the CPI "tends to report somewhat higher inflation" than does the PCE index. Historical data indicate that the CPI-based rate is about one-half of a percentage point higher than the PCE-based rate. According to Haubrich and Millington, from 2000 through April 2014 the average annual rate of inflation as measured with the CPI was 2.4% compared with 1.9% as measured with the PCE index.

As of 2014 the Federal Reserve had a target annual inflation rate of 2% for the U.S. economy. In "Federal Reserve Formally Targets 2% U.S. Inflation" (Forbes.com, January 25, 2012), Steve Schaefer explains that the Federal Reserve officially announced the target rate in January 2012. The nation's central bank indirectly affects inflation rates by manipulating interest rates. Low interest rates tend to increase demand for goods and services, which can push prices upward. By contrast, high interest rates tend to dampen demand, which can push prices downward.

HEALTH CARE SPENDING AND PRICES

Chapter 1 examines the nation's health spending which totaled $2.8 trillion in 2012. These expenditures included amounts paid by consumers, businesses, and the government for health insurance, medical goods and services, investments in the medical system, research, public health, and other purposes. As shown in Figure 1.14 in Chapter 1, national health expenditures grew from 5% of GDP in 1960 to 17.2% of GDP in 2012. It is difficult to assign a specific proportion of the nation's health spending to consumers. In many cases employers and government programs, such as Medicare and Medicaid, cover some of the costs.

The BEA includes health care expenditures in the PCE, as shown in Table 3.6. These amounts include payments made on behalf of households by businesses and the government. Health care PCE fall into two broad categories: health and net health insurance (i.e., premiums minus benefits). In 2013 nearly $2.4 trillion was spent on health care goods and services (excluding health insurance). Hospital services ($893 billion) and outpatient physician services ($456.3 billion) were the most expensive components. As noted earlier, the PCE proportion devoted to health care services more than doubled from 7.9% in 1973 to 16.7% in 2013. (See Figure 3.2.) The BEA calculates that $145.1 billion was spent on net health insurance during 2013.

Because the PCE includes expenditures made on behalf of households by businesses and the government, the PCE index is very useful for measuring medical inflation rates. Table 3.7 shows PCE index values for selected years between 1960 and 2013. Note that the index value is 100 for the reference year of 2009. The index for health changed from about 75.7 in 2000 to approximately 108.2 in 2013, a difference of 32.5. Dividing the latter value by 75.7 and multiplying by 100 provides an inflation rate of 42.9%; thus, on average, prices for health care goods and services (excluding health insurance) increased by 42.9% between 2000 and 2013.

CPI values reflect prices experienced by urban households only. As noted earlier and shown in Table 3.5, the medical care inflation rate based on the CPI-U

TABLE 3.6

Personal consumption expenditures for health care and health insurance, 2013

[Billions of dollars]

	2013
Health	2,372.1
Medical products, appliances, and equipment	451.8
Pharmaceutical and other medical products[a]	388.6
Pharmaceutical products	383.9
Other medical products	4.7
Therapeutic appliances and equipment	63.2
Outpatient services	864.5
Physician services[b]	456.3
Dental services	109.3
Paramedical services	298.9
Home health care	89.4
Medical laboratories	33.8
Other professional medical services[c]	175.7
Hospital and nursing home services	1,055.8
Hospitals[d]	893
Nursing homes	162.8
Net health insurance	145.1
Medical care and hospitalization[e]	122.9
Income loss[f]	2.9
Workers' compensation[g]	19.3

[a]Excludes drug preparations and related products dispensed by physicians, hospitals, and other medical services.
[b]Consists of offices of physicians, health maintenance organization medical centers, and freestanding ambulatory surgical and emergency centers.
[c]Includes podiatrists, chiropractors, mental health practitioners (except physicians), physical, occupational and speech therapists, audiologists, all other health practitioners, ambulance services, kidney dialysis centers, and blood and organ bank services.
[d]Consists of nonprofit hospitals, proprietary hospitals, and government hospitals. Consists of primary sales of these hospitals for personal consumption. Expenses of nonprofit hospitals are included in the expenditures of nonprofit institutions serving households (NPISHs).
[e]Consists of premiums less benefits for health, hospitalization, and accidental death and dismemberment insurance.
[f]Consists of premiums less benefits for income loss insurance.
[g]Consists of premiums plus premium supplements less normal losses and dividends paid to policyholders for privately administered workers' compensation.

SOURCE: Adapted from "Table 2.5.5. Personal Consumption Expenditures by Function," in *National Income and Product Accounts Tables*, U.S. Department of Commerce, Bureau of Economic Analysis, August 5, 2014, http://www.bea.gov/iTable/iTable.cfm?ReqID=9&step=1#reqid=9&step=3&isuri=1&903=74 (accessed August 11, 2014)

outpaced the "all items" rate for 2006 through 2013. Table 3.8 provides a breakdown of the annual rates for medical care components. Inpatient and outpatient hospital services had the highest rates in most years. In 2013 they had inflation rates of 4.4% and 3.8%, respectively. Health insurance prices showed significant variability between 2006 and 2013. In 2013 the health insurance inflation rate was 0.9%. Historical data indicate that medical care inflation as measured by the CPI-U has outpaced "all items" inflation in most years dating back to 1945. (See Table 3.9.)

Increased spending and prices for health care have financially stressed American consumers. In *Report on the Economic Well-Being of U.S. Households in 2013* (July 2014, http://www.federalreserve.gov/econresdata/2013-report-economic-well-being-us-households-201407.pdf), the Federal Reserve reports the results obtained in 2013 from its Survey of Household Economics and

TABLE 3.7

Price indexes for personal consumption expenditures for health care and health insurance, selected years, 1960–2013

[Index numbers, 2009 = 100]

	1960	1970	1980	1990	2000	2010	2011	2012	2013
Health	7.794	11.197	23.730	52.136	75.733	102.610	104.736	106.837	108.187
Medical products, appliances, and equipment	17.966	17.782	28.546	56.583	77.351	103.013	106.203	109.253	109.795
Pharmaceutical and other medical products[a]	17.217	16.288	25.739	54.107	75.459	103.628	107.272	110.681	111.234
Pharmaceutical products	17.090	16.142	25.511	53.867	75.257	103.699	107.386	110.824	111.379
Other medical products	28.265	28.017	44.126	75.937	94.669	97.895	98.162	99.395	99.816
Therapeutic appliances and equipment	21.762	25.713	43.909	70.752	89.228	99.369	99.977	101.038	101.515
Outpatient services	8.499	13.177	28.495	56.920	80.725	102.197	103.617	104.867	105.711
Physician services[b]	8.225	12.969	28.728	60.404	84.908	102.446	104.011	105.235	105.427
Dental services	6.989	10.182	20.514	40.377	66.623	102.743	105.095	107.528	111.207
Paramedical services	9.980	15.712	34.167	60.228	80.871	101.591	102.428	103.276	104.097
Home health care	—	—	—	63.446	86.797	101.129	101.125	101.944	101.814
Medical laboratories	—	—	—	66.811	85.885	99.960	100.329	100.378	98.986
Other professional medical services[c]	—	—	—	57.681	77.175	102.191	103.566	104.601	106.380
Hospital and nursing home services	5.474	8.624	19.441	47.078	70.826	102.786	105.056	107.472	109.590
Hospitals[d]	5.154	8.116	18.550	47.670	70.988	102.949	105.181	107.784	110.142
Nursing homes	7.744	12.233	25.741	44.991	70.130	101.969	104.431	105.889	106.737
Net health insurance	5.822	9.053	16.546	41.707	72.048	104.314	107.876	110.179	111.744
Medical care and hospitalization[e]	3.724	6.698	13.458	40.665	67.556	105.394	109.730	112.502	114.364
Income loss[f]	61.992	76.009	112.264	289.964	80.282	101.642	104.850	107.030	108.596
Workers' compensation[g]	10.525	16.195	25.217	35.853	97.003	98.472	97.111	96.621	96.535

[a]Excludes drug preparations and related products dispensed by physicians, hospitals, and other medical services.
[b]Consists of offices of physicians, health maintenance organization medical centers, and freestanding ambulatory surgical and emergency centers.
[c]Includes podiatrists, chiropractors, mental health practitioners (except physicians), physical, occupational and speech therapists, audiologists, all other health practitioners, ambulance services, kidney dialysis centers, and blood and organ bank services.
[d]Consists of nonprofit hospitals, proprietary hospitals, and government hospitals. Consists of primary sales of these hospitals for personal consumption. Expenses of nonprofit hospitals are included in the expenditures of nonprofit institutions serving households (NPISHs).
[e]Consists of premiums less benefits for health, hospitalization, and accidental death and dismemberment insurance.
[f]Consists of premiums less benefits for income loss insurance.
[g]Consists of premiums plus premium supplements less normal losses and dividends paid to policyholders for privately administered workers' compensation.

SOURCE: Adapted from "Table 2.5.4. Price Indexes for Personal Consumption Expenditures by Function," in *National Income and Product Accounts Tables*, U.S. Department of Commerce, Bureau of Economic Analysis, August 5, 2014, http://www.bea.gov/iTable/iTable.cfm?ReqID=9&step=1#reqid=9&step=3&isuri=1&903=74 (accessed August 12, 2014)

Decisionmaking. The total sample population of 4,134 people is considered representative of the U.S. population. The survey participants were asked if they had not obtained needed medical care in the previous year because they could not afford it. As shown in Table 3.10 more than a quarter (25.7%) of the respondents said they had foregone dental care and 17.6% had skipped seeing a doctor because of the costs. Nearly 15% had not purchased needed prescription medicines because they could not afford them.

The Affordable Care Act

Concern about rising health care spending was one of the drivers behind passage in 2010 of the Patient Protection and Affordable Care Act (ACA), which is described in Chapter 1 and commonly known as Obamacare. In *Economic Report of the President: Together with the Annual Report of the Council of Economic Advisers* (March 2014, http://www.whitehouse.gov/sites/default/files/docs/full_2014_economic_report_of_the_president.pdf), the Council of Economic Advisers (CEA) notes that changes in health care spending reflect changes in both prices and utilization (i.e., demand). Increased prices are reflected in the medical care inflation rates discussed earlier. However, medical care has greatly changed over time due to technological advances. The CEA uses as an example an appendectomy, which is the surgical removal of the appendix. The CEA states "an appendectomy done in 1990 and an appendectomy done in 2010 might be treated as the 'same item' in a health care price index, but it is likely that the 2010 version of the procedure reflects substantial improvements in surgical technique relative to its 1990 counterpart." Such improvements account for some of the increases seen in prices and spending. Many other factors, including an aging population, are also at play, as discussed in Chapter 1.

When the ACA was passed in 2010, medical inflation had already slowed its pace of growth. Reduced spending during the Great Recession (December 2007–June 2009) was one factor. As shown in Table 3.9 medical care

TABLE 3.8

Percentage change in the urban Consumer Price Index (CPI-U) for medical care expenditure categories, December 2006–December 2013, December 2013–June 2014

Item and group	Percent change from previous December								
	December								
	2006	2007	2008	2009	2010	2011	2012	2013	June 2014
Expenditure category									
Medical care	3.6	5.2	2.6	3.4	3.3	3.5	3.2	2.0	1.9
Medical care commodities	1.8	2.7	1.6	3.3	2.9	3.2	1.7	0.3	2.8
Medicinal drugs[a]	—	—	—	—	3.1	3.4	1.6	0.4	2.9
Prescription drugs	1.9	3.3	1.5	4.4	4.1	4.1	1.9	0.8	3.6
Nonprescription drugs[a]	—	—	—	—	−1.0	0.1	0.7	−1.3	0.5
Medical equipment and supplies[a]	—	—	—	—	−0.1	−0.4	1.9	−1.6	0.3
Medical care services	4.1	5.9	3.0	3.4	3.4	3.6	3.7	2.5	1.7
Professional services	2.6	4.2	3.0	2.5	2.7	2.2	1.9	2.1	1.1
Physicians' services[b]	1.7	4.1	2.9	2.5	3.4	2.7	2.0	1.9	0.7
Dental services[b]	5.0	5.8	3.7	3.2	2.7	2.2	2.8	2.8	1.5
Eyeglasses and eye care[c]	2.0	1.5	0.3	1.7	0.3	0.7	0.3	1.2	2.3
Services by other medical professionals[b,c]	3.1	3.1	3.8	1.8	1.8	1.3	0.8	2.1	0.9
Hospital and related services	6.1	8.1	5.4	7.1	6.7	5.3	4.6	3.9	3.2
Hospital services[b,d]	6.2	8.3	5.9	7.7	7.6	5.8	4.9	4.2	3.5
Inpatient hospital services[b,d,e]	6.8	7.6	5.7	7.7	9.2	6.2	4.2	4.4	4.4
Outpatient hospital services[b,c,e]	5.2	9.9	5.6	8.2	5.1	5.0	5.5	3.8	2.9
Nursing homes and adult day services[b,d]	5.0	4.8	3.2	3.6	3.1	2.9	3.6	3.0	1.8
Care of invalids and elderly at home[f]	3.1	3.4	1.6	1.6	1.5	1.9	0.9	0.3	1.5
Health insurance[f]	6.4	8.8	−3.5	−3.0	−4.0	6.1	9.9	0.9	−0.2

—Data not available.
[a]Indexes on a December 2009 = 100 base.
[b]This index series was calculated using a Laspeyres estimator.
[c]Indexes on a December 1986 = 100 base.
[d]Indexes on a December 1996 = 100 base.
[e]Special index based on a substantially smaller sample.
[f]Indexes on a December 2005 = 100 base.
All other item stratum index series were calculated using a geometric means estimator.
Note: Index applies to a month as a whole, not to any specific date.

SOURCE: Adapted from "Table 26. Historical Consumer Price Index for All Urban Consumers (CPI-U): U.S. City Average, by Commodity and Service Group and Detailed Expenditure Categories," in *CPI Detailed Report (Tables 1–29) June 2014*, U.S. Department of Labor, Bureau of Labor Statistics, July 22, 2014, http://www.bls.gov/cpi/cpid1406.pdf (accessed August 9, 2014)

---

prices as measured by the CPI-U increased by 5.2% in 2007 and then increased by only 2.6% in 2008. The rates over subsequent years were also historically low. In 2013 medical care inflation increased by 2%, the smallest annual increase since 1949.

A similar trend is seen in the PCE index for health. In *Economic Report of the President*, the CEA provides the following annual growth rates for health care goods and services and net health insurance based on PCE data:

- 1960–2010: 5.4%
- 2000–2007: 3.3%
- 2007–2010: 2.8%
- 2010–2013: 1.7%

The CEA notes that the year-over-year rate for December 2013 was about 1%, the lowest rate since 1963.

There is great debate about why medical inflation rates remained relatively low in the years following the recession. In *Economic Report of the President*, the CEA claims "some already-implemented features of the Affordable Care Act, including reductions in overpayments to Medicare providers and private insurers as well as payment reforms that incentivize better patient outcomes, are contributing to this trend." (ACA provisions affecting government spending on health care are examined in Chapter 9.) Other analysts, particularly ACA critics, believe the trend simply reflects lingering recessionary effects. In "How the ACA May Affect Health Costs" (WSJ.com, February 23, 2014), Louise Radnofsky notes that skeptics say "people and providers cut back on spending during the economic downturn and it has yet to bounce back."

Additional factors also played a role. In "Obamacare Wins U.S. Bond Converts as Slowing Costs Tame CPI" (Bloomberg.com, April 15, 2014), Daniel Kruger mentions cuts in government payments to Medicare providers that were dictated by the Budget Control Act of 2011. These cuts went into effect in 2013 as part of a broad budgetary reduction called the sequestration. Kruger also points out that "Americans gained access to more generic drugs." Generic drugs are cheaper than "branded" drugs that can monopolize the market until their patent protection expires. For example, Lipitor, a very popular cholesterol-controlling drug, saw its patent expire in 2011. Cheaper generic equivalents then entered the marketplace.

TABLE 3.9

Percentage change in the urban Consumer Price Index (CPI-U) for medical care, 1945–2013

December to December	All items	Medical care*
1945	2.2	2.6
1946	18.1	8.3
1947	8.8	6.9
1948	3.0	5.8
1949	−2.1	1.4
1950	5.9	3.4
1951	6.0	5.8
1952	0.8	4.3
1953	0.7	3.5
1954	−0.7	2.3
1955	0.4	3.3
1956	3.0	3.2
1957	2.9	4.7
1958	1.8	4.5
1959	1.7	3.8
1960	1.4	3.2
1961	0.7	3.1
1962	1.3	2.2
1963	1.6	2.5
1964	1.0	2.1
1965	1.9	2.8
1966	3.5	6.7
1967	3.0	6.3
1968	4.7	6.2
1969	6.2	6.2
1970	5.6	7.4
1971	3.3	4.6
1972	3.4	3.3
1973	8.7	5.3
1974	12.3	12.6
1975	6.9	9.8
1976	4.9	10.0
1977	6.7	8.9
1978	9.0	8.8
1979	13.3	10.1
1980	12.5	9.9
1981	8.9	12.5
1982	3.8	11.0
1983	3.8	6.4
1984	3.9	6.1
1985	3.8	6.8
1986	1.1	7.7
1987	4.4	5.8
1988	4.4	6.9
1989	4.6	8.5
1990	6.1	9.6
1991	3.1	7.9
1992	2.9	6.6
1993	2.7	5.4
1994	2.7	4.9
1995	2.5	3.9
1996	3.3	3.0
1997	1.7	2.8
1998	1.6	3.4
1999	2.7	3.7
2000	3.4	4.2
2001	1.6	4.7
2002	2.4	5.0
2003	1.9	3.7
2004	3.3	4.2
2005	3.4	4.3
2006	2.5	3.6
2007	4.1	5.2
2008	0.1	2.6
2009	2.7	3.4
2010	1.5	3.3
2011	3.0	3.5
2012	1.7	3.2
2013	1.5	2.0

TABLE 3.9

Percentage change in the urban Consumer Price Index (CPI-U) for medical care, 1945–2013 [CONTINUED]

*Commodities and services.
Note: Changes from December to December are based on unadjusted indexes. Series reflect changes in composition and renaming beginning in 1998, and formula and methodology changes in 1999.

SOURCE: Adapted from "Table B-10. Changes in Consumer Price Indexes, 1945–2013," in *Economic Report of the President: Together with the Annual Report of the Council of Economic Advisers*, Executive Office of the President of the United States, March 2014, http://www.whitehouse.gov/sites/default/files/docs/full_2014_economic_report_of_the_president.pdf (accessed August 11, 2014)

TABLE 3.10

Survey respondents who needed health care but did not get it because of cost, 2013

[Percent, except as noted]

Response	Rate
Prescription medicine	14.9
To see a doctor	17.6
Mental health care or counseling	6.7
Dental care	25.7
To see a specialist	11.4
Follow-up care	10.3
Number of respondents	4,134

SOURCE: "Table C.48. During the past 12 months, was there a time when you needed any of the following, but didn't get it because you couldn't afford it?" in *Report on the Economic Well-Being of U.S. Households in 2013*, Board of Governors of the Federal Reserve System, July 2014, http://www.federalreserve.gov/econresdata/2013-report-economic-well-being-us-households-201407.pdf (accessed August 9, 2014)

ACA IMPACTS ON CONSUMERS. Many ACA provisions that directly affect consumers went into effect at the beginning of 2014. This includes the individual mandate, which requires most legal residents of the United States to either obtain health insurance or pay a penalty. As shown in Table 1.4 in Chapter 1, the law prohibits insurers from denying coverage to people with preexisting conditions. In addition, lower-income applicants for health insurance can benefit from government subsidies that reduce their insurance premiums and out-of-pocket payments.

In *Personal Income and Outlays: January 2014* (March 3, 2014, http://www.bea.gov/newsreleases/national/pi/2014/pdf/pi0114.pdf), the BEA notes large boosts in personal income and outlays during January 2014 precipitated by the ACA. For example, Medicaid benefits increased by $19.2 billion due to expanded coverage options under the new law. ACA subsidies and tax credits added another $14.7 billion to income. Personal outlays also increased. The BEA states, "Measures of health care services were boosted $29 billion to reflect the effect of the ACA on demand for these services."

The RAND Corporation is a nonprofit organization that conducts research on issues of interest to policy makers. In "Effects of the Affordable Care Act on Consumer Health Care Spending and Risk of Catastrophic Health Costs" (October 2013, http://www.rand.org/pubs/research_reports/RR383.html), Sarah A. Nowak et al. report on their analysis of the potential effects of the ACA on health care spending by three groups of consumers:

- An estimated 11.5 million people who were previously uninsured and will transition by 2016 to Medicaid under the ACA expansion: Nowak et al. predict, "Total out-of-pocket spending for consumers in this group will fall dramatically." In addition, these people will not pay health insurance premiums.

- An estimated 16.5 million people who were previously uninsured and will transition by 2016 to ACA-regulated individual health insurance policies: These consumers are expected to experience a decrease in out-of-pocket spending. However, they are projected to see an increase in total health care spending (i.e., premiums plus out-of-pocket expenses) because "these people, who previously opted not to [buy insurance] or were ineligible to buy insurance, will be paying premiums for the first time."

- An estimated 8.5 million people who will transition by 2016 from pre-ACA individual health insurance policies to ACA-regulated individual health insurance policies: These consumers are expected to experience a decrease in out-of-pocket spending. Those with incomes less than 400% of the federal poverty level (that is, less than $95,400 for a family of four in 2014) are projected to see a decrease in total health care spending. Those with incomes greater than 400% of the federal poverty level are expected to experience an increase in total health care spending.

CHAPTER 4
PERSONAL DEBT

Beautiful credit! The foundation of modern society.
—Mark Twain and Charles Dudley Warner, *The Gilded Age: A Tale of To-day* (1873)

He who goes a borrowing, goes a sorrowing.
—Benjamin Franklin, *Poor Richard's Almanack* (1757)

Personal debt has both good and bad effects on the U.S. economy. Americans borrow money to buy houses, cars, and other consumer goods. They also take out loans to pay for vacations, investments, and educational expenses. All of this spending helps businesses and boosts the nation's gross domestic product (GDP; the total market value of final goods and services that are produced within an economy in a given year). As long as debt is handled prudently, it can be a positive economic force. However, some Americans take on too much debt and get into financial difficulties. Debt becomes a problem when people must devote large amounts of their disposable income (after-tax income or take-home pay) to repaying loans instead of spending or investing their money.

During the late 1990s and the first half of the first decade of the 21st century Americans took on massive amounts of mortgage debt as a housing boom swept the nation. When the boom ended, many people found themselves unable (or unwilling) to repay their loans. This produced a widespread financial crisis that affected the economy as a whole. The United States sank into the so-called Great Recession, an economic slowdown that officially lasted from December 2007 to June 2009. The latter month represents the time when the economy quit contracting and began expanding (recovering). Americans' debt declined dramatically during the Great Recession and for several years afterward, but in 2012 it began to increase again.

CATEGORIES OF DEBT

Economists divide personal debt into two broad categories: investment debt and consumer debt. Money that is borrowed to buy houses and real estate is considered investment debt. Because most property appreciates (increases in value) over time, the debt assumed to finance its purchase will likely be a wise investment. Likewise, money borrowed to start a business or pay for a college education can bring financial benefits. This assumes that the investment was a wise one and the short-term costs of the debt can be borne. By contrast, consumer debt is purely for consumption purposes. The money is spent to gain immediate access to goods and services that will not appreciate in value (and will likely lose value) over time.

Credit falls into two other categories: nonrevolving credit and revolving credit. Nonrevolving loans require regular payments of amounts that ensure that the original debt (the principal) plus interest will be paid off in a particular amount of time. They are also known as closed-end loans and are commonly used to finance the purchase of real estate, cars, and boats or to pay for educational expenses. Nonrevolving loans feature predictable payment amounts and schedules that are laid out in amortization tables. The term *amortize* is derived from the Latin term *mort*, which means "to kill or deaden." An amortization schedule details how a loan will be gradually eliminated (killed off) over a set period. Revolving debt is a different kind of arrangement in which the debtor is allowed to borrow against a predetermined total amount of credit and is billed for the outstanding principal plus interest. The loans typically require regularly scheduled minimum payments but not a set period for repaying the entire amount due. Credit card loans are the primary example of revolving debt.

Loans can also be secured or unsecured. A secured loan is one in which the borrower puts up an asset called collateral to lessen the financial risk of the lender. If the borrower defaults (fails to pay back the loan), the lender can seize the collateral and sell it to recoup some or all of the money that was lent. Mortgages on homes and property and loans on cars, boats, motor homes, and other goods of high value are typically secured loans. In all these cases the collateral can be legally repossessed by the lenders. Unsecured loans are not backed by collateral. They are granted solely on the good financial reputation of the borrower. Credit card debts and debts owed to medical practitioners and hospitals are the major types of unsecured debts.

INTEREST RATES

One of the chief factors affecting the amount of debt that people assume is the amount of interest charged on loans. Banks and other financial institutions charge interest to make money on lending money. The interest rate charged must be low enough to tempt potential borrowers, but high enough to make a profit for lenders. In general, commercial lenders base their interest rates on the rates that are charged by the Federal Reserve. Lower interest rates encourage consumers to borrow money.

The Federal Reserve makes short-term loans to banks at an interest rate called the discount rate. If the Federal Reserve raises or lowers the discount rate, then banks adjust the federal funds rate, which is the rate they charge each other for loans. This affects the prime rate, the interest rate banks charge their best customers (typically large corporations), which in turn affects the rates on other loans. Figure 1.9 in Chapter 1 shows the bank prime loan rate from January 1949 to July 2014. The rate varied widely over time from about 2% in 1949 to more than 20% during the early to mid-1980s. Since 1990 the prime rate has consistently remained below 10% and has dipped below 5%. From January 2009 through July 2014 the rate was stationary at 3.25%.

When a loan is granted, the creditor sets terms that specify whether the interest rate to be paid will be fixed or variable. A fixed interest rate remains constant throughout the life of the loan. A variable rate changes and is typically tied to a publicly published interest rate, such as the prime rate. For example, a loan can be made with the stipulation that the interest rate charged each month will be one percentage point higher than the prime rate. As the prime rate changes, so will the interest rate on the loan and the borrower's monthly payments.

Table 4.1 shows the tremendous difference between loans with differing interest rates and repayment periods. A $100,000 mortgage with a 30-year fixed interest rate of 5% will result in $193,300 being paid over the lifetime of the loan. The same $100,000 loan at 10% interest will cost $315,900. Loans for new cars typically have a repayment period of three to five years. A 7% fixed interest car loan for $20,000 paid over three years results in a total payment of $22,200. The same loan spread over five years will end up costing $23,800. Thus, the shorter loan period results in a much lower overall payout for the car. However, the trade-off to the borrower for a shorter loan period will be higher monthly payments. Borrowers must consider the financial consequences of monthly payments and interest rates to get a loan they can afford in the short and long term.

TABLE 4.1

Interest payments for particular loans

Interest rate	Years of loan	Amount borrowed (principal)	Total interest paid*	Total principal + interest paid*
5%	30	$100,000	$93,300	$193,300
10%	30	$100,000	$215,900	$315,900
15%	30	$100,000	$355,200	$455,200
5%	15	$100,000	$42,300	$142,300
10%	15	$100,000	$93,400	$193,400
15%	15	$100,000	$151,900	$251,900
7%	3	$20,000	$2,200	$22,200
7%	4	$20,000	$3,000	$23,000
7%	5	$20,000	$3,800	$23,800

*Rounded to nearest $100

SOURCE: Created by Kim Masters Evans for Gale, © 2010

CREDIT REPORTS AND SCORES

A credit score is a numerical rating of a person's creditworthiness based on his or her past history of managing credit. According to Malgorzata Wozniacka and Snigdha Sen, in "Credit Scores: What You Should Know about Your Own" (November 23, 2004, http://www.pbs.org/wgbh/pages/frontline/shows/credit/more/scores.html), credit scoring began during the late 1950s with companies called credit bureaus that collected information about the credit history and general reputation of individuals. In 1971 Congress passed the Fair Credit Reporting Act. Wozniacka and Sen note that the law "established a framework for fair information practices to protect privacy and promote accuracy in credit reporting."

Several companies compile credit scores in the United States. The three major U.S. credit bureaus are Equifax, Experian, and TransUnion. Each company has its own scoring system and maintains a record called a credit report for each individual in its database. Anyone who has ever taken out a loan of any type (including credit cards) from a commercial entity, such as a bank, department store, or auto dealership, most likely has a credit report on file. Lenders regularly update the credit bureaus about the loan payment histories of borrowers, such as whether or not the borrowers make their loan payments on time and in full. Other sources that can influence credit scores are utilities, medical service providers, landlords and leasing agents, and companies with which consumers establish contracts to pay for goods or services, such as cell phone companies.

Fair Isaac Company (now known as FICO) is a financial company that was founded during the 1950s. Over the decades it has developed statistical models that are widely used by the credit bureaus to determine credit scores. A credit score calculated using FICO software is called a FICO score. The FICO method is highly regarded in the credit industry. Thus, many major financial institutions rely on FICO scores as reliable measures of the creditworthiness of customers. According to FICO, in "What's in My FICO Score" (2014, http://www.myfico.com/CreditEducation/WhatsInYourScore.aspx), a FICO score generally depends on the following elements:

- Payment history—35%
- Amounts owed—30%
- Length of credit history—15%
- New credit (i.e., recently opened accounts)—10%
- Types of credit used—10%

In "What's Not in My FICO Score" (2014, http://www.myfico.com/CreditEducation/WhatsNotInYourScore.aspx), FICO notes that federal law prohibits the use of information on a person's race, religion, color, national origin, sex, marital status, or age in credit scores.

There are some general guidelines about what FICO scores mean to lenders. In "FICO Credit Scoring" (2014, http://www.mbda.gov/blogger/financial-education/fico-credit-scoring), the U.S. Department of Commerce's Minority Business Development Agency indicates that FICO scores range from 375 to 900 points, with higher scores indicating better creditworthiness. People with higher scores are more likely to be granted credit and to receive better credit terms (e.g., lower interest rates) than people with lower scores. Typically, a score of at least 680 is required to get the best (or prime) credit terms. People with scores lower than about 620 are said to have "subprime" (less than prime) creditworthiness. If they are granted credit, they may be charged a high interest rate by the lender because they are "risky" borrowers who are less likely than "prime" borrowers to repay a loan.

BAD DEBT AND COLLECTIONS

Debtors are typically required to make monthly payments on loans. Missed payments, particularly for several months in a row, greatly raise the chances that the debtor will default (fail to fulfill the obligation to repay the loan). Creditors report missed (and late) payments to the credit reporting agencies. Missed and late payments negatively impact credit scores. Creditors typically allow debtors only a limited time (e.g., 3 months) to bring current any missed payments. After that time the creditors consider the debtor in default and the debt to be "uncollectible" or "bad." Businesses can deduct such debts (which are called charge-offs) on their income and tax statements.

Once a loan is deemed to be in default, a creditor can take certain actions, including suing the debtor. However, each state has a statute-of-limitations that restricts the number of years in which a lawsuit can be brought. Any collateral (e.g., a home or vehicle) used to secure a debt can be seized. This process is called foreclosure when mortgaged property is involved. Cars, boats, and other types of collateral are said to be "repossessed." However, even after a foreclosure or repossession a creditor can still sue a debtor until the statute of limitations date is reached.

Some creditors operate in-house collections departments that try to find debtors in default and convince them to pay past-due amounts. Third-parties, such as collection agencies or attorneys, may also be hired for this purpose. The amounts involved in collection efforts are massive. In "The Impact of Third-Party Debt Collection on the U.S. National and State Economies in 2013" (July 2014, http://www.acainternational.org/files.aspx?p=/images/21594/theimpactofthird-partydebtcollectiononthenationalandstateeconomies2014.pdf), the professional services firm Ernst & Young indicates that in 2013 more than 1 billion consumer accounts were placed with third-party agencies by creditors. The total value of these accounts was $756 billion. Overall, third-party debt collectors were able to recoup $55 billion in payments during 2013 for the original creditors.

It should be noted that creditors can also sell "bad" debts to debt-buying companies. These firms receive information about the debtors and then try to track them down and convince them to pay. In these cases any payments are kept by the debt-buying companies. In "The Structure and Practices of the Debt Buying Industry" (January 2013, http://www.ftc.gov/sites/default/files/documents/reports/structure-and-practices-debt-buying-industry/debtbuyingreport.pdf), the Federal Trade Commission (FTC) describes its study of nine large debt buying companies, which it estimates bought about three-fourths of the debt sold in the United States in 2008. The FTC notes that over a three-year period these companies purchased nearly 90 million consumer accounts with a total value (debt owed) of $143 billion.

DEBT DATA

The Federal Reserve is the central bank of the United States. It is overseen by a board of governors and includes regional banks throughout the country. Within the U.S. government, the Federal Reserve System is the primary source of publicly available data about the debt of American households.

Total Liabilities

The Federal Reserve publishes debt statistics in *Federal Reserve Statistical Release: Z.1 Financial Accounts of the United States* for households and nonprofit organizations. As shown in Table 4.2, total liabilities were nearly $13.8 trillion at the end of the first quarter of 2014. Almost all ($13.1 trillion, or 95% of the total) of the total liabilities were in the form of credit market instruments, mainly mortgages ($9.3 trillion) and consumer credit ($3.1 trillion). Consumer credit includes nonmortgage loans, such as credit cards, student loans, and auto loans.

Table 4.3 shows the percent change in credit market debt for households and nonprofit organizations from 1979 through the first quarter of 2014. The highest growth year since the mid-1980s was 2003, when total credit market debt grew by 11.9% from the year before. The growth slowed considerably over subsequent years and went negative from 2009 through 2011 before rebounding. During the first quarter of 2014 total credit market debt increased by 2%. The breakdown by debt type in Table 4.3 shows that home mortgage debt decreased by 0.9% that quarter, and consumer credit increased by 6.6%.

Another economic indicator compiled by the Federal Reserve is the household debt service ratio (DSR). The DSR is the ratio of household debt payments to disposable personal income. It indicates the estimated fraction of disposable income that is devoted to payments on outstanding mortgage and consumer debt.

As shown in Figure 4.1, the DSR generally increased between 1994 and 2007. It peaked above 13% in late 2007 and then declined dramatically. By the first quarter of 2014 the DSR was about 10%. Thus, Americans as a whole were spending about 10% of their disposable income on their debt.

Aggregated Household Data

The Federal Reserve Bank of New York (FRBNY) tracks certain kinds of debt held by U.S. households (excluding nonprofit organizations). In *Quarterly Report on Household Debt and Credit* (August 2014, http://www.newyorkfed.org/householdcredit/2014-q2/data/pdf/HHDC_2014Q2.pdf), the FRBNY notes that it performs its analyses by examining the Equifax credit reports for a nationally representative sample of individuals and households. The data are then aggregated to represent the nation as a whole.

The FRBNY provides the following debt balances as of the end of the second quarter of 2014. (Note that the component percentages do not sum to 100% due to rounding.)

- Mortgages—$8.1 trillion or 70% of total. This category includes first mortgages and closed-end home equity installment loans.

- Student loans—$1.1 trillion or 10% of total. This category includes loans provided by government and private entities.

- Auto loans—$905 billion or 8% of total. This category includes loans provided by banking institutions, automobile dealers, and automobile financing companies.

- Credit card debt—$669 billion or 6% of total. This category includes only credit cards issued by banks or other financial institutions.

TABLE 4.2

Household debt at end of period, 2008 through the first quarter of 2014

[Billions of dollars; amounts outstanding end of period, not seasonally adjusted.]

	2008	2009	2010	2011	2012	2013	2014 first quarter
Total liabilities	14,278.2	14,049.5	13,766.5	13,566.0	13,626.8	13,768.2	13,784.8
Credit market instruments	13,849.6	13,546.3	13,214.8	13,052.9	13,044.2	13,146.1	13,147.5
Home mortgages*	10,579.0	10,417.6	9,912.7	9,697.5	9,481.7	9,386.2	9,349.1
Consumer credit	2,650.6	2,552.8	2,647.4	2,755.9	2,923.6	3,097.4	3,103.6
Municipal securities	259.5	265.4	263.2	255.5	241.0	227.8	227.6
Depository institution loans n.e.c.	26.4	−15.9	61.0	11.5	62.6	92.7	124.4
Other loans and advances	133.2	133.7	136.1	138.1	139.3	141.3	141.3
Commercial mortgages	200.9	192.6	194.3	194.3	195.9	200.8	201.6
Security credit	164.8	203.0	278.2	238.9	303.7	339.2	351.7
Trade payables	236.7	278.2	248.8	250.0	254.0	255.0	256.0
Deferred and unpaid life insurance premiums	27.0	22.1	24.7	24.3	24.9	27.9	29.6

*Includes loans made under home equity lines of credit and home equity loans secured by junior liens.
Note: n.e.c. = not elsewhere classified. Sector includes domestic hedge funds, private equity funds, and personal trusts.

SOURCE: Adapted from "L.100. Households and Nonprofit Organizations," in *Federal Reserve Statistical Release: Z.1 Financial Accounts of the United States: Flow of Funds, Balance Sheets, and Integrated Macroeconomic Accounts—First Quarter 2014,* Board of Governors of the Federal Reserve System, June 5, 2014, http://www.federalreserve.gov/releases/z1/current/z1.pdf (accessed August 14, 2014)

TABLE 4.3

Percentage growth in household debt, 1979 through the first quarter of 2014

[In percent; quarterly figures are seasonally adjusted annual rates. Data shown are on an end-of-period basis.]

	Domestic nonfinancial sectors		
	Total	Households home mortgage	Consumer credit
1979	14.8	16.4	13.9
1980	8.2	10.9	1.0
1981	7.3	7.2	5.5
1982	5.6	4.8	5.0
1983	11.0	10.2	12.1
1984	13.0	11.4	18.4
1985	16.1	14.6	15.9
1986	11.4	13.7	9.1
1987	10.4	13.4	4.8
1988	9.9	11.8	6.7
1989	9.1	10.9	6.3
1990	7.1	8.8	1.9
1991	5.1	7.0	−1.1
1992	5.4	6.5	1.1
1993	6.1	5.5	7.4
1994	7.7	5.6	15.2
1995	7.1	4.9	14.4
1996	6.7	6.2	9.0
1997	5.9	6.1	5.5
1998	7.5	8.0	7.2
1999	8.0	9.4	7.8
2000	9.0	8.7	11.4
2001	9.6	10.6	8.6
2002	10.7	13.3	5.6
2003	11.9	14.5	5.3
2004	11.1	13.5	5.6
2005	11.2	13.4	4.5
2006	10.2	11.2	5.2
2007	7.0	7.4	6.1
2008	1.1	0.9	1.3
2009	−0.1	0.6	−3.9
2010	−1.1	−1.8	−1.0
2011	−0.2	−0.7	4.1
2012	1.5	−0.8	6.2
2013	1.5	−0.1	5.9
2014—first quarter	2.0	−0.9	6.6

SOURCE: Adapted from "D.1. Credit Market Debt Growth by Sector," in *Federal Reserve Statistical Release: Z.1 Financial Accounts of the United States: Flow of Funds, Balance Sheets, and Integrated Macroeconomic Accounts—First Quarter 2014*, Board of Governors of the Federal Reserve System, June 5, 2014, http://www.federalreserve.gov/releases/z1/current/z1.pdf (accessed August 14, 2014)

- Home equity revolving loans—$521 billion or 4% of total.

- Other loans—$323 billion or 3% of total. This category includes personal loans and loans provided for retail purchases (e.g., credit cards issued by department stores).

The total balance for these debt types was $11.6 trillion.

The FRBNY also monitors the delinquency status of the debt it tracks. The vast majority (93.8%) of the $11.6 trillion household debt described above was current (i.e., the payments were up to date) as of June 30, 2014. From 2003 through the first quarter of 2007 the payments-current percentage exceeded 95% and crept as high as 96.2%. It then began a downfall dropping to a low point of 88.1% at year-end 2009. Thus, 11.1% of household debt was delinquent at that time. The financial strains of the housing crisis and the Great Recession (December 2007 to June 2009) are blamed for the delinquency spike. Improving economic conditions over subsequent years saw the delinquency rate decrease. As of June 30, 2014, the FRBNY notes that 6.2% of the debt it tracks (or $721 billion) was delinquent as follows:

- 30–59 days late (not more than two payments past due)—1.2%

- 60–89 days late (not more than three payments past due)—0.5%

- 90–119 days late (not more than four payments past due)—0.3%

- 120+ days late (five or more payments past due) or in collections (e.g., by a collection agency)—1.9%

- Severely derogatory (defined by the FRBNY as "any of the previous states combined with reports of a repossession, charge off to bad debt or foreclosure")—2.3%

Family-Level Data

The Survey of Consumer Finances is conducted every three years by the Federal Reserve in cooperation with the Internal Revenue Service to collect detailed financial information on American families. As of December 2014, the most recent survey results available were from the 2013 survey. According to Jesse Bricker et al. of the Federal Reserve, in *Changes in U.S. Family Finances from 2010 to 2013: Evidence from the Survey of Consumer Finances* (September 4, 2014, http://www.federalreserve.gov/econresdata/scf/scfindex.htm), more than 6,000 interviews were conducted as part of the 2013 survey. The resulting data are believed to be representative of 122.5 million U.S. families.

Bricker et al. note that 74.5% of U.S. families held any type of debt in 2013. Nearly half (48.1%) held debt secured by residential property, mainly primary residences (42.9%). The largest fractions holding other types of debt were 38.1% with credit card balances, 30.9% with vehicle loans, and 20% with education loans. The average total debt amount in 2013 was $122,300 per family. Average amounts for specific debt types were:

- Debt secured by primary residence—$156,700 per family

- Education loans—$28,900 per family

- Vehicle loans—$14,600 per family

- Credit card balances—$5,700 per family

Note that these averages consider only families holding each type of debt, not the entire sample population.

FIGURE 4.1

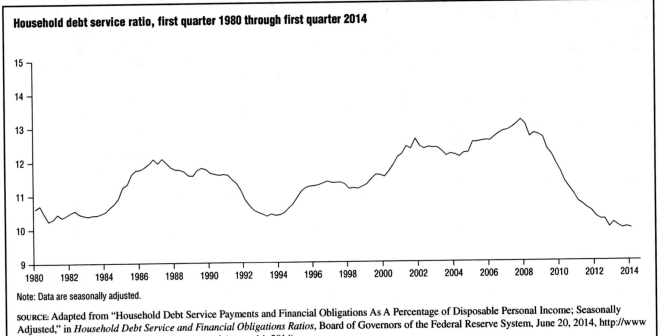

Household debt service ratio, first quarter 1980 through first quarter 2014

Note: Data are seasonally adjusted.

SOURCE: Adapted from "Household Debt Service Payments and Financial Obligations As A Percentage of Disposable Personal Income; Seasonally Adjusted," in *Household Debt Service and Financial Obligations Ratios*, Board of Governors of the Federal Reserve System, June 20, 2014, http://www.federalreserve.gov/releases/housedebt/ (accessed August 14, 2014)

MORTGAGES

For most Americans a mortgage is the largest personal debt they will ever incur. Mortgage debt is a form of investment debt because real estate usually increases in value. Thus, assuming mortgage debt is generally considered to be a sensible economic move, as long as the payments are well matched to the borrower's income and ability to pay. During the second half of the first decade of the 21st century the U.S. housing market suffered a deep financial crisis. In many cases real estate did not appreciate in value, but depreciated (decreased in value). The causes and consequences of this crisis will be explained later in the chapter. First, it is necessary to understand some basic information about mortgages and how the mortgage market operates.

A mortgage represents a lien, or binding charge, against a piece of property for the payment of a debt. In other words, the loan is granted on the condition that the property can be claimed by the lender (creditor) in the event the borrower defaults. If the loan is satisfactorily paid, full ownership of the property is granted to the borrower. Lenders use a process called underwriting to assess the creditworthiness of potential mortgage borrowers.

Mortgage Underwriting Standards

A home mortgage is a very large loan, usually totaling $100,000 or more, for which the repayment period may be several decades. Mortgage agreements are complex legal documents that bind lenders and borrowers to do certain things. Lenders typically charge borrowers fees (e.g., application fees, home appraisal fees, and so on) to originate a mortgage. In addition, lenders that hold the mortgages they originate will earn money from the interest payments that the borrowers make each month.

If a borrower quits paying on a mortgage, the lender can take possession of the home through a legal process called foreclosure and resell the home to someone else. However, this requires a significant investment of time and resources by the lender. In addition, foreclosed homes are typically sold at a loss. Thus, it is in a lender's best interest to issue mortgages to borrowers who are highly likely to repay the loans. As a result, lenders use specific underwriting standards to screen potential borrowers. There are four important factors that lenders take into consideration: the borrower's credit score, documentation of the borrower's income and assets, the borrower's debt ratio, and the loan-to-value ratio of the mortgage. Credit scoring was described earlier in this chapter; the remaining factors are described below.

DOCUMENTATION OF INCOME AND ASSETS. Lenders typically require mortgage applicants to provide documentation of their income and assets. Many workers receive regular paychecks from their employer. These applicants provide the lender with paperwork showing the amount and timing of their salary or wages. The length of time the applicant has had the same job is also important to the lender. Some people are self-employed. They typically have to provide the lender with copies of their income tax returns or other documents for proof of income.

Borrowers must also supply written documentation of their assets. These assets usually include cash in bank accounts, savings bonds, stocks or other investments with value, and real estate or vehicles for which the loans have already been paid off.

BORROWER'S DEBT RATIO. A borrower's debt ratio can be calculated in different ways. One ratio results from dividing total assets (or total income) by total debt. Another example is a ratio of monthly debt payments to all monthly obligations. For example, a borrower's monthly obligations might include utility payments, car and student loan payments, and contract payments, such as for cell phone service. A borrower who already has substantial monthly obligations will find it difficult to also make mortgage payments.

LOAN-TO-VALUE RATIO. A loan-to-value (LTV) ratio is calculated by dividing the amount of a mortgage loan by the value of the home to be purchased. For example, a borrower seeking an $80,000 mortgage to buy a $100,000 home would have an LTV ratio of 0.8. The borrower would have to pay in cash the difference between the purchase price and the mortgage amount (in this case $20,000). This is called a down payment on the home purchase. Historically, mortgage lenders have been reluctant to lend the full purchase price of a home. A borrower who defaults on a mortgage and loses the home through foreclosure also loses the down payment that he or she made when the home was purchased. Thus, a down payment has long been considered by lenders to be a good sign that the borrower will faithfully make the mortgage payments.

Risk Classifications

Lenders use the results of the underwriting process to classify mortgages into different risk categories, depending on the creditworthiness of the borrower and the likelihood that the person will repay the loan.

As noted earlier, prime loans feature better (lower) interest rates than subprime loans. Nonprime loans are for people who do not qualify for prime mortgages for various reasons (poor or short credit history, lack of assets, low income or inability to prove income, and so on). Nonprime loan holders are more likely than prime loan holders to default on their loans. As a result, lenders charge higher interest rates on nonprime loans because of the greater risk that is associated with them.

Some lenders further classify nonprime loans as near-prime or subprime. In *The Rise in Mortgage Defaults* (November 2008, http://www.federalreserve.gov/pubs/feds/2008/200859/200859pap.pdf), Christopher J. Mayer, Karen M. Pence, and Shane M. Sherlund point out that these classifications are not strictly or uniformly defined in the mortgage industry. However, near-prime mortgages are generally for borrowers with "minor credit quality issues" or those "who are unable or unwilling to provide full documentation of assets or income." People buying a home as an investment, rather than to live in, might also be considered near-prime borrowers because investors are more likely than owner-occupiers to default on mortgage loans. (Investors buy homes with the expectation that the properties will appreciate in value. If this happens, they can resell the homes in a relatively short period for more than they originally paid and make a profit.) Subprime mortgages are for borrowers posing the greatest risk of nonpayment. This classification is typically assigned due to lender concerns about the borrower's credit history, income, assets, debt ratios, and/or lack of documentation.

Mortgage Interest Rates: Fixed and Adjustable

A fixed-rate mortgage charges a set interest rate over the entire lifetime of the loan, typically 30 years. Figure 4.2 shows the average annual interest rate charged on a 30-year fixed mortgage from April 1971 to July 2014, when the rate was 4.1%. Comparison to Figure 1.9 in Chapter 1 shows that fixed mortgage rates mirror the ups and downs of the prime rate. One feature of a fixed-rate mortgage is that the monthly payment remains the same throughout the lifetime of the loan.

Creative financing terms introduced by creditors since the late 1990s have led to many alternatives to the conventional 30-year fixed-rate mortgage. One alternative is a shorter loan period, for example, 15 years instead of 30 years. Another option is an adjustable-rate mortgage (ARM). ARMs feature variable interest rates (and consequently variable monthly payments) over the lifetime of the loan. An ARM rate is typically tied to a published benchmark rate called an index rate. Low index rates during the first half of the first decade of the 21st century enticed many home buyers to take on ARMs instead of fixed-rate mortgages. Lenders may offer discount (or teaser) rates that are even lower than the index rate during the early months or years of the ARM repayment period. This translates into extra low monthly payments for an initial period, followed by much higher payments as the ARM matures.

Some creditors offer mortgages that allow homeowners to make interest-only payments for a short initial portion of the loan period. This is followed by a longer period of much higher monthly payments. A similar product is the payment-option mortgage, which allows homeowners to make small minimum payments for an initial short period. Short-term payment schedules requiring one large balloon payment are also offered in some mortgage products.

Mortgage arrangements with changeable monthly payments and balloon payments can pose a financial problem for homeowners who overestimate their ability

FIGURE 4.2

Interest rate on conventional 30-year fixed mortgage, April 1971–July 2014

SOURCE: Adapted from "Contract Rate on 30-Year, Fixed-Rate Conventional Home Mortgage Commitments," in *Selected Interest Rates (Daily)—H.15: Historical Data*, Board of Governors of the Federal Reserve System, August 11, 2014, http://www.federalreserve.gov/releases/h15/data.htm (accessed August 14, 2014)

to meet the costs of the mortgage. Figure 4.3 illustrates how monthly payments can vary significantly between different types of mortgages on a $200,000 home. The buyer assuming a fixed-rate 30-year mortgage at 6% interest pays $1,199.10 per month for the entire lifetime of the loan. The 5/1 ARM is a common ARM arrangement in which the initial interest rate remains fixed for five years and then begins to fluctuate with the index rate. In this example, the buyer pays a discounted rate of 4% during the first five years of the ARM. This translates to a monthly mortgage payment of $954.83. In year six the monthly payment is tied to a 6% ARM rate, and the monthly payment jumps to $1,165.51. In year seven the ARM rate increases to 7%; consequently, the monthly payment increases to $1,389.51. Two other types of mortgages depicted in Figure 4.3 (a 5/1 ARM with interest-only payments and a payment-option mortgage) both feature large increases in monthly payments after the initial low-rate period.

Refinancing Mortgages

Most home mortgages cover long periods (up to 30 years). Interest rates, however, can change dramatically in the short term, rising and falling in response to macroeconomic factors. Home buyers who assume mortgages during times of high interest rates can ask creditors to refinance (adjust the mortgage terms) when interest rates go down. Basically, refinancing entails drawing up a new mortgage contract on a property. Because mortgage contracts are complicated legal documents, creditors usually charge fees to refinance mortgages. Thus, homeowners must weigh the long-term benefits of a reduced interest rate against the expense of refinancing fees.

During the first decade of the 21st century interest rates trended downward, making mortgage refinancing popular. This was particularly true for consumers who had purchased homes during the 1980s, when interest rates were extremely high by historical standards.

Refinancing can result in lower monthly payments for the homeowner because of the lower interest rate and because refinancing is commonly performed after several years of payments have been made on the original loan. This frees up the borrowers' money for consumer spending, investing, or saving. However, refinances that are conducted with a long payment period will keep the homeowner in mortgage debt for a longer period than originally anticipated. Some homeowners opt for a

FIGURE 4.3

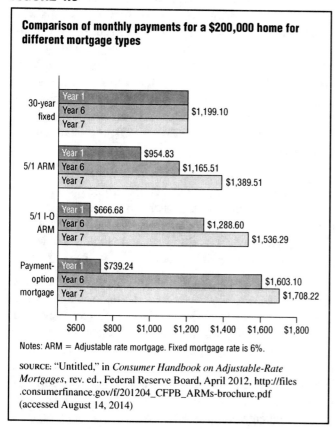

Comparison of monthly payments for a $200,000 home for different mortgage types

Notes: ARM = Adjustable rate mortgage. Fixed mortgage rate is 6%.

SOURCE: "Untitled," in *Consumer Handbook on Adjustable-Rate Mortgages*, rev. ed., Federal Reserve Board, April 2012, http://files.consumerfinance.gov/f/201204_CFPB_ARMs-brochure.pdf (accessed August 14, 2014)

shorter loan payback period when they refinance. For example, consider a homeowner who has been paying for five years on a fixed-rate 30-year mortgage. There are 25 years left in the repayment period. Refinancing at a much lower interest rate with a new 15-year payback period may not decrease the monthly payment, but it will reduce by 10 years the amount of time the homeowner will be in mortgage debt.

Home Equity Loans

Real estate tends to appreciate in value. Thus, a property can increase in value above the amount that was originally borrowed to pay for it. For example, imagine a homeowner who bought a house in 2000 for $100,000 with a 30-year fixed-rate mortgage. After making mortgage payments for several years, the homeowner discovers that the principal due on the loan has dropped to $90,000, but the property has increased in value to $140,000. The difference between the amount of principal owed (the outstanding loan balance) and the value of the property is $50,000 and is called home equity. Home equity is an asset that can be borrowed against. Basically, homeowners can liquefy (turn into cash) the equity they have built up in their homes.

During the 1990s and the first half of the first decade of the 21st century, U.S. homeowners took advantage of rising property values and low interest rates to liquefy billions of dollars in equity from their homes. Alan Greenspan and James Kennedy calculate in *Sources and Uses of Equity Extracted from Homes* (March 2007, http://www.federalreserve.gov/pubs/feds/2007/200720/200720pap.pdf) that the cash extracted from home equity between 1991 and 2005 averaged $530 billion per year. Approximately two-thirds of this total was from sales of existing homes. The remaining one-third was due to equity-based loans. Greenspan and Kennedy estimate that the vast majority (87%) of the liquefied equity from home sales was used by Americans toward the purchase of their next home. The remaining 13% was spent or invested. They believe that one-third of the cash from equity-based loans was applied toward nonmortgage debt (e.g., credit card debt). Another one-third was likely spent on home improvements, and the remainder was spent on other purchases or invested.

Economists generally approve of the use of home equity loans for home improvements. This type of spending is considered to be an investment because it adds value to the home. Many homeowners choose to use home equity loans to repay other debt. Because mortgage loans typically have lower interest rates than other loans, this exchange is beneficial. In addition, the interest paid on mortgage loans is tax deductible for most Americans, whereas interest paid on other types of loans is not deductible. Thus, conversion of "bad" types of debt (such as credit cards) to mortgage debt has favorable consequences.

However, some economists worry that homeowners who use home equity loans to pay off bad kinds of debt may succumb to temptation and run up bad debt again. This could put them in a dire financial situation. They will no longer have their home equity to fall back on if their new debts become more than they can afford, and they might have to default on their loans. Home equity loans, like all mortgage loans, are secured by property. Thus, defaulting on a home equity loan can result in the loss of the home by the owner.

Foreclosures, Defaults, and Delinquencies

Foreclosure is a legal process in which a lender takes possession of the collateral (i.e., the home) of a borrower who has stopped paying on a mortgage loan. The borrower is said to be in default, meaning that the borrower has failed to abide by the legally binding mortgage agreement. Typically, lenders grant borrowers up to 90 days to "catch up" on late mortgage payments. After that 90-day period, the foreclosure process begins. Mortgage loans on which borrowers have not made a payment for at least 90 days are said to be "seriously delinquent."

The Secondary Market for Mortgages

Mortgage loans represent an investment for lenders. Mortgages will provide income well into the future as the loan payments (including interest) are paid by the borrowers. As a result, mortgages are commodities that are purchased by investors. This is known as the secondary mortgage market. In general, individual mortgages are bundled together and sold on the stock market. This process is called securitization. The packages are known as mortgage-backed securities or mortgage-based securities. Both are abbreviated MBS.

The original lenders are often eager to sell mortgages on the secondary market to obtain cash that they can use to make new loans. Companies purchase MBS products because mortgages have historically been considered relatively safe investments that will provide regular income in the future. Of course, the secondary buyers are trusting that the original lenders used good underwriting practices and lent money only to people who are very likely to keep making their mortgage payments. If borrowers default on mortgages that are within MBS packages, the MBS investors suffer a financial loss.

The Federal Government's Role in Mortgages

There are two types of mortgages in common use: conventional mortgages and government-underwritten mortgages. Conventional mortgages are loans made by nongovernmental businesses, such as banks and finance companies. Government-underwritten mortgages are insured by a federal, state, or local government agency.

Because high rates of homeownership are considered good for the U.S. economy, the government has taken an active role in the mortgage market. Mortgage terms have changed dramatically since the early 1930s. At that time home buyers could borrow only up to half of a property's market value (i.e., an LTV ratio of 0.5). A typical repayment plan included three to five years of regular payments and then one large balloon payment of the remaining balance. According to the Federal Housing Administration (FHA; 2014, http://www.hud.gov/offices/hsg/fhahistory.cfm), these terms discouraged many potential homeowners. As a result, the homeownership rate stood at about 40%.

During the 1930s the federal government introduced a variety of initiatives to boost a housing industry that was devastated by the Great Depression (1929–1939) and increase homeownership. These efforts were focused on encouraging the supply side of the mortgage industry. They benefited consumers by enhancing the availability and flexibility of home mortgages. For example, amortization schedules covering 15 years or more became common and balloon payments were eliminated; both of these changes made it much easier for consumers to afford houses. Following World War II (1939–1945) the Veterans Administration (VA; now known as the U.S. Department of Veterans Affairs) began offering mortgages with favorable terms to returning veterans. Postwar economic prosperity and relatively low interest rates led to a housing boom. According to Robert R. Callis and Melissa Kresin of the U.S. Census Bureau, in "Residential Vacancies and Homeownership in the Second Quarter 2014" (July 29, 2014, http://www.census.gov/housing/hvs/files/qtr214/q214press.pdf), the nation's homeownership rate was about 65% during the mid-1990s. It climbed to a record high of 69.4% in 2004 and then began to decline. As of the second quarter of 2014, the homeownership rate was 64.8%.

Besides the VA, there are several other agencies and organizations that operate under government control or mandate to increase homeownership among Americans, including the FHA, the Federal National Mortgage Association (Fannie Mae), and the Federal Home Loan Mortgage Corporation (Freddie Mac).

FEDERAL HOUSING ADMINISTRATION. The FHA was created in 1934 and later placed under the oversight of the U.S. Department of Housing and Urban Development. The FHA provides mortgage insurance on loans that are made by FHA-approved lenders to buyers of single- and multi-family homes. FHA-insured loans require less cash down payment from the home buyer than most conventional loans. The insurance provides assurances to lenders that the government will cover losses resulting from homeowners who default on their loans.

FEDERAL NATIONAL MORTGAGE ASSOCIATION. Fannie Mae was created in 1938. It began buying FHA-insured mortgages from banks and other lenders, bundled the mortgages together, and then sold the mortgage packages as investments on the stock market. In essence, Fannie Mae created the secondary market for home mortgages. Lenders benefit because they receive immediate money that can be lent to new customers, and home buyers benefit from the increased availability of mortgage loans. The mortgage packages are attractive to investors because the mortgages are backed by FHA insurance. In 1968 Fannie Mae became a private organization and expanded its portfolio beyond FHA-insured mortgages to private-label MBS (i.e., MBS products securitized by nongovernmental businesses, such as banks and financial institutions).

FEDERAL HOME LOAN MORTGAGE CORPORATION. Freddie Mac was created by the federal government in 1970 to prevent Fannie Mae's monopolization of the mortgage market. Like Fannie Mae, Freddie Mac is a private organization operating under a government charter and buys and sells home mortgages on the secondary market. Both Fannie Mae and Freddie Mac are shareholder-owned corporations.

The Housing Market Booms and Busts

As noted in Chapter 1, markets sometimes boom (become overly inflated in value) and then bust (lose value suddenly and dramatically). What frustrates investors and analysts alike is that the evidence of a boom is not obvious until after a bust occurs. In other words, most people do not see a bust coming because they are caught up in the excitement of making money and expect the financial windfall to continue indefinitely.

The historically low interest rates during the first half of the first decade of the 21st century spurred demand in the real estate market. For example, sales of newly built single-family homes increased steadily from the late 1990s to 2005. (See Figure 4.4.)

High demand pushes prices upward. The Federal Housing Finance Agency uses an index called the house price index (HPI) to track home prices for mortgages that are acquired by Fannie Mae and Freddie Mac. (It should be noted that the HPI covers purchase-only mortgages and not refinanced loans.) The HPI for January 1991 is arbitrarily set at 100. According to the Federal Housing Finance Agency, as presented in Figure 4.5, the HPI increased dramatically through mid-2007, when it peaked above 220 and then declined over time to about 180. By June 2014 the HPI had increased (appreciated) to approximately 210.

The financial company Standard & Poor's (S&P) maintains home price indexes known as the S&P/Case-Shiller indexes. They are named after the economists Karl E. Case and Robert J. Shiller. One of the indexes, the 20-city composite, measures home price changes by compiling and weighting data for 20 major metropolitan areas around the country. In the press release "For the Past Year Home Prices Have Generally Moved Sideways According to the S&P/Case-Shiller Home Price Indices" (July 27, 2010, http://www.standardandpoors.com), S&P reports that the 20-city composite index reached its highest point (just over 200) in 2006. The index is benchmarked to a value of 100 that is assigned to January 2000. Thus, the index more than doubled between 2000 and 2006.

During the first half of the first decade of the 21st century homeowner-occupiers and investors became excited about appreciating home prices. The homeowner-occupiers saw their homes quickly increase in value, allowing them to borrow against the rising equity.

FIGURE 4.4

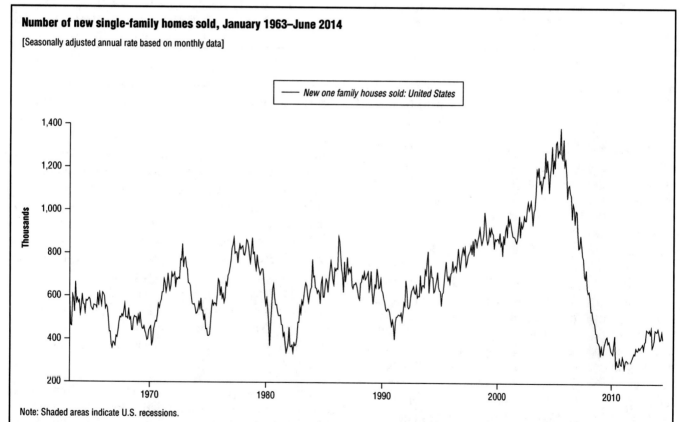

Number of new single-family homes sold, January 1963–June 2014

[Seasonally adjusted annual rate based on monthly data]

Note: Shaded areas indicate U.S. recessions.

SOURCE: "New One Family Houses Sold: United States," in *FRED® Economic Data*, Federal Reserve Bank of St. Louis, July 24, 2014, http://research.stlouisfed.org/fredgraph.pdf?hires=1&type=application/pdf&chart_type=line&recession_bars=on&log_scales=&bgcolor=%23e1e9f0&graph_bgcolor=%23ffffff&fo=verdana&ts=12&tts=12&txtcolor=%23444444&show_legend=yes&show_axis_titles=yes&drp=0&cosd=1963-01-01&coed=2014-06-01&width=670&height=445&stacking=&range=&mode=fred&id=HSN1F&transformation=lin&nd=&ost=-99999&oet=99999&scale=left&line_color=%234572a7&line_style=solid&lw=2&mark_type=none&mw=1&mma=0&fml=a&fgst=lin&fq=Monthly&fam=avg&vintage_date=&revision_date= (accessed August 15, 2014)

FIGURE 4.5

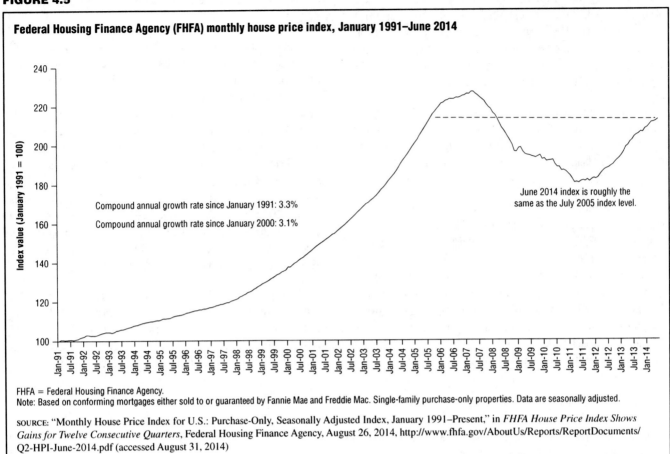

Federal Housing Finance Agency (FHFA) monthly house price index, January 1991–June 2014

FHFA = Federal Housing Finance Agency.
Note: Based on conforming mortgages either sold to or guaranteed by Fannie Mae and Freddie Mac. Single-family purchase-only properties. Data are seasonally adjusted.

SOURCE: "Monthly House Price Index for U.S.: Purchase-Only, Seasonally Adjusted Index, January 1991–Present," in *FHFA House Price Index Shows Gains for Twelve Consecutive Quarters*, Federal Housing Finance Agency, August 26, 2014, http://www.fhfa.gov/AboutUs/Reports/ReportDocuments/Q2-HPI-June-2014.pdf (accessed August 31, 2014)

Investors reaped handsome profits by selling homes relatively quickly for more than what they paid for them. This process became known as "flipping houses."

Huge demand from potential homeowner-occupiers and investors prompted many lenders to relax their underwriting standards and extend loans to borrowers they might have previously rejected. In addition, many lenders greatly expanded their use of creative financing terms, such as ARMs that featured low initial monthly payments. Subprime ARMs became particularly popular. Eric Petroff indicates in "Who Is to Blame for the Subprime Crisis?" (September 5, 2007, http://www.investopedia.com/articles/07/subprime-blame.asp) that only $35 billion in subprime mortgage loans were initiated in 1994. By 2002 that number had risen to $213 billion. Over the next three years the subprime market skyrocketed. In 2005 subprime lenders originated $665 billion of the risky mortgage loans. The White House notes in *Economic Report of the President* (February 2008, http://www.gpoaccess.gov/eop/2008/2008_erp.pdf) that the percentage of mortgage originations that were subprime increased from 5% in 2001 to over 20% in 2006.

Mortgage originators also greatly relaxed their LTV standards during the housing boom. As noted earlier, a low LTV ratio (e.g., less than 0.8 to 0.9) has historically been associated with buyers who are more invested in their properties and less likely to default. Mayer, Pence, and Sherlund find that the median LTV ratio (half of the LTV ratios were higher and half were lower) for subprime mortgages actually increased from 90% in 2003 to 100% in 2005. In other words, many subprime borrowers made no down payment when they purchased their home. In addition, subprime borrowers were regularly put into nontraditional mortgage arrangements, such as ARMs and loans that offered interest-only payments for a short period. These loans featured balloon payments and/or sharp increases in monthly payments at some point during the loan period. The variable interest rates were often tied to economic indexes that can change suddenly and drastically. Mayer, Pence, and Sherlund conclude that "the riskiest borrowers were matched with the most complicated products."

Both the borrowers and lenders of subprime loans expected homes to keep appreciating in value. For homeowners, this appreciation would allow them to refinance their loan and tap into home equity to offset the financial burden of the coming higher monthly payments. Likewise, lenders believed that home appreciation would offset the risk they were taking in lending money to people with poor credit histories or other financial problems.

Thus, like all booms, the housing boom was driven by high expectations that future events would take a favorable path.

Investment companies also drove the housing boom by enthusiastically buying mortgage products on the secondary market. This provided cash to lending companies to underwrite more mortgages. Meanwhile, the stock of banks and other financial institutions providing mortgages greatly rose in value, as investors became excited about the increasing loan portfolios.

DEFAULTING BEGINS TO GROW. According to Mayer, Pence, and Sherlund, the mortgage market first showed signs of trouble in mid-2005, when large lenders noticed a disturbing increase in the percentage of their subprime loans that had become "seriously delinquent." Between 2000 and 2004 an average of 1.5% of subprime loans were in default only one year after the loans were originated. In other words, the borrowers stopped making their mortgage payments within months of taking out the loans. Analysts refer to these events as "early payment defaults." By the end of 2005 early payment defaults for subprime loans had climbed to 2.5% and continued to grow, reaching 5% at year-end 2006 and 8% at year-end 2007.

Early payment defaults are particularly troubling because they happen only months after underwriting occurs. Mayer, Pence, and Sherlund note that these loans "may have been underwritten so poorly that borrowers were unable to afford the monthly payments almost from the moment of origination." They also point to evidence suggesting "fraudulent practices by both borrowers and mortgage brokers." These practices may have kept buyers on the secondary mortgage market from realizing the actual riskiness of the investments they were buying. Regardless, Mayer, Pence, and Sherlund maintain that "some of the deterioration in underwriting characteristics should have been apparent to investors in mortgage-backed securities."

In *Economic Report of the President*, the White House notes that between 2004 and 2006 the default rate among homeowners with subprime ARMs was about 6%. By late 2007 the rate had skyrocketed to more than 15%. The default rate for prime mortgage loans also increased during this period, from about 1% to nearly 4%.

Maura Reynolds reports in "Loan Troubles Hit New Heights" (LATimes.com, March 7, 2008) that the national average foreclosure rate at the end of the fourth quarter of 2007 was the highest in recorded history. Just over 2% of mortgages were in foreclosure at that time. The foreclosure rate was highest in states that had experienced the greatest housing boom (and hence the greatest housing bust): Arizona, California, Florida, and Nevada.

THE ECONOMY SUFFERS. The unusually high default rates caused mounting financial losses for original lenders and MBS investors. The White House states in *Economic Report of the President* that "there were significant disruptions in [the] financial markets in the summer of 2007." In fact, it was the beginning of a vicious downward spiral. As banks, financial institutions, and investment companies reported huge losses, their stock plummeted in value. Some of the companies failed, and many others were in danger of failing.

By this time many banks and financial institutions had stopped originating new mortgages because they were fearful of experiencing more losses. Sales of newly built single-family homes dropped considerably. (See Figure 4.4.) Home prices declined at a record pace, as seen in Figure 4.5. Figure 4.6 shows the median value for existing homes in the United States between 2005 and 2011. The data are based on owners' estimates collected during the American Community Survey conducted by the U.S. Census Bureau. The median home value peaked in late 2007 and early 2008 at around $193,000 and then plummeted. The decline in housing values left many homeowners owing more on their mortgage than their home was worth in the marketplace. This is called being "upside down" or "underwater" in one's mortgage.

FIGURE 4.6

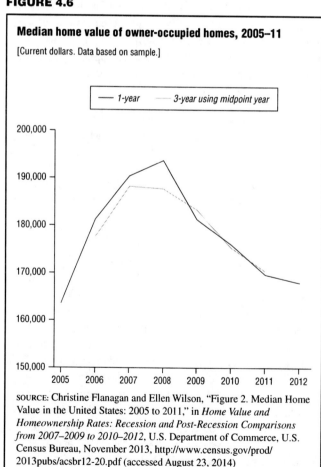

Median home value of owner-occupied homes, 2005–11
[Current dollars. Data based on sample.]

SOURCE: Christine Flanagan and Ellen Wilson, "Figure 2. Median Home Value in the United States: 2005 to 2011," in *Home Value and Homeownership Rates: Recession and Post-Recession Comparisons from 2007–2009 to 2010–2012*, U.S. Department of Commerce, U.S. Census Bureau, November 2013, http://www.census.gov/prod/2013pubs/acsbr12-20.pdf (accessed August 23, 2014)

Mayer, Pence, and Sherlund report that by mid-2008 more than half of the borrowers in Arizona, California, Florida, and Nevada had negative equity (i.e., they were upside down in their mortgages). More than one-third of borrowers in Indiana, Michigan, and Ohio and approximately 10% of borrowers in the rest of the country were upside down in their mortgages.

TROUBLE FOR FANNIE MAE AND FREDDIE MAC. The housing market bust put tremendous financial pressure on Fannie Mae and Freddie Mac. Like many other MBS holders, the two companies suffered huge financial losses due to defaulted loans. By mid-2008 investors were concerned that Fannie Mae and Freddie Mac would be unable to raise sufficient money to cover these losses.

Katie Benner reports in "The $5 Trillion Mess" (CNNMoney.com, July 14, 2008) that shares in Fannie Mae and Freddie Mac dropped in value by 65% and 75%, respectively, during the first half of 2008. At that time the two companies held or guaranteed approximately $5 trillion in mortgage loans.

In July 2008 Congress passed the Housing and Economic Recovery Act. The act created the Federal Housing Finance Agency (FHFA) and gave it the power to put Fannie Mae and/or Freddie Mac into conservatorship. Mark Jickling of the Congressional Research Service explains in "Fannie Mae and Freddie Mac in Conservatorship" (September 15, 2008, http://fpc.state.gov/documents/organization/110097.pdf) that the FHFA was given "full powers to control the assets and operations of the firms." A conservatorship for both companies was officially established in September 2008.

Jickling notes that the U.S. Department of the Treasury began investing money into the firms by buying shares of their stock and "newly issued" Fannie Mae and Freddie Mac MBS packages. In addition, the two firms were allowed to get short-term loans from the Treasury Department using MBS as collateral.

Under the Housing and Economic Recovery Act, the FHFA must submit an annual report to Congress about the status and financial soundness of Fannie Mae and Freddie Mac. In *Report to Congress, 2009* (May 25, 2010, http://www.fhfa.gov/AboutUs/Reports/Pages/FHFA-2009-Annual-Report-to-Congress-.aspx), the FHFA notes that Fannie Mae's annual net income was positive in 2005 and 2006. In 2007 the company suffered a loss of about $2 billion. In 2008 and 2009 its annual losses skyrocketed between $60 billion and $70 billion per year. Freddie Mac's financial performance followed a similar pattern. The company had positive net income in 2005 and 2006 and then reported losses of $2 billion in 2007, $50 billion in 2008, and $22 billion in 2009. At year-end 2009 Fannie Mae and Freddie Mac held or guaranteed nearly half (47%) of the country's outstanding residential mortgage debt.

In *Report to Congress, 2013* (June 13, 2014, http://www.fhfa.gov/AboutUs/Reports/ReportDocuments/FHFA_2013_Report_to_Congress.pdf), the FHFA reports that the enterprises have taken a variety of actions since being put into conservatorship to conserve their assets and prevent future losses. These safeguards include higher FICO credit scores for potential borrowers than were required before 2008 and lower average LTV ratios for newly issued mortgages. In addition, the FHFA notes, "Higher-risk mortgages, such as no-income documentation or interest-only mortgages, have largely been eliminated from the Enterprises' new guarantees." The FHFA also states, "Since conservatorship, the Enterprises have completed 3.1 million foreclosure alternative actions, including over 1.6 million permanent loan modifications."

In the fact sheet "FHFA's Update on Private Label Securities Actions" (September 12, 2014, http://www.fhfa.gov/Media/PublicAffairs/Pages/FHFAs-Update-on-Private-Label-Securities-Actions.aspx), the FHFA notes that in 2011 it filed lawsuits against more than a dozen financial institutions that had sold MBS to Fannie Mae or Freddie Mac. The institutions were accused of violating securities law and committing fraud in some cases. As of August 2014, the FHFA had obtained settlement agreements totaling $17.7 billion. (See Table 4.4.) In addition, Wells Fargo Bank had agreed to pay over $335 million in

TABLE 4.4

Settled and pending federal litigation cases regarding private-label securities (PLS) as of August 2014

PLS litigation settlements

1. General Electric Company	$6.25 million
2. CitiGroup Inc.	$250 million
3. UBS Americas, Inc. (Union Bank of Switzerland)	$885 million
4. J.P. Morgan Chase & Co.	$4 billion
5. Deutsche Bank AG	$1.925 billion
6. Ally Financial, Inc.	$475 million
7. Morgan Stanley	$1.25 billion
8. SG Americas (Societe Generale)	$122 million
9. Credit Suisse Holdings (USA) Inc.	$885 million
10. Bank of America Corp.	
11. Merrill Lynch & Co.	$5.83 billion
12. Countrywide Financial Corporation	
13. Barclays Bank PLC	$280 million
14. First Horizon National Corp.	$110 million
15. RBS Securities, Inc. (in Ally action)	$99.5 million
16. Goldman Sachs & Co.	$1.2 billion

Non-litigation PLS settlements

Wells Fargo Bank, N.A.	$335.23 million

Remaining PLS cases

Southern District of New York cases:

HSBC North America Holdings, Inc. (Hong Kong Shanghai Banking Corp.)
Nomura Holding America, Inc.

District of Connecticut case:

The Royal Bank of Scotland Group, PLC

SOURCE: "PLS Litigation Settlements," in *FHFA's Update on Private Label Securities Actions: 2013 and 2014 Settlements and Remaining Cases*, Federal Housing Finance Agency, August 22, 2014, http://www.fhfa.gov/Media/PublicAffairs/Pages/FHFAs-Update-on-Private-Label-Securities-Actions.aspx (accessed August 24, 2014)

a nonlitigation settlement. Additional cases were pending against three other financial institutions.

GOVERNMENT PROGRAMS. The Housing and Economic Recovery Act is only one of many actions that have been taken by the federal government to try to boost the ailing housing market. After taking office in January 2009, President Barack Obama (1961–) pushed for major reforms of the financial markets and set up programs that were specifically designed to address the housing market crisis. The Making Home Affordable (MHA) program (http://www.makinghomeaffordable.gov) helps individual homeowners restructure their mortgages, while the Hardest Hit Fund (2012, http://www.treasury.gov/initiatives/financial-stability/TARP-Programs/housing/hhf/Pages/Program-Purpose-and-Overview.aspx) funnels money to state housing finance agencies "to develop locally-tailored foreclosure prevention solutions in areas that have been hard hit by home price declines and high unemployment." Both programs were established by the Treasury Department's Office of Financial Stability under the Troubled Asset Relief Program (TARP), a broad-scaled financial program that was started during the administration of President George W. Bush (1946–) and authorized by Congress through the Emergency Economic Stabilization Act of 2008. The TARP program is described in detail in Chapter 6.

MHA program components include the Home Affordable Modification Program, which allows homeowners who are struggling to make their monthly payments and who face foreseeable and/or imminent default to have their mortgages restructured to be more affordable. Another component is the Home Affordable Refinance Program (HARP), which provides homeowners with mortgages that are owned or guaranteed by Fannie Mae or Freddie Mac an opportunity to refinance the loans to make the monthly payments more affordable.

The MHA program got off to a slow start with lenders and borrowers blaming each other for various problems. The article "Obama Administration to Extend and Expand Foreclosure Relief Program" (Associated Press, January 27, 2012) notes that "homeowners have complained that they were disqualified after banks lost their documents and failed to return phone calls. Banks have blamed homeowners for failing to submit needed paperwork." The Obama administration modified and extended the program, and participation by homeowners and lenders began to increase. In "Making Home Affordable: Program Performance Report through April 2014" (June 13, 2014, http://www.treasury.gov/initiatives/financial-stability/reports/Documents/April%202014%20MHA%20Report%20final.pdf), the Treasury Department reports that more than 2 million "homeowner assistance actions" had been taken under the program as of April 2014. This included more than 1.3 million homeowners who had received a Home Affordable Modification Program permanent mortgage modification. In "Mortgage Program Pans Out" (WSJ.com, December 31, 2014), Nick Timiraos notes that as of December 2013 nearly 3 million homeowners had refinanced their loans through the HARP program.

President Obama has tried to persuade Congress to approve other programs for homeowner assistance that would be funded by fees or taxes imposed on large banks; however, as of December 2014, these efforts had proved unsuccessful.

Federal and state officials have successfully sued large home lenders over alleged misconduct during the housing crisis. In February 2012 a $25 billion settlement was reached with Ally Financial (formerly known as GMAC), Bank of America, Citigroup, JPMorgan Chase, and Wells Fargo over accusations they had engaged in improper foreclosure procedures. The case involved a practice called "robo-signing," in which loan servicers allegedly signed documents that authorized foreclosures to proceed without checking to see if all the proper paperwork (i.e., proof of ownership changes) had been filed as mortgages were sold between companies and bundled into securities. Chris Isidore and Jennifer Liberto note in "Mortgage Deal Could Bring Billions in Relief" (CNNMoney.com, February 15, 2012) that most of the settlement money was slated to help homeowners who were underwater in their mortgages. The deal covered all states, except Oklahoma, which negotiated a separate $18.6 billion settlement for its residents who were allegedly victims of improper foreclosures.

THE HOUSING MARKET SLOWLY RECOVERS. Figure 4.7 shows an economic indicator commonly called "housing starts" (i.e., the number of new privately owned housing units that began construction). Note that monthly data have been seasonally adjusted and annualized. Housing starts bottomed out in early 2009 at a seasonally adjusted annual rate of about 500,000 units. The number began to climb, reaching nearly 900,000 units in June 2014. In other words it was expected in June 2014 that housing starts for the year would total almost 900,000 units. Sales of new homes have also slowly recovered, from a seasonally adjusted annual rate of fewer than 300,000 units in 2010 to just over 400,000 units as of June 2014. (See Figure 4.4.)

House prices have also been creeping upward. As shown in Figure 4.5, the FHFA HPI began climbing in early 2011. Table 4.5 provides a breakdown by state for home appreciation rates as of June 30, 2014. The national five-year average was 8.25%. The District of Columbia, North Dakota, and California experienced the highest five-year appreciation in home values with rates ranging from 30% to 40%, considerably higher than the national average. Nearly all states reported upward movement in their one-year and one-quarter

FIGURE 4.7

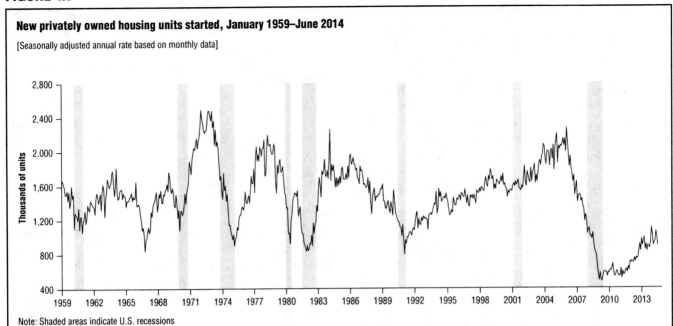

New privately owned housing units started, January 1959–June 2014

[Seasonally adjusted annual rate based on monthly data]

Note: Shaded areas indicate U.S. recessions

SOURCE: "Housing Starts: Total: New Privately Owned Housing Units Started," in *FRED® Economic Data*, Federal Reserve Bank of St. Louis, July 17, 2014, http://research.stlouisfed.org/fredgraph.pdf?hires=1&type=application/pdf&chart_type=line&recession_bars=on&log_scales=&bgcolor=%23e1e9f0&graph_bgcolor=%23ffffff&fo=verdana&ts=12&tts=12&txtcolor=%23444444&show_legend=yes&show_axis_titles=yes&drp=0&cosd=1959-01-01&coed=2014-06-01&width=670&height=445&stacking=&range=&mode=fred&id=HOUST&transformation=lin&nd=&ost=-99999&oet=99999&scale=left&line_color=%234572a7&line_style=solid&lw=2&mark_type=none&mw=1&mma=0&fml=a&fgst=lin&fq=Monthly&fam=avg&vintage_date=&revision_date= (accessed August 15, 2014)

rates. The national one-year average was 5.25%. According to the Federal Reserve Bank of St. Louis (2014, http://research.stlouisfed.org/fred2/series/SPCS20RSA), the 20-city composite S&P/Case-Shiller Home Price Index plunged to about 140 in 2009 (from its record-high value above 200 in late 2006). The index hovered around 140 through 2012 and then began to increase. It was about 170 through much of 2014.

Meanwhile, delinquency and foreclosure data show improvements in the housing market. The Mortgage Bankers Association provides regular reports on delinquency and foreclosure rates based on surveys it conducts. In "Delinquency and Foreclosure Rates Decrease in Second Quarter" (August 7, 2014, http://www.mbaa.org/NewsandMedia/PressCenter/89113.htm), the association reports that 6% of all mortgage loans on one-to-four-unit residential properties were past due on payments at the end of the second quarter of 2014. It was the lowest rate since the fourth quarter of 2007. Likewise, the percentage of mortgages in the foreclosure process was 2.5% at the end of the second quarter of 2014. This was the lowest rate since the first quarter of 2008.

CONSUMER CREDIT

The Federal Reserve defines the term *consumer credit* as credit extended to individuals that does not include loans secured by real estate. In other words, mortgages are excluded. Creditors providing consumer loans are primarily banks, finance companies, and credit unions. The federal government is also a major source of consumer credit because it finances loans for educational purposes (i.e., student loans). Creditors often base the interest rates they charge on a particular benchmark or index, such as the prime rate. As index rates move up and down, the interest rates offered for consumer loans often change also.

Consumer loans are typically for much lower amounts than mortgage loans and undergo a much less rigid underwriting process. They also have a higher risk of default than mortgage loans. As such, consumer loans generally have higher interest rates than mortgage loans. Applicants with good credit histories (as indicated by their credit scores and credit reports) are able to obtain larger amounts of consumer credit and lower interest rates than applicants with poorer credit histories.

Consumer credit includes nonrevolving and revolving loans. As noted earlier, nonrevolving loans have an amortization schedule over a set period (e.g., four years) that includes fixed payment amounts (e.g., $350 per month). If the borrower makes payments in accordance with the schedule, the loan is completely paid off at the end of the period. This is possible because nonrevolving loans have fixed interest rates. In other words, the same interest rate applies throughout the payment period. Nonrevolving

TABLE 4.5

Federal Housing Finance Agency (FHFA) house price appreciation, by state, period ending June 30, 2014

State	Rank*	1-Yr.	Qtr.	5-Yr.	Since 1991 Q1
Nevada (NV)	1	14.80%	0.87%	18.07%	68.95%
California (CA)	2	11.38%	1.33%	30.16%	112.60%
District of Columbia (DC)	3	10.74%	−1.94%	40.12%	349.76%
North Dakota (ND)	4	8.68%	1.72%	30.79%	185.08%
Arizona (AZ)	5	8.39%	0.61%	16.30%	135.91%
Florida (FL)	6	8.32%	1.25%	12.34%	115.40%
Georgia (GA)	7	8.13%	0.45%	7.12%	85.00%
Colorado (CO)	8	7.76%	0.69%	19.92%	219.86%
Michigan (MI)	9	7.18%	1.46%	13.45%	75.41%
Oregon (OR)	10	6.98%	1.15%	6.61%	207.11%
Washington (WA)	11	6.97%	2.43%	2.36%	148.61%
Idaho (ID)	12	6.56%	1.25%	−1.84%	129.62%
Texas (TX)	13	6.46%	0.17%	17.55%	121.85%
South Dakota (SD)	14	5.94%	0.70%	11.24%	150.33%
Minnesota (MN)	15	5.89%	1.81%	6.59%	134.50%
USA		**5.25%**	**0.81%**	**8.25%**	**108.94%**
Utah (UT)	16	5.24%	0.64%	10.17%	197.44%
West Virginia (WV)	17	4.99%	0.72%	7.45%	104.80%
Massachusetts (MA)	18	4.95%	2.50%	7.44%	136.09%
Tennessee (TN)	19	4.85%	0.63%	7.44%	104.67%
South Carolina (SC)	20	4.60%	0.90%	0.74%	93.28%
North Carolina (NC)	21	4.35%	1.91%	−0.02%	94.46%
Ohio (OH)	22	4.09%	0.38%	4.20%	65.10%
Kansas (KS)	23	4.03%	0.08%	5.58%	104.01%
Montana (MT)	24	4.03%	0.88%	7.39%	224.92%
Nebraska (NE)	25	3.99%	1.21%	9.80%	112.66%
Indiana (IN)	26	3.71%	0.53%	6.56%	70.58%
Wyoming (WY)	27	3.69%	−0.54%	5.91%	209.52%
Maryland (MD)	28	3.60%	1.22%	2.66%	127.05%
New Hampshire (NH)	29	3.40%	−0.50%	−0.43%	104.26%
Illinois (IL)	30	3.11%	1.06%	−3.33%	80.09%
Vermont (VT)	31	3.06%	6.18%	2.29%	118.06%
Wisconsin (WI)	32	2.77%	1.01%	−1.34%	112.91%
Missouri (MO)	33	2.68%	−0.14%	1.21%	93.69%
New Jersey (NJ)	34	2.62%	0.80%	−4.25%	117.37%
Rhode Island (RI)	35	2.57%	−0.49%	−2.66%	87.45%
Pennsylvania (PA)	36	2.40%	1.43%	3.11%	96.44%
Iowa (IA)	37	2.35%	0.11%	7.36%	109.17%
Kentucky (KY)	38	2.28%	0.34%	4.54%	97.26%
New York (NY)	39	2.28%	0.64%	1.31%	110.42%
Louisiana (LA)	40	2.22%	0.89%	6.79%	145.34%
Alabama (AL)	41	1.70%	−0.94%	−2.81%	87.21%
Maine (ME)	42	1.60%	1.02%	−1.45%	109.16%
Hawaii (HI)	43	1.40%	−4.55%	10.01%	101.01%
Oklahoma (OK)	44	1.38%	−0.77%	6.73%	107.03%
New Mexico (NM)	45	1.37%	1.26%	−6.10%	113.11%
Arkansas (AR)	46	1.28%	0.74%	3.47%	90.96%
Virginia (VA)	47	1.24%	−0.23%	4.33%	125.32%
Delaware (DE)	48	1.13%	2.41%	−9.14%	86.68%
Alaska (AK)	49	0.34%	0.91%	7.68%	133.24%
Connecticut (CT)	50	0.29%	−1.18%	−5.96%	66.58%
Mississippi (MS)	51	−0.51%	−0.72%	−0.48%	80.48%

*Rankings based on annual percentage change.
Note: Based on conforming mortgages either sold to or guaranteed by Fannie Mae and Freddie Mac. Single-family purchase-only properties.

SOURCE: "House Price Appreciation by State: Percent Change in House Prices," in *FHFA House Price Index Shows Gains for Twelve Consecutive Quarters*, Federal Housing Finance Agency, August 26, 2014, http://www.fhfa.gov/AboutUs/Reports/ReportDocuments/Q2-HPI-June-2014.pdf (accessed August 31, 2014)

---

loans are commonly obtained for vehicles, boats, vacations, and educational expenses.

Revolving debt is quite different because it does not have an amortization payment schedule. The creditor grants the borrower a total amount of credit at a particular interest rate that may be fixed for a limited time before beginning to vary. Each month the borrower is billed for the outstanding balance, which includes principal plus interest. The borrower can pay off the entire balance or a lesser amount down to the minimum payment required by the creditor. Payment of any amount less than the entire balance will result in additional finance charges on the remaining balance. This is an example of compound interest (interest charged on an amount that already includes built-up interest charges). The major example of revolving debt is credit card debt. Some home equity loans and personal loans also use the revolving model.

Total Consumer Debt

The nation's total consumer debt is calculated and tracked by the Federal Reserve based on data obtained from creditors. Table 4.6 provides a detailed breakdown of outstanding consumer debt by creditor type as of June 2014. Depository institutions, such as commercial banks, held nearly $1.3 trillion (40% of the total), the federal government held $785.6 billion (25% of the total), and finance companies held $672 billion (21% of the total). The remaining balance was held by credit unions and other types of creditors.

Figure 4.8 shows that the amount of outstanding consumer credit grew relatively slowly during the 1970s and 1980s and then skyrocketed beginning in the 1990s. The total peaked at nearly $2.7 trillion in mid-2008 and then declined sharply. At first, this drop was hailed as proof that Americans had retreated from the heavy reliance on credit that characterized the years leading up to the Great Recession. However, the change was short lived. As shown in Figure 4.8, the amount of consumer credit began rising again, reaching $3.2 trillion in June 2014.

Consumer credit during the 1970s and 1980s consisted almost entirely of nonrevolving debt. During the 1990s the breakdown began to change with revolving debt accounting for an ever-increasing proportion of consumer debt. Of the $3.2 trillion in consumer debt in June 2014, nonrevolving debt accounted for $2.3 trillion (73% of the total) and revolving debt accounted for $837 billion (27% of the total). (See Table 4.6.)

As noted earlier, in *Quarterly Report on Household Debt and Credit* the FRBNY found that student loans, auto loans, and credit card debt were the three largest types of nonmortgage debt held by U.S. households at the end of the second quarter of 2014. Each of these debt types will be examined in more detail.

Student Loans

Student loans are loans obtained to pay for educational expenses, primarily at the college level. Although

TABLE 4.6

Consumer credit outstanding, by major holders and major types of credit, June 2014

[Billions of dollars]

	June 2014 (preliminary)
Total	3,168.5
Major holders	
Depository institutions	1,281.4
Finance companies	672.0
Credit unions	284.3
Federal government[a]	785.6
Nonprofit and educational institutions[b]	56.5
Nonfinancial business	42.6
Pools of securitized assets[c, d]	46.2
Major types of credit, by holder	
Revolving	837.0
Depository institutions	681.3
Finance companies	61.9
Credit unions	43.3
Federal government[a]	N/A
Nonprofit and educational institutions[b]	N/A
Nonfinancial business	22.2
Pools of securitized assets[c, d]	28.3
Nonrevolving	2,331.5
Depository institutions	600.1
Finance companies	610.1
Credit unions	241.0
Federal government[a]	785.6
Nonprofit and educational institutions[b]	56.5
Nonfinancial business	20.4
Pools of securitized assets[c, d]	17.8
Memo	
Student loans[e]	1,274.6
Motor vehicle loans[f]	918.1

[a]Includes student loans originated by the Department of Education under the Federal Direct Loan Program and the Perkins Loan Program, as well as Federal Family Education Program loans that the government purchased under the Ensuring Continued Access to Student Loans Act.
[b]Includes student loans originated under the Federal Family Education Loan Program and held by educational institutions and nonprofit organizations that are affiliated with state governments.
[c]Outstanding balances of pools upon which securities have been issued; these balances are no longer carried on the balance sheets of the loan originators.
[d]The shift of consumer credit from pools of securitized assets to other categories is largely due to financial institutions' implementation of the FAS 166/167 accounting rules.
[e]Includes student loans originated under the Federal Family Education Loan Program and the Direct Loan Program; Perkins loans; and private student loans without government guarantees. This memo item includes loan balances that are not included in the nonrevolving credit balances. Data for this memo item are released for each quarter-end month.
[f]Includes motor vehicle loans owned and securitized by depository institutions, finance companies, credit unions, and nonfinancial business. Includes loans for passenger cars and other vehicles such as minivans, vans, sport-utility vehicles, pickup trucks, and similar light trucks for personal use. Loans for boats, motorcycles and recreational vehicles are not included. Data for this memo item are released for each quarter-end month.
N/A = not applicable.

SOURCE: Adapted from "Consumer Credit Outstanding (Levels)," in *Federal Reserve Statistical Release G.19: Consumer Credit, June 2014*, Board of Governors of the Federal Reserve System, August 7, 2014, http://www.federalreserve.gov/releases/g19/current/g19.pdf (accessed August 15, 2014)

they are called consumer loans, student loans are not for consumption purposes. They fund the advancement of skill and knowledge in individuals, likely increasing the potential for higher future income. Thus, student loans are considered a type of "human investment." Because the federal government encourages higher education, it plays a major role in ensuring that student loans are available. As shown in Table 4.6, the federal government held $785.6 billion in consumer credit in June 2014. According to the Federal Reserve, this credit consisted of loans that were originated by the U.S. Department of Education under the William D. Ford Federal Direct Loan Program, the Perkins Loan Program, and the Federal Family Education Loan program, which consisted of loans that the government purchased from depository institutions and finance companies. Student loans are also available through the private sector, such as from commercial banks.

The FRBNY indicates that student loan debt was an estimated $1.1 trillion as of June 30, 2014. This amount had skyrocketed from the first quarter of 2003, when it was only $241 billion. Student loan debt has grown to be the largest single type of consumer debt held by American households. According to the FRBNY, in *Quarterly Report on Household Debt and Credit*, 10.9% of student loan balances were 90 or more days delinquent as of June 30, 2014. This was the highest rate for all the loan types considered. While the rate was down slightly from its peak of 11.8% in 2013, it had grown substantially since 2003 when it was only 6.1%.

Concern about student loan delinquency rates has been building for years. In fact, some analysts believe that student loan debt constitutes an economic "bubble" that will burst, much as the housing loan bubble burst. This would have devastating financial effects on the individuals involved and put a financial burden on the federal government (and hence U.S. taxpayers) in the event of massive defaults on student loans.

According to the U.S. Department of Education (ED), in "Collections" (2014, https://studentaid.ed.gov/repay-loans/default/collections), once a borrower defaults on a federal student loan the entire balance—principal and interest—become due at once. The debt can be turned over to a collection agency. Unlike a private company, the ED can take certain actions without a court order against debtors in default. Specifically, it can collect the debt by withholding money from a debtor's wages, income tax refund, or other federal payments.

Auto Loans

Vehicle loans are typically nonrevolving loans that are secured by the vehicle being purchased. In other words, the creditor can repossess the vehicle if the loan is in default. As a result of this collateral, the interest rates on vehicle loans tend to be lower than on other types of consumer loans. According to the Federal Reserve, in "Consumer Credit: June 2014" (August 7, 2014, http://www.federalreserve.gov/releases/g19/current/g19.pdf), as of May 2014 the average interest rate on a 48-month (four-year) loan from a commercial bank for the purchase

FIGURE 4.8

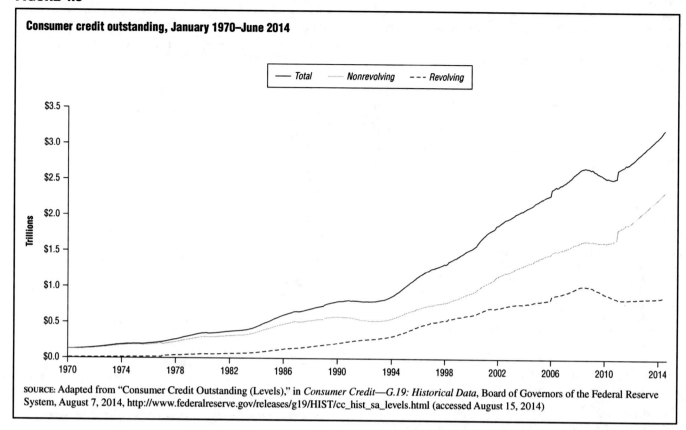

Consumer credit outstanding, January 1970–June 2014

SOURCE: Adapted from "Consumer Credit Outstanding (Levels)," in *Consumer Credit—G.19: Historical Data*, Board of Governors of the Federal Reserve System, August 7, 2014, http://www.federalreserve.gov/releases/g19/HIST/cc_hist_sa_levels.html (accessed August 15, 2014)

of a new car was 4.5%. This was down from an average rate of 6.7% in 2009.

As shown in Table 4.6, outstanding loans on motor vehicles totaled $918.1 billion as of June 2014. This includes cars, trucks, vans, and sport utility vehicles (SUVs) for personal use, but excludes boats, motorcycles, and recreational vehicles.

Credit Cards

On average, credit cards have higher interest rates than other types of loans, including mortgages, vehicle loans, and student loans. Most credit card loans are unsecured. The higher risk factor for the creditor and the huge demand for credit cards contribute to the high interest rates that are charged. In "Consumer Credit: June 2014," the Federal Reserve assesses all credit card accounts and the credit card accounts that were assessed interest during the period. Note that credit card owners with zero balances do not pay interest. In addition, some credit card issuers offer introductory or special deals (usually for a set period) during which no interest is due. According to the Federal Reserve, the average interest rate charged by commercial banks in May 2014 on all credit card accounts was 11.8%. The rate was slightly higher (12.7%) when considering only accounts that were assessed interest. These rates were down from the averages of 13.4% and 14.3%, respectively, reported in 2009.

When credit cards were first introduced, it was common for creditors to require 5% or more of the balance as a minimum monthly payment. Minimum payment requirements were gradually reduced to 2% by most credit card issuers. Low required minimum payments, high interest rates, and the effect of compounding interest make it difficult for many consumers to pay off credit card debt. The Bankruptcy Abuse Prevention and Consumer Protection Act of 2005 requires creditors to tell borrowers how long it will take to pay off their credit card debt if only minimum payments are made. The Credit CARD Act of 2009 puts limits on when credit card issuers can raise interest rates and bans "unfair or deceptive" billing procedures.

CREDIT CARD USAGE AND HABITS. In *Quarterly Report on Household Debt and Credit*, the FRBNY estimates there were 405.9 million household credit card accounts as of June 30, 2014 (the end of the second quarter of 2014), and the total balance owed on them was $669 billion. Thus, the average amount owed per account was $1,648.

The Gallup Organization regularly polls Americans on their credit card habits. In *Americans Rely Less on Credit Cards Than in Previous Years* (April 24, 2014, http://www.gallup.com/poll/168668/americans-rely-less-credit-cards-previous-years.aspx), Art Swift notes that a poll conducted in April 2014 found that 29% of respondents

TABLE 4.7

Poll respondents categorize their credit card payments, selected years, 2001–14

HOW DO YOU GENERALLY PAY YOUR CREDIT CARD(S) EACH MONTH?

	Always pay full amount %	Usually pay full amount %	Usually leave balances %	Usually pay minimum amount %	Pay less than minimum amount %
Apr 3–6, 2014	48	16	20	12	1
Apr 6–9, 2008	43	17	25	12	1
Apr 10–13, 2006	42	17	27	11	2
Apr 5–8, 2004	37	16	31	12	2
Apr 8–11, 2002	41	14	28	13	2
Apr 6–8, 2001	42	16	29	11	1

Note: Based on 765 credit card owners.

SOURCE: Art Swift, "Credit Cards and Balances," in *Americans Rely Less on Credit Cards Than in Previous Years*, The Gallup Organization, April 24, 2014, http://www.gallup.com/poll/168668/americans-rely-less-credit-cards-previous-years.aspx (accessed August 18, 2014). Copyright © 2014 Gallup, Inc. All rights reserved. The content is used with permission; however, Gallup retains all rights of republication.

TABLE 4.8

Credit card debt of poll respondents, selected years, 2006–14

WHAT IS THE TOTAL AMOUNT OF MONEY YOU OWE ON ALL OF YOUR CREDIT CARDS, THAT YOU DO NOT INTEND TO PAY OFF THIS MONTH?

	Apr 10–13, 2006 %	Apr 6–9, 2008 %	Apr 3–6, 2014 %
National adults			
None/do not own credit card	21	22	29
None/own credit card	29	33	30
$0 to $1,000	12	10	10
$1,001 to $5,000	17	13	11
$5,001 to $10,000	8	7	4
More than $10,000	6	7	6
No opinion	7	8	9
Mean (including zero)	$2,947	$2,941	$2,426
Credit card owners			
None/own credit card	40	42	45
$0 to $1,000	17	14	15
$1,001 to $5,000	20	17	17
$5,001 to $10,000	9	10	6
More than $10,000	7	9	8
No opinion	6	8	8
Mean (including zero)	$3,426	$3,848	$3,573

SOURCE: Art Swift, "Credit Card Debt in the U.S.," in *Americans Rely Less on Credit Cards Than in Previous Years*, The Gallup Organization, April 24, 2014, http://www.gallup.com/poll/168668/americans-rely-less-credit-cards-previous-years.aspx (accessed August 18, 2014). Copyright © 2014 Gallup, Inc. All rights reserved. The content is used with permission; however, Gallup retains all rights of republication.

did not own a credit card. This was the highest percentage recorded by Gallup since 2001. Of the persons who did own at least one credit card in 2014, the largest portion (33%) had one or two credit cards, while 18% had three or four credit cards. Among credit card owners, the average was 3.7 credit cards. As shown in Table 4.7, nearly half (48%) of the credit card owners said they "always" pay the full amount due each month. Another 16% said they "usually" pay the full amount due. One percent of those asked said they pay less than the minimum amount due each month.

According to Swift, the average balance for the credit card owners in April 2014 was $3,573. Gallup occasionally asks poll participants about their intentions to carry (not pay off) their credit card balances. The results are shown in Table 4.8. The April 2014 poll revealed that 45% of credit card owners intended to pay off their total balance that month. Another 15% planned to carry a balance of $1,000 or less, while 17% intended to carry a balance between $1,001 and $5,000. Smaller percentages planned to carry even higher balances.

The many variables involved in credit card usage and payment habits make it very difficult to calculate the amount of credit card debt held by the average American. In "Average Credit Card Debt Statistics" (September 23, 2014, http://www.creditcards.com/credit-card-news/average-credit_card_debt-1276.php), Fred O. Williams provides estimates obtained from various sources of $1,098 to $7,743 per card, per person, or per household and notes, "The complications stem from different ways of defining credit cards, different ways of counting the card-carrying population, and the different ways people use their cards." For example, some sources do not include the balances on unused cards (i.e., cards with long-time zero balances), cards which are paid off each month, or cards issued by stores and other nonfinancial institutions.

PERSONAL BANKRUPTCIES

The word *bankrupt* is derived from the Italian phrase *banca rotta*, which means "bench broken," referring to the benches or tables that were used by merchants in outdoor markets in 16th-century Italy. Bankruptcy is a state of financial ruin. Under U.S. law, people with more debts than they can reasonably hope to repay can file for personal bankruptcy. This results in a legally binding agreement between debtors and the federal government worked out in a federal bankruptcy court. The agreement calls for the debtors to pay as much as they can with whatever assets they have, and after a predetermined time (usually a number of years) the debtors begin again with new credit. Depending on state law, certain belongings may be kept through the bankruptcy.

The American Bankruptcy Institute explains in "Frequently Asked Questions" (2014, http://bankruptcyresources.org/faq-page) that an official declaration of bankruptcy benefits individuals in the short term because it puts a stop to all collection efforts by creditors. An "automatic stay" goes into effect that prevents creditors from calling, writing, or suing debtors who are covered by a bankruptcy plan.

FIGURE 4.9

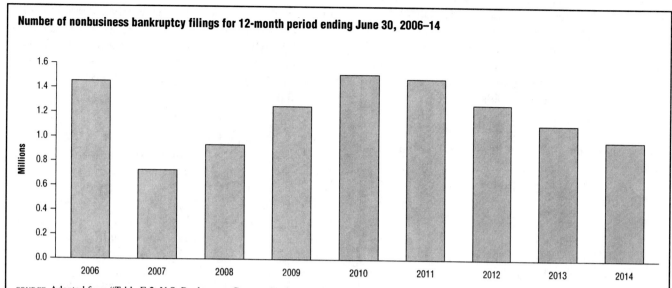

Number of nonbusiness bankruptcy filings for 12-month period ending June 30, 2006–14

SOURCE: Adapted from "Table F-2. U.S. Bankruptcy Courts—Business and Nonbusiness Cases Commenced, by Chapter of the Bankruptcy Code, during the 12 Month Period Ending June 30, 2014," in *Bankruptcy Statistics: 12 Month Period Ending June 30, 2014*, The U.S. Courts, 2014, http://www.uscourts.gov/Statistics/BankruptcyStatistics/12-month-period-ending-june.aspx (accessed August 14, 2014)

Figure 4.9 shows the number of nonbusiness (personal) bankruptcy cases per 12-month period ending June 30 of each year between 2006 and 2014. For the 12-month period ending June 30, 2014, there were nearly 1 million personal bankruptcy filings. This value was down from the previous five years, but higher than levels reported for 2007 and 2008.

There are three types (or chapters) of personal bankruptcy under which individuals may file:

- Chapter 7—a liquidation plan is developed in which the debtor turns over certain assets that are sold and used to pay creditors.

- Chapter 13—a payment plan is developed under which debtors receiving regular income repay their creditors. Liquidation of assets is not required in most cases.

- Chapter 11—while similar to Chapter 13, Chapter 11 is reserved for individuals with substantial debts and assets.

According to the U.S. Courts (October 2014, http://www.uscourts.gov/Statistics/BankruptcyStatistics/2014-bankruptcy-filings.aspx), 623,349 personal bankruptcy cases were filed under Chapter 7 during the 12-month period ending September 30, 2014. This represented 67% of the total 935,420 personal bankruptcy filings. Another 310,914 (33%) filings were under Chapter 13 and the remaining 1,154 (0.1%) were under Chapter 11.

MEDICAL DEBT AND BANKRUPTCIES

Medical debt is debt incurred for medical expenses. Even people with health insurance plans can owe money to hospitals, clinics, and physician offices for expenses that are not covered by the plans. These include copays, deductibles, and coinsurance, which are defined in Chapter 1. It is difficult to determine the total amount of medical debt in the United States because it is spread across various loan types. People may charge medical expenses on their credit cards or take out personal or home equity loans to pay them. Some people, however, owe debts directly to medical businesses. As noted earlier, Ernst & Young report in "The Impact of Third-Party Debt Collection on the U.S. National and State Economies in 2013" that third-party debt collectors recouped $55 billion for creditors during 2013. More than one-third (38%) of this amount was for health care debt, which was the largest single type of debt recouped that year.

Surveys conducted by public and private organizations reveal that medical debt is a significant problem for some Americans. The Centers for Disease Control and Prevention (CDC) is an agency within the U.S. Department of Health and Human Services. The CDC regularly conducts a National Health Interview Survey (NHIS) in which it quizzes U.S. residents about health-related matters. In "Financial Burden of Medical Care: A Family Perspective" (*NCHS Data Brief*, no. 142, January 2014, http://www.cdc.gov/nchs/data/databriefs/db142.pdf), Robin A. Cohen and Whitney K. Kirzinger report that the 2012 NHIS included 43,345 families considered representative of the civilian noninstitutionalized population of the United States (i.e., people not in the military and not in institutions, such as prisons or long-term care facilities). More than a quarter (25.4%) of the families indicated they had experienced "problems paying medical bills in the past 12 months." Of this group, 8.9% said they currently had medical bills that "they were unable to pay at all." Overall, 21.4% of the families noted they currently had medical bills that they were paying over time. According to Cohen and Kirzinger, there were significant rate differences between families. For example, 26% of families in the lowest income group had experienced problems paying their medical bills in the previous 12 months, compared with only 6.4% in the highest income group. More families with children had payment difficulties than did childless families. Likewise, families in which some or all members lacked health insurance were more likely to report problems paying their medical bills than did families in which all members were insured.

The Commonwealth Fund is a private organization that conducts research on the health care system. Every two years it reports the results of its Biennial Health Insurance Survey. In "Insuring the Future: Current Trends in Health Coverage and the Effects of Implementing the Affordable Care Act, Findings from the Commonwealth Fund Biennial Health Insurance Survey, 2012" (April 2013, http://www.commonwealthfund.org/~/media/files/publications/fund-report/2013/apr/1681_collins_insuring_future_biennial_survey_2012_final.pdf), Sara R. Collins et al. indicate that the 2012 survey included 3,393 respondents aged 19 to 64 years. They are believed representative of approximately 183.9 million U.S. adults in this age range. Overall, 68% of the respondents reported one or more problems with medical bills or debt during the previous two years. This included 42% whose credit rating had been lowered, 37% who had used all of their savings to pay medical bills, 27% who had taken on credit card debt, 7% who had taken out a mortgage on their home or another type of loan, and 6% who had declared bankruptcy because of medical debt.

Some researchers believe that medical debt is a major cause of bankruptcy in the United States. In June 2009 the *American Journal of Medicine* published "Medical Bankruptcy in the United States, 2007: Results of a National Study" (http://www.pnhp.org/new_bankruptcy_study/Bankruptcy-2009.pdf) by David U. Himmelstein et al. In this often-quoted study, the researchers indicate that in 2007 they obtained self-reported information from 2,314 U.S. debtors who had declared bankruptcy that year. Just over 1,000 of the respondents were interviewed by telephone about their experiences. Himmelstein et al. note that in 29% of the bankruptcies "the debtor said medical bills were reason for bankruptcy." By taking into account additional criteria, including lost income due to illness and cases in which people mortgaged their homes to pay medical bills, the researchers conclude that 62.1% of all bankruptcies in 2007 were precipitated by medical factors. The latter figure has been widely reported in the media and is hotly disputed. Some analysts believe it is too broad and that the 29% figure is more appropriate, because it is directly supported by the survey respondents.

CHAPTER 5
THE AMERICAN WORKER

When we are all in the business working together, we all ought to have some share in the profits—by way of a good wage, or salary, or added compensation.

—Henry Ford, *My Life and Work* (1922)

The American workforce plays a major role in the U.S. economy. Workers produce goods and provide services, the consumption of which drives growth in the nation's gross domestic product (the total market value of final goods and services that are produced within an economy in a given year). Nevertheless, there is an age-old struggle between employers and employees over compensation. Businesses must compensate workers with pay and benefits that are high enough to attract and keep motivated employees, but not so high as to damage the profitability and growth of the business itself. On a macroeconomic scale, gainful employment of large numbers of workers is important to the overall health of the U.S. economy. As noted in Chapter 2, the so-called Great Recession (which lasted from December 2007 to June 2009) led to a huge increase in the nation's unemployment rate. High unemployment lingered even after the recession ended and greatly affected the economic well-being of the American people.

EMPLOYMENT AND UNEMPLOYMENT

The U.S. Bureau of Labor Statistics (BLS) publishes the monthly news release "The Employment Situation," which the media commonly calls the "monthly jobs report." The BLS tracks employment and unemployment using two surveys. The Current Employment Statistics program includes monthly surveys of approximately 144,000 nonfarm businesses and government agencies with about 554,000 worksites around the country to obtain detailed information on employment, work hours, and payroll. The BLS refers to these data as "establishment data." The agency tracks the number of unemployed using data collected by the U.S. Census Bureau from approximately 60,000 households as part of the Current Population Survey. The BLS refers to these data as "household data."

The BLS defines the civilian labor force as including all civilian noninstitutionalized people (i.e., people not in the military and not in institutions, such as prisons or long-term care facilities) aged 16 years and older who have a job or are actively looking for a job. People are considered to be employed during a given week if they meet any of the following criteria:

- They performed any work that week for pay or profit.
- They worked without pay for at least 15 hours that week in a family operated enterprise.
- They had a job but could not work that week due to illness, vacation, personal obligations, leave of absence, bad weather, or labor disputes.

People considered not to be in the labor force are those who do not have a job and are not looking for a job. This category includes many students, retirees, stay-at-home parents, the mentally and physically challenged, and people in prison and other institutions, as well as those who are not employed but have become discouraged from looking for work. The unemployed are counted as those who do not have a job but have actively looked for a job during the previous four weeks and are available for work. Also included are people who did not work during a given week due to temporary layoffs.

As shown in Table 5.1, the civilian labor force consisted of 155.4 million people in 2013 or 63.2% of the civilian noninstituionalized population aged years 16 and older. Another 90.3 million people were considered not to be in the labor force. The percentage of people in the U.S. labor force has declined since the late 1990s, when it peaked at 67.1%. A variety of factors are believed to be behind the drop. In "Explaining the Decline in the U.S.

TABLE 5.1

Employment status of the civilian noninstitutional population, 1980–2013

[Numbers in thousands]

Year	Civilian noninstitutional population	Civilian labor force							Not in labor force
				Employed				Unemployed		
		Total	Percent of population	Total	Percent of population	Agriculture	Nonagricultural industries	Number	Percent of labor force	
				Persons 16 years of age and over						
1980	167,745	106,940	63.8	99,303	59.2	3,364	95,938	7,637	7.1	60,806
1981	170,130	108,670	63.9	100,397	59.0	3,368	97,030	8,273	7.6	61,460
1982	172,271	110,204	64.0	99,526	57.8	3,401	96,125	10,678	9.7	62,067
1983	174,215	111,550	64.0	100,834	57.9	3,383	97,450	10,717	9.6	62,665
1984	176,383	113,544	64.4	105,005	59.5	3,321	101,685	8,539	7.5	62,839
1985	178,206	115,461	64.8	107,150	60.1	3,179	103,971	8,312	7.2	62,744
1986	180,587	117,834	65.3	109,597	60.7	3,163	106,434	8,237	7.0	62,752
1987	182,753	119,865	65.6	112,440	61.5	3,208	109,232	7,425	6.2	62,888
1988	184,613	121,669	65.9	114,968	62.3	3,169	111,800	6,701	5.5	62,944
1989	186,393	123,869	66.5	117,342	63.0	3,199	114,142	6,528	5.3	62,523
1990	189,164	125,840	66.5	118,793	62.8	3,223	115,570	7,047	5.6	63,324
1991	190,925	126,346	66.2	117,718	61.7	3,269	114,449	8,628	6.8	64,578
1992	192,805	128,105	66.4	118,492	61.5	3,247	115,245	9,613	7.5	64,700
1993	194,838	129,200	66.3	120,259	61.7	3,115	117,144	8,940	6.9	65,638
1994	196,814	131,056	66.6	123,060	62.5	3,409	119,651	7,996	6.1	65,758
1995	198,584	132,304	66.6	124,900	62.9	3,440	121,460	7,404	5.6	66,280
1996	200,591	133,943	66.8	126,708	63.2	3,443	123,264	7,236	5.4	66,647
1997	203,133	136,297	67.1	129,558	63.8	3,399	126,159	6,739	4.9	66,837
1998	205,220	137,673	67.1	131,463	64.1	3,378	128,085	6,210	4.5	67,547
1999	207,753	139,368	67.1	133,488	64.3	3,281	130,207	5,880	4.2	68,385
2000	212,577	142,583	67.1	136,891	64.4	2,464	134,427	5,692	4.0	69,994
2001	215,092	143,734	66.8	136,933	63.7	2,299	134,635	6,801	4.7	71,359
2002	217,570	144,863	66.6	136,485	62.7	2,311	134,174	8,378	5.8	72,707
2003	221,168	146,510	66.2	137,736	62.3	2,275	135,461	8,774	6.0	74,658
2004	223,357	147,401	66.0	139,252	62.3	2,232	137,020	8,149	5.5	75,956
2005	226,082	149,320	66.0	141,730	62.7	2,197	139,532	7,591	5.1	76,762
2006	228,815	151,428	66.2	144,427	63.1	2,206	142,221	7,001	4.6	77,387
2007	231,867	153,124	66.0	146,047	63.0	2,095	143,952	7,078	4.6	78,743
2008	233,788	154,287	66.0	145,362	62.2	2,168	143,194	8,924	5.8	79,501
2009	235,801	154,142	65.4	139,877	59.3	2,103	137,775	14,265	9.3	81,659
2010	237,830	153,889	64.7	139,064	58.5	2,206	136,858	14,825	9.6	83,941
2011	239,618	153,617	64.1	139,869	58.4	2,254	137,615	13,747	8.9	86,001
2012	243,284	154,975	63.7	142,469	58.6	2,186	140,283	12,506	8.1	88,310
2013	245,679	155,389	63.2	143,929	58.6	2,130	141,799	11,460	7.4	90,290

Note: Revisions to population controls and other changes can affect the comparability of labor force levels over time. In recent years, for example, updated population controls have been introduced annually with the release of January data. Persons 16 years of age and over. Data are annual averages.

SOURCE: Adapted from "1. Employment Status of the Civilian Noninstitutional Population, 1943 to Date," in *Labor Force Statistics from the Current Population Survey*, U.S. Department of Labor, Bureau of Labor Statistics, February 26, 2014, http://www.bls.gov/cps/cpsaat01.htm (accessed August 29, 2014)

Labor Force Participation Rate" (March 2012, http://chicagofed.org/digital_assets/publications/chicago_fed_letter/2012/cflmarch2012_296.pdf), Daniel Aaronson, Jonathan Davis, and Luojia Huthe of the Federal Reserve Bank of Chicago estimate that nearly half of the decline is due to aging Americans leaving the work force. The first of the baby boomers (i.e., people born between 1946 and 1964) entered their 50s during the late 1990s. Growing retirement rates among this generation are expected to shrink the labor force through the 2020s. In addition, Aaronson, Davis, and Huthe note, "There has been a long-running downward shift in teen work activity—which picked up speed during the latter half" of the first decade of the 21st century.

Industries and Jobs

The federal government broadly characterizes jobs as being in the goods-providing or service-providing categories. Goods-providing industries include businesses engaged in manufacturing, construction, mining, and natural resources. Service-providing industries include businesses whose main function is to provide a professional or trade service, rather than a product. Service-providing industries are extremely diverse and include businesses involved in retail and wholesale trade, professional and business services, education and health services, leisure and hospitality, government, and many other services.

Since the mid-20th century the service-providing industries have grown to dominate the U.S. economy as employment in goods-producing industries has declined. As shown in Figure 5.1, goods-producing industries employed nearly 25 million people in 1980. This value rose and fell over the following two decades, but declined sharply during the first decade of the 21st century. By 2006 approximately 22.5 million people were employed

FIGURE 5.1

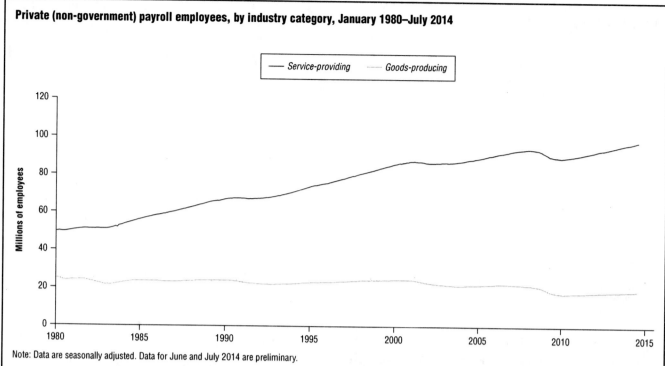

Private (non-government) payroll employees, by industry category, January 1980–July 2014

Note: Data are seasonally adjusted. Data for June and July 2014 are preliminary.

SOURCE: Adapted from "Series ID CES0600000001. Goods Producing," in *Employment, Hours, and Earnings from the Current Employment Statistics Survey (National)*, U.S. Department of Labor, Bureau of Labor Statistics, August 18, 2014, http://data.bls.gov/timeseries/CES0600000001 (accessed August 29, 2014), and "Series ID CES0800000001. Private Service-Providing," in *Employment, Hours, and Earnings from the Current Employment Statistics Survey (National)*, U.S. Department of Labor, Bureau of Labor Statistics, August 18, 2014, http://data.bls.gov/timeseries/CES0800000001 (accessed August 29, 2014)

in goods-producing industries. The Great Recession, however, was associated with deep employment losses in these industries. Employment plummeted to fewer than 18 million in 2010 before beginning to rise. It was at 19.1 million as of July 2014. By comparison, the number of people employed in service-providing industries (private and government) has soared. Figure 5.1 shows that private service-providing employment rose from about 50 million in 1980 to nearly 94 million in early 2008. After a slight decline to just under 90 million in 2009, service employment rebounded to nearly 98 million in July 2014.

Figure 5.2 shows historical trends in private and government employment. Private employment has increased dramatically from approximately 75 million in 1980 to more than 117 million in July 2014. Government employment grew from about 16 million in 1980 to nearly 22 million in July 2014.

Table 5.2 provides a breakdown of employment by industry type as of July 2014. The data are based on payroll records kept by employers and are seasonally adjusted. Overall, 139 million payroll workers were employed in nonfarm occupations in July 2014. The industries with the largest numbers of workers were professional and business services (19.3 million), retail trade (15.4 million), health care (14.7 million), local government (14.1 million), and manufacturing (12.2 million).

Unemployment Rates

Table 5.1 indicates that the civilian labor force in 2013 included 143.9 million employed and 11.5 million unemployed, giving an overall unemployment rate of 7.4%. This rate was down from 2010, when unemployment peaked at 9.6%. Figure 5.3 shows annual unemployment rates dating back to 1925. Rates above 8% have been uncommon since the 1930s, during the height of the Great Depression (1929–1939).

Figure 5.4 shows monthly unemployment rates from January 1994 through July 2014. Note that the values have been seasonally adjusted. As explained in Chapter 2, seasonal adjustment is used by economists to render data more useful for analysis. The U.S. economy shows regular seasonal patterns in employment. For example, retail employment increases every December due to holiday shopping, and then declines in January. Likewise, educator employment increases each fall and declines each summer. Economists use statistical tools to adjust monthly or quarterly employment (and unemployment) data so they can see underlying trends and determine if changes from month to month or from quarter to quarter are actually significant.

From January 1994 through October 2008 the unemployment rate varied from about 4% to 6.5%. (See

FIGURE 5.2

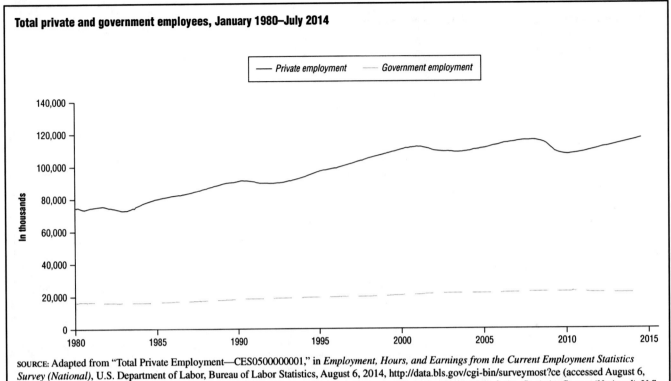

Total private and government employees, January 1980–July 2014

SOURCE: Adapted from "Total Private Employment—CES0500000001," in *Employment, Hours, and Earnings from the Current Employment Statistics Survey (National)*, U.S. Department of Labor, Bureau of Labor Statistics, August 6, 2014, http://data.bls.gov/cgi-bin/surveymost?ce (accessed August 6, 2014), and "Series ID CES9000000001: Government," in *Employment, Hours, and Earnings from the Current Population Statistics Survey (National)*, U.S. Department of Labor, Bureau of Labor Statistics, August 19, 2014, http://data.bls.gov/cgi-bin/surveymost?ce (accessed August 19, 2014)

Figure 5.4.) This is considered a generally healthy range for the U.S. economy. In "What Is the Lowest Level of Unemployment That the U.S. Economy Can Sustain?" (June 18, 2014, http://www.federalreserve.gov/faqs/economy_14424.htm), the Federal Reserve Board states, "The lowest level of unemployment that the economy can sustain is difficult to determine and has probably changed over time due to changes in the composition of the labor force, and changes in how employers search for workers and how workers search for jobs." The Federal Reserve Board indicates, however, that a "normal" unemployment rate for the late 2010s "in the absence of shocks to the economy" would be 5% to 6%. As shown in Figure 5.4 and Figure 5.3, the unemployment rate climbed historically high during late 2008 and remained above 7% through late 2013. By July 2014 it had declined to 6.2%.

WORKER CATEGORIES. Table 5.3 shows unemployment levels and rates by industry and class of worker as of July 2014. Note that the data are not seasonally adjusted. The largest numbers of unemployed workers by industry were: wholesale and retail trade (1.3 million), leisure and hospitality (1.2 million), and professional and business services (1.1 million). The industries with the highest unemployment rates were leisure and hospitality (8.1%), construction (7.5%), and professional and business services (6.7%). The mining, quarrying, and oil and gas extraction sector had, by far, the lowest unemployment rate in July 2014 at 3%.

The unemployment rate in July 2014 varied widely according to certain demographic factors. Teenagers aged 16 to 19 years were unemployed at a rate of 20.2%. (See Table 5.4.) Racial and ethnic differences were also significant; African American workers had an unemployment rate of 11.4%, and Hispanic workers had a rate of 7.8%. These values were higher than the 5.3% unemployment rate reported for white workers. The rate for Asian American workers was only 4.5%.

Educational attainment played a major role in unemployment rates in July 2014. (See Table 5.4.) The rate for all workers aged 25 years and older was 5%. Among those with less than a high school diploma the rate was much higher, at 9.6%. High school graduates with no college had an unemployment rate of 6.1%. Workers who had taken some college courses or who had earned an associate's degree had a rate of 5.3%. Those with a bachelor's degree and higher had the lowest unemployment rate, at 3.1%.

Table 5.4 also provides information about the nearly 9.7 million people who were unemployed as of July 2014. Most had been unemployed for 26 weeks or less. Nearly 3.2 million (32.6%) had been unemployed for 27 weeks or more (i.e., roughly six months or more).

As noted earlier the BLS does not include in the labor force people who have become discouraged from looking for work. In its "Economic News Release"

TABLE 5.2

Employees on nonfarm payrolls by industry sector, July 2014

[In thousands]

	July 2014
Total nonfarm	**139,004**
Total private	**117,082**
Goods-producing	19,117
Mining and logging	916
Logging	55.20
Mining	860.80
Construction	6,041
Manufacturing	12,160
Durable goods	7,695
Nondurable goods	4,465
Service-providing	119,887
Private service-providing	97,965
Trade, transportation, and utilities	26,438
Wholesale trade	5,876.40
Retail trade	15,386.40
Transportation and warehousing	4,622.40
Utilities	552.40
Information	2,666
Financial activities[a]	7,951
Professional and business services	19,269
Education and health services	21,483
Educational services	3,393.90
Health care and social assistance	18,089.30
Health care[b]	14,730.80
Social assistance	3,358.50
Leisure and hospitality	14,647
Other services	5,511
Repair and maintenance	1,214.60
Personal and laundry services	1,368.20
Membership associations and organizations	2,928.30
Government	21,922
Federal	2,714
State government	5,066
Local government	14,142

[a]Excludes nonoffice commissioned real estate sales agents.
[b]Includes ambulatory health care services, hospitals, and nursing and residential care facilities.
Note: Data are seasonally adjusted and preliminary.

SOURCE: Adapted from "Table B-1a. Employees on Nonfarm Payrolls by Industry Sector and Selected Industry Detail, Seasonally Adjusted," in *Current Employment Statistics—CES (National)*, U.S. Department of Labor, Bureau of Labor Statistics, August 1, 2014, http://www.bls.gov/web/empsit/ceseeb1a.htm (accessed August 19, 2014)

(August 8, 2014, http://www.bls.gov/news.release/empsit.t15.htm), the agency explains that "discouraged" workers are those who "have given a job-market related reason for not currently looking for work." There were 741,000 discouraged workers in July 2014. (See Table 5.4.) This category swelled immensely during and after the Great Recession, as shown in Table 5.5. The number of discouraged workers peaked in 2010 at nearly 1.2 million, far above the values common during the years leading up to the recession. Discouraged workers are a subset of people the BLS considers "marginally attached to the labor force." In "Economic News Release," the agency explains in that these workers "currently are neither working nor looking for work but indicate that they want and are available for a job and have looked for work sometime in the past 12 months." In total, nearly 2.2 million workers were in this category as of July 2014.

(See Table 5.4.) The BLS notes that people other than discouraged workers had noneconomic reasons for their marginal attachment to the labor force. These reasons included family responsibilities, ill health, and transportation problems.

A labor market is said to be slack when available workers outnumber jobs. By contrast a tight labor market features more available jobs than workers, and employers compete with each other for employees. During and after the Great Recession the U.S. labor market was very slack. Some workers desperate for employment accepted part-time work, even though they wanted and were available for full-time work. As shown in Table 5.4, more than 7.5 million people were working part time for economic reasons in July 2014. The BLS notes that these reasons included work slowdowns, unfavorable business conditions, the inability of workers to find full-time work, or seasonal declines in employee demand.

Differing Measures of Unemployment

The unemployment rate that is highlighted in "The Employment Situation" and reported by the media is the number of unemployed as a percent of the civilian labor force. This is the official unemployment rate or the U-3 rate.

As shown in Table 5.6, the BLS also calculates other unemployment rates (U-1, U-2, U-4, U-5, and U-6) using different population subsets. Some critics argue that the U-4 rate is actually more demonstrative than the U-3 rate of overall unemployment because the U-4 rate includes discouraged workers. The U-4 unemployment rate in July 2014 was 6.6%, slightly higher than the official unemployment rate of 6.2%. During and soon after the recession when the number of discouraged workers was historically high, some analysts argued that the U-4 rate more aptly described the nation's unemployment situation.

The U-5 and U-6 unemployment rates are even more inclusive than the U-4 rate. As shown in Table 5.6, the U-5 rate includes all persons marginally attached to the labor force and was 7.5% in July 2014. The last category of people not included in the official unemployment rate is "total employed part time for economic reasons." Including the marginally attached and the so-called involuntary part-time workers in the unemployment rate provides a U-6 unemployment rate of 12.2%. (See Table 5.6.) This rate is substantially higher than the 6.2% official unemployment rate for July 2014.

Proponents of the U-4, U-5, and/or U-6 unemployment rates argue that the U-3 rate (the official unemployment rate) is too restrictive and does not provide an accurate estimate of the unemployed population. The Great Recession witnessed huge jumps in the numbers of marginally attached people and involuntary part-time workers. Because these people are not considered to be

FIGURE 5.3

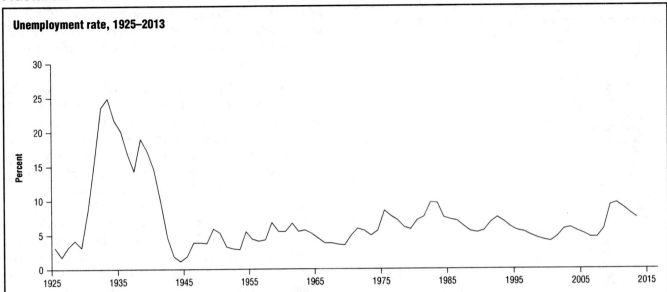

Unemployment rate, 1925–2013

Note: Rate calculated through 1947 for people aged 14 years and older. After that, the rate is based on people aged 16 years and above.

SOURCE: Adapted from "Employment Status of the Civilian Noninstitutional Population," in *Labor Force Statistics from the Current Population Survey*, U.S. Department of Labor, Bureau of Labor Statistics, February 26, 2014, http://data.bls.gov/cgi-bin/surveymost?bls (accessed August 7, 2014), and "Series D85–86. Unemployment: 1890 to 1970," in *Historical Statistics of the United States, Colonial Times to 1970, Bicentennial Edition, Part 1*, U.S. Department of Commerce, U.S. Census Bureau, September 1975, http://www2.census.gov/prod2/statcomp/documents/CT1970p1-05.pdf (accessed August 7, 2014)

FIGURE 5.4

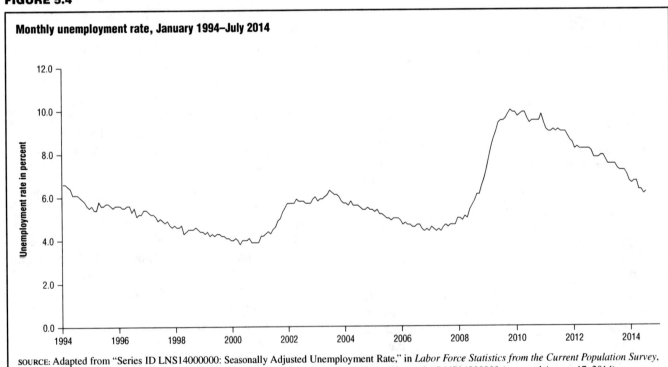

Monthly unemployment rate, January 1994–July 2014

SOURCE: Adapted from "Series ID LNS14000000: Seasonally Adjusted Unemployment Rate," in *Labor Force Statistics from the Current Population Survey*, U.S. Department of Labor, Bureau of Labor Statistics, August 17, 2014, http://data.bls.gov/timeseries/LNS14000000 (accessed August 17, 2014)

part of the labor force, they are basically ignored by the U-3 unemployment rate. However, it must be acknowledged that at least some of the controversy over the U-3 rate is driven by politics, rather than by concern for accurate data reporting. In "Is the Real Unemployment Rate 40 Percent?" (HuffingtonPost.com, July 6, 2012), Eli Lehrer notes that during the administration of President George W. Bush (1946–), who was a conservative, liberal

TABLE 5.3

Unemployment by industry and class of worker, July 2014

Industry and class of worker	Number of unemployed persons (in thousands) July 2014	Unemployment rates July 2014
Total, 16 years and over*	10,307	6.5
Nonagricultural private wage and salary workers	7,366	5.9
Mining, quarrying, and oil and gas extraction	34	3.0
Construction	666	7.5
Manufacturing	825	5.2
Durable goods	510	5.2
Nondurable goods	315	5.4
Wholesale and retail trade	1,283	6.2
Transportation and utilities	372	6.0
Information	128	4.5
Financial activities	341	3.6
Professional and business services	1,066	6.7
Education and health services	1,051	4.8
Leisure and hospitality	1,171	8.1
Other services	430	6.3
Agriculture and related private wage and salary workers	93	5.7
Government workers	937	4.6
Self-employed workers, unincorporated, and unpaid family workers	432	4.4

*Persons with no previous work experience and persons whose last job was in the U.S. armed forces are included in the unemployed total.
Note: Updated population controls are introduced annually with the release of January data. Data are not seasonally adjusted.

SOURCE: Adapted from "Table A-14. Unemployed Persons by Industry and Class of Worker, Not Seasonally Adjusted," in *The Employment Situation—July 2014*, U.S. Department of Labor, Bureau of Labor Statistics, August 1, 2014, http://www.bls.gov/news.release/pdf/empsit.pdf (accessed August 8, 2014)

TABLE 5.4

Employment and unemployment data, July 2014

[Numbers in thousands]

Category	July 2014
Employment status	
Civilian noninstitutional population	248,023
Civilian labor force	156,023
Participation rate	62.9
Employed	146,352
Employment-population ratio	59.0
Unemployed	9,671
Unemployment rate	6.2
Not in labor force	92,001
Unemployment rates	
Total, 16 years and over	6.2
Adult men (20 years and over)	5.7
Adult women (20 years and over)	5.7
Teenagers (16 to 19 years)	20.2
White	5.3
Black or African American	11.4
Asian (not seasonally adjusted)	4.5
Hispanic or Latino ethnicity	7.8
Total, 25 years and over	5.0
Less than a high school diploma	9.6
High school graduates, no college	6.1
Some college or associate degree	5.3
Bachelor's degree and higher	3.1
Reason for unemployment	
Job losers and persons who completed temporary jobs	4,859
Job leavers	862
Reentrants	2,848
New entrants	1,087
Duration of unemployment	
Less than 5 weeks	2,587
5 to 14 weeks	2,431
15 to 26 weeks	1,412
27 weeks and over	3,155
Employed persons at work part time	
Part time for economic reasons	7,511
Slack work or business conditions	4,609
Could only find part-time work	2,519
Part time for noneconomic reasons	19,662
Persons not in the labor force (not seasonally adjusted)	
Marginally attached to the labor force	2,178
Discouraged workers	741

Note: Persons whose ethnicity is identified as Hispanic or Latino may be of any race. Data are seasonally adjusted. Detail for the seasonally adjusted data shown in this table will not necessarily add to totals because of the independent seasonal adjustment of the various series. Updated population controls are introduced annually with the release of January data.

SOURCE: Adapted from "Summary Table A. Household Data, Seasonally Adjusted," in *The Employment Situation—July 2014*, U.S. Department of Labor, Bureau of Labor Statistics, August 1, 2014, http://www.bls.gov/news.release/pdf/empsit.pdf (accessed August 8, 2014)

---

commentators used alternative unemployment rates to "cast doubt on his economic policies." Lehrer points out that the same thing was happening in mid-2012, with conservative commentators using the broader unemployment rates, such as the U-6 rate, to argue that the economic policies of President Barack Obama (1961–), a liberal, were failing.

Unemployment Insurance Claims

Another indicator used to measure unemployment in the country is derived from the government's unemployment insurance (UI) program. Every state operates a UI trust fund that is managed in concert with the federal government. The UI trust funds are funded by taxes levied on employers and in some states on employees. Payments from the UI trust funds are made temporarily, typically up to 26 weeks, to unemployed people who meet specific criteria set by each state. The U.S. Department of Labor explains in "State Unemployment Insurance Benefits" (May 23, 2013, http://workforcesecurity.doleta.gov/unemploy/uifactsheet.asp) that workers are eligible to receive UI benefits (i.e., UI trust fund payments) if they become unemployed "through no fault of their own" and meet other state-defined criteria for time worked and wages earned before becoming unemployed.

Figure 5.5 shows the number of new state UI claims filed by applicants on a weekly basis from January 1, 2005, to July 19, 2014. The number peaked at 665,000 in early 2009. Initial UI claims then fell over subsequent years, dropping to 279,000 claims per week in mid-July 2014.

TABLE 5.5

Discouraged workers, 1996–2013

[Number in thousands]

Year	Monthly average
1996	397
1997	343
1998	331
1999	273
2000	262
2001	321
2002	369
2003	457
2004	466
2005	436
2006	381
2007	369
2008	462
2009	778
2010	1,173
2011	989
2012	909
2013	861

Note: Data are not seasonally adjusted.

SOURCE: Adapted from "Series ID LNU05026645: Discouraged," in Labor Force Statistics from the Current Population Survey, U.S. Department of Labor, Bureau of Labor Statistics, August 19, 2014, http://data.bls.gov/timeseries/LNU05026645 (accessed August 19, 2014)

TABLE 5.6

Alternative measures of labor underutilization, July 2014

[Percent]

Measure	Seasonally adjusted July 2014
U-1 Persons unemployed 15 weeks or longer, as a percent of the civilian labor force	2.9
U-2 Job losers and persons who completed temporary jobs, as a percent of the civilian labor force	3.1
U-3 Total unemployed, as a percent of the civilian labor force (official unemployment rate)	6.2
U-4 Total unemployed plus discouraged workers, as a percent of the civilian labor force plus discouraged workers	6.6
U-5 Total unemployed, plus discouraged workers, plus all other persons marginally attached to the labor force, as a percent of the civilian labor force plus all persons marginally attached to the labor force	7.5
U-6 Total unemployed, plus all persons marginally attached to the labor force, plus total employed part time for economic reasons, as a percent of the civilian labor force plus all persons marginally attached to the labor force	12.2

Note: Persons marginally attached to the labor force are those who currently are neither working nor looking for work but indicate that they want and are available for a job and have looked for work sometime in the past 12 months. Discouraged workers, a subset of the marginally attached, have given a job-market related reason for not currently looking for work. Persons employed part time for economic reasons are those who want and are available for full-time work but have had to settle for a part-time schedule. Updated population controls are introduced annually with the release of January data. Data are seasonally adjusted.

SOURCE: Adapted from "Table A-15. Alternative Measures of Labor Underutilization," in The Employment Situation—July 2014, U.S. Department of Labor, Bureau of Labor Statistics, August 1, 2014, http://www.bls.gov/news.release/pdf/empsit.pdf (accessed August 8, 2014)

Employment Perceptions and Predictions

The Gallup Organization conducts regular polls that question Americans about their perceptions of the job situation. In *In U.S., 28% Say Now Is a Good Time to Find a Quality Job* (March 19, 2014, http://www.gallup.com/poll/167990/say-good-time-find-quality-job.aspx), Frank Newport of the Gallup Organization notes that in January 2007 (before the Great Recession began) 48% of respondents said it was a "good time" to find a quality job. (See Figure 5.6.) This value plummeted to 8% during 2009 and remained below 20% well into 2012. In March 2014 more than a quarter of respondents (28%) said it was a "good time" to find a quality job, indicating slowly growing optimism about the job market.

In December 2013 the BLS published employment projections, by industry sector, through 2022 using employment data from 2012. Total employment is expected to grow to nearly 161 million in 2022. (See Table 5.7.) The largest job gains are projected for the health care and social assistance sector (up 5 million) and professional and business services sector (up 3.5 million). The BLS predicts the construction sector and the health care and social assistance sector will each experience an annual job growth rate of 2.6% through 2022. They are followed by the educational services sector (1.9%) and the professional and business services sector (1.8%). The BLS projects that certain sectors will lose jobs through 2022. Employment of agriculture self-employed and unpaid family workers is expected to decline 2.8%. Job losses are also predicted for the federal government sector (down 1.6%), the utilities sector and the agriculture, forestry, fishing, and hunting sector (both down 1.1%), the manufacturing sector (down 0.5%), and the information sector (down 0.2%).

COMPENSATION OF AMERICAN WORKERS

As explained in Chapter 3, the U.S. Department of Commerce's Bureau of Economic Analysis tracks personal income as part of its compilation of the National Income and Product Accounts. Table 5.8 provides a breakdown for one component of personal income (compensation of employees) for 2012, 2013, and the first two quarters of 2014. Note that the quarterly data are seasonally adjusted at annual rates.

Employee compensation consists of pay and supplements. Pay includes wages (which is the term used primarily for pay made on an hourly, weekly, or monthly basis) and salaries (which is pay calculated yearly). In addition, many employers provide supplemental amounts in addition to wages and salaries for the benefit of their employees. These amounts are not paid directly to the employees but are made on their behalf. For example, some employers pay a portion of the premiums for health insurance plans for their employees. Some employers make contributions to employee pension plans, which are basically savings plans for retirement. Employers are required by law to make payments to certain government

FIGURE 5.5

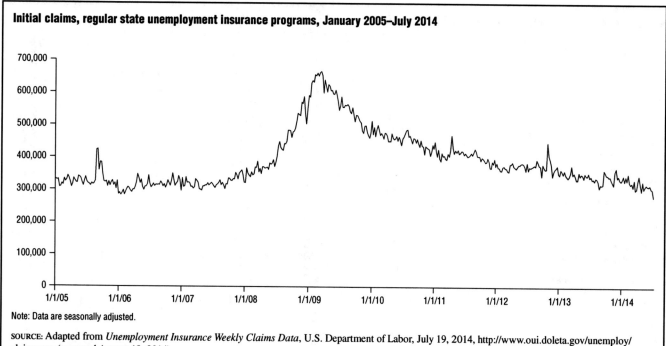

Initial claims, regular state unemployment insurance programs, January 2005–July 2014

Note: Data are seasonally adjusted.

SOURCE: Adapted from *Unemployment Insurance Weekly Claims Data*, U.S. Department of Labor, July 19, 2014, http://www.oui.doleta.gov/unemploy/claims.asp (accessed August 18, 2014)

FIGURE 5.6

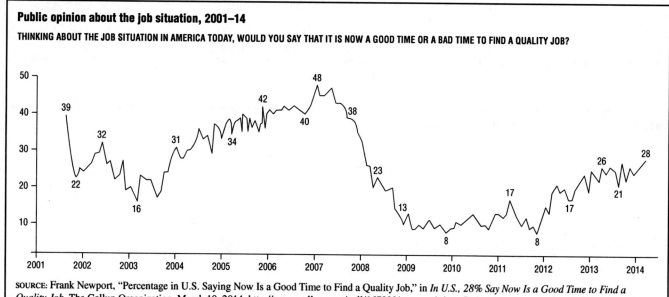

Public opinion about the job situation, 2001–14

THINKING ABOUT THE JOB SITUATION IN AMERICA TODAY, WOULD YOU SAY THAT IT IS NOW A GOOD TIME OR A BAD TIME TO FIND A QUALITY JOB?

SOURCE: Frank Newport, "Percentage in U.S. Saying Now Is a Good Time to Find a Quality Job," in *In U.S., 28% Say Now Is a Good Time to Find a Quality Job*, The Gallup Organization, March 19, 2014, http://www.gallup.com/poll/167990/say-good-time-find-quality-job.aspx (accessed August 17, 2014). Copyright © 2014 Gallup, Inc. All rights reserved. The content is used with permission; however, Gallup retains all rights of republication.

social insurance funds on behalf of their employees. These funds are for programs such as Social Security, Medicare, and unemployment insurance.

As shown in Table 5.8, employee compensation totaled $8.6 trillion in 2012 and $8.8 trillion in 2013. As of the second quarter of 2014, total compensation for 2014 was expected to be $9.2 trillion. This value includes nearly $7.5 trillion in wages and salaries and almost $1.8 trillion in supplements.

Employee Pay: Wages and Salaries

The BLS obtains employee pay data from households through the Current Population Survey and from businesses and government agencies through the Current Employment Statistics program. Establishment data from the latter survey are published monthly for workers on private (i.e., nongovernmental) payrolls. This includes both full-time and part-time employees. In "The Employment Situation—November 2014" (December 5, 2014,

TABLE 5.7

Employment by major industry sector, 2012 and projected for 2022

Industry sector	Thousands of jobs 2012	Thousands of jobs 2022	Change 2012–2022	Percent distribution 2012	Percent distribution 2022	Annual growth rate (percent) 2012–2022
Total[a]	145,355.8	160,983.7	15,627.9	100.0	100.0	1.0
Nonagriculture wage and salary[b]	134,427.6	149,751.3	15,323.7	92.5	93.0	1.1
Goods-producing, excluding agriculture	18,360.3	19,554.2	1,193.9	12.6	12.1	0.6
Mining	800.5	921.7	121.2	0.6	0.6	1.4
Construction	5,640.9	7,263.0	1,622.1	3.9	4.5	2.6
Manufacturing	11,918.9	11,369.4	−549.5	8.2	7.1	−0.5
Services-providing	116,067.3	130,197.1	14,129.8	79.9	80.9	1.2
Utilities	554.2	497.8	−56.4	0.4	0.3	−1.1
Wholesale trade	5,672.8	6,143.2	470.4	3.9	3.8	0.8
Retail trade	14,875.3	15,966.2	1,090.9	10.2	9.9	0.7
Transportation and warehousing	4,414.7	4,742.0	327.3	3.0	2.9	0.7
Information	2,677.6	2,612.4	−65.2	1.8	1.6	−0.2
Financial activities	7,786.3	8,537.3	751.0	5.4	5.3	0.9
Professional and business services	17,930.2	21,413.0	3,482.8	12.3	13.3	1.8
Educational services	3,346.9	4,022.2	675.3	2.3	2.5	1.9
Health care and social assistance	16,971.8	21,965.9	4,994.1	11.7	13.6	2.6
Leisure and hospitality	13,745.8	15,035.0	1,289.2	9.5	9.3	0.9
Other services	6,174.5	6,823.4	648.9	4.2	4.2	1.0
Federal government	2,814.0	2,406.5	−407.5	1.9	1.5	−1.6
State and local government	19,103.2	20,032.2	929.0	13.1	12.4	0.5
Agriculture, forestry, fishing, and hunting[c]	2,112.7	1,889.2	−223.5	1.5	1.2	−1.1
Agriculture wage and salary	1,306.9	1,281.8	−25.1	0.9	0.8	−0.2
Agriculture self-employed and unpaid family workers	805.8	607.4	−198.4	0.6	0.4	−2.8
Nonagriculture self-employed and unpaid family workers	8,815.5	9,343.2	527.7	6.1	5.8	0.6

[a]Employment data for wage and salary workers are from the Bureau of Labor Statistics Current Employment Statistics survey, which counts jobs, whereas self-employed, unpaid family workers, and agriculture, forestry, fishing, and hunting are from the Current Population Survey (household survey), which counts workers.
[b]Includes wage and salary data from the Current Employment Statistics survey, except private households, which is from the Current Population Survey. Logging workers are excluded.
[c]Includes agriculture, forestry, fishing, and hunting data from the Current Population Survey, except logging, which is from Current Employment Statistics survey. Government wage and salary workers are excluded.

SOURCE: Adapted from "Table 3. Employment by Major Industry Sector, 2002, 2012, and Projected 2022," in *Employment Projections—2012–2022*, U.S. Department of Labor, Bureau of Labor Statistics, December 19, 2013, http://www.bls.gov/news.release/pdf/ecopro.pdf (accessed August 19, 2014)

TABLE 5.8

Employee compensation, 2012–13, and first and second quarters 2014

[Billions of dollars]

	2012	2013	Seasonally adjusted at annual rates 2012 I	Seasonally adjusted at annual rates 2012 II
Compensation of employees	8,606.5	8,844.8	9,100.2	9,235.0
Wages and salaries	6,932.1	7,124.7	7,343.6	7,462.2
Private industries	5,733.8	5,916.6	6,129.1	6,242.8
Goods-producing industries	1,157.1	1,195.3	1,237.8	1,259.2
Manufacturing	734.4	747.6	768.1	747.0
Services-producing industries	4,576.7	4,721.3	4,891.3	4,983.6
Trade, transportation, and utilities	1,094.0	1,121.3	1,148.7	1,171.6
Other services-producing industries	3,482.7	3,600.0	3,742.6	3,812.0
Government	1,198.3	1,208.1	1,214.5	1,219.4
Supplements to wages and salaries	1,674.4	1,720.1	1,756.6	1,772.8
Employer contributions for employee pension and insurance funds*	1,160.5	1,193.9	1,213.6	1,222.0
Employer contributions for government social insurance	513.9	526.1	542.9	550.8

*Includes actual employer contributions and actuarially imputed employer contributions to reflect benefits accrued by defined benefit pension plan participants through service to employers in the current period.

SOURCE: Adapted from "Table 2. Personal Income and Its Disposition (Years and Quarters)," in *Personal Income and Outlays: June 2014, Revised Estimates: 1999 through May 2014*, U.S. Department of Commerce, Bureau of Economic Analysis, August 1, 2014, http://www.bea.gov/newsreleases/national/pi/2014/pdf/pi0614.pdf (accessed August 19, 2014)

http://www.bls.gov/news.release/pdf/empsit.pdf), the BLS notes that in November 2014 the average pay for all employees was $24.66 per hour or $853.24 per week. In addition, the BLS reports the average pay for production and nonsupervisory (P&NS) employees. This subset includes only production employees in the mining, logging, and manufacturing industries; construction employees in the construction industry; and nonsupervisory employees in the service-providing industries. The BLS indicates that approximately 80% of all employees on private nonfarm payrolls are P&NS employees. In November 2014 the average pay for P&NS employees was $20.74 per hour or $701.01 per week. This subset includes many workers toward the lower end of the income scale. As such, economic data for P&NS workers are often used by politicians in reference to America's "working class" or "middle class," even those these terms do not have strict definitions.

Table 5.9 lists average weekly hours, hourly earnings, and weekly earnings for P&NS workers for 1970 through 2013 and for January 2014. The table includes earnings in current dollars (which are not inflation-adjusted) and in real (inflation-adjusted) dollars. As explained in Chapter 3, the consumer price index (CPI) can be used to adjust current-dollar values to real-dollar values. The real earnings values listed in Table 5.9 were calculated using the CPI-W, which is the CPI for urban wage earners and clerical workers. The distinction between current-dollar and real-dollar earnings is important. Prices typically

TABLE 5.9

Average hours and earnings for production and nonsupervisory employees in private industry, 1970–2013 and January 2014

[Monthly data seasonally adjusted]

| | | Average hourly earnings | | Average weekly earnings, total private | | | |
| | | Total private | | Level | | Percent change from year earlier | |
Year or month	Average weekly hours Total private	Current dollars	1982–84 dollars*	Current dollars	1982–84 dollars*	Current dollars	1982–84 dollars*
1970	37.0	$3.40	$8.72	$125.79	$322.54	4.2	−1.4
1971	36.7	3.63	8.92	133.22	327.32	5.9	1.5
1972	36.9	3.90	9.26	143.87	341.73	8.0	4.4
1973	36.9	4.14	9.26	152.59	341.36	6.1	−0.1
1974	36.4	4.43	8.93	161.61	325.83	5.9	−4.5
1975	36.0	4.73	8.74	170.29	314.77	5.4	−3.4
1976	36.1	5.06	8.85	182.65	319.32	7.3	1.4
1977	35.9	5.44	8.93	195.58	321.15	7.1	0.6
1978	35.8	5.88	8.96	210.29	320.56	7.5	−0.2
1979	35.6	6.34	8.67	225.69	308.74	7.3	−3.7
1980	35.2	6.85	8.26	241.07	290.80	6.8	−5.8
1981	35.2	7.44	8.14	261.53	286.14	8.5	−1.6
1982	34.7	7.87	8.12	273.10	281.84	4.4	−1.5
1983	34.9	8.20	8.22	286.43	287.00	4.9	1.8
1984	35.1	8.49	8.22	298.26	288.73	4.1	0.6
1985	34.9	8.74	8.18	304.62	284.96	2.1	−1.3
1986	34.7	8.93	8.22	309.78	285.25	1.7	0.1
1987	34.7	9.14	8.12	317.39	282.12	2.5	−1.1
1988	34.6	9.44	8.07	326.48	279.04	2.9	−1.1
1989	34.5	9.80	7.99	338.34	275.97	3.6	−1.1
1990	34.3	10.20	7.91	349.63	271.03	3.3	−1.8
1991	34.1	10.51	7.83	358.46	266.91	2.5	−1.5
1992	34.2	10.77	7.79	368.20	266.43	2.7	−0.2
1993	34.3	11.05	7.78	378.89	266.64	2.9	0.1
1994	34.5	11.34	7.79	391.17	268.66	3.2	0.8
1995	34.3	11.65	7.78	400.04	267.05	2.3	−0.6
1996	34.3	12.04	7.81	413.25	268.17	3.3	0.4
1997	34.5	12.51	7.94	431.86	274.02	4.5	2.2
1998	34.5	13.01	8.15	448.59	280.90	3.9	2.5
1999	34.3	13.49	8.27	463.15	283.79	3.2	1.0
2000	34.3	14.02	8.30	480.99	284.78	3.9	0.3
2001	34.0	14.54	8.38	493.74	284.58	2.7	−0.1
2002	33.9	14.97	8.51	506.60	288.00	2.6	1.2
2003	33.7	15.37	8.55	517.82	288.00	2.2	0.0
2004	33.7	15.69	8.50	528.89	286.66	2.1	−0.5
2005	33.8	16.12	8.44	544.05	284.84	2.9	−0.6
2006	33.9	16.75	8.50	567.39	287.87	4.3	1.1
2007	33.8	17.42	8.59	589.27	290.61	3.9	1.0
2008	33.6	18.07	8.56	607.53	287.86	3.1	−0.9
2009	33.1	18.61	8.88	616.01	293.86	1.4	2.1
2010	33.4	19.05	8.90	636.25	297.36	3.3	1.2
2011	33.6	19.44	8.77	653.19	294.79	2.7	−0.9
2012	33.7	19.74	8.73	665.82	294.31	1.9	−0.2
2013	33.7	20.13	8.78	677.67	295.51	1.8	0.4
2014 Jan	33.5	20.39	8.82	683.07	295.40	1.9	0.3

*Current dollars divided by the consumer price index for urban wage earners and clerical workers on a 1982–84 = 100 base.

SOURCE: Adapted from "Table B-15. Hours and Earnings in Private Nonagricultural Industries, 1970–2014," in *Economic Report of the President: Together with the Annual Report of the Council of Economic Advisers*, Executive Office of the President of the United States, March 2014, http://www.whitehouse.gov/sites/default/files/docs/full_2014_economic_report_of_the_president.pdf (accessed August 11, 2014)

increase over time, as evidenced by the inflation rates shown in Figure 3.4 in Chapter 3. In theory, earnings must increase each year by at least as much as the inflation rate in order for workers to maintain the same purchasing power. This is commonly phrased as "wages must keep up with inflation."

As shown in Table 5.9, current-dollar wages (hourly and weekly) increased every year from 1970 through 2013. By contrast, the real-dollar amounts experienced a variety of positive, negative, and neutral changes. For example, in 2012 average weekly real earnings declined by 0.2%. P&NS employees lost purchasing power that year even though their earnings increased by 1.9% in current dollars. In 2013 average weekly real earnings increased by 0.4%. From 1996 through 2000, P&NS employees saw their real weekly earnings rise for five years in a row. However, since that time the changes have been mixed. After the Great Recession ended in 2009, the economy slowly began to recover. As will be explained in subsequent chapters, corporate profits and stock market indexes (which are indicators of business performance) showed great improvements. Some analysts complain that employees have not benefited as much from the economic upturn because businesses have not used their profits to raise wages to keep up with inflation. In "This Is Why Wages Have Risen So Slowly" (WashingtonPost.com, August 20, 2014), the economist Jared Bernstein asks, "In an economic recovery that's more than five years old, isn't it about time we saw some real wage gains?"

THE MINIMUM WAGE. The Fair Labor Standards Act (FLSA) offers protection for full- and part-time workers in private and government jobs and covers minimum wages, overtime pay, employer record keeping, and child labor. The FLSA also established the standard 40-hour workweek. Local fire and police employees are typically not covered by the FLSA. It was passed in 1938 and has been amended many times over the years.

The FLSA established a federal hourly minimum wage that U.S. employers must honor for many nonsupervisory, nonfarm, private sector, and government employees. Most states have their own minimum wage as well. In states with minimum wages that differ from the federal minimum wage, the employer must pay the higher of the two. As of December 2014, the federal minimum wage was $7.25 per hour, a level maintained since 2009.

There are many exceptions to the minimum wage law. Employers may apply for subminimum wage certificates for disabled workers, full-time students, workers under the age of 20 years who are in their first 90 days of employment, workers who receive tips, and student-learners (usually high school students). Lawmakers reason that exempting employers from paying the minimum wage to certain workers (e.g., the disabled and students) encourages them to hire more of those workers who may otherwise be at a disadvantage. Employers may not, however, displace other workers to hire those subject to the subminimum wage. Other workers exempt from the minimum wage include certain professional and administrative employees, certain workers in the fishing industry, certain seasonal employees, babysitters, and certain farm workers.

The BLS tracks the number of hourly wage earners in the United States based on the results of the Current Population Survey. In "Characteristics of Minimum Wage Workers: 2013" (March 2014, http://www.bls.gov/cps/minwage2013.pdf), the BLS indicates that there were 1.5 million workers paid the federal minimum wage during 2013. Another 1.8 million workers earned less than the minimum wage. The total number of workers earning minimum wage or less accounted for 4.3% of the 75.9 million people earning hourly wages that year.

The minimum wage policy is not without controversy. Advocates for low-income workers believe the minimum wage should be increased regularly to keep up with the effects of inflation. In *Economic Report of the President: Together with the Annual Report of the Council of Economic Advisers* (March 2014, http://www.whitehouse.gov/sites/default/files/docs/full_2014_economic_report_of_the_president.pdf), President Barack Obama (1961–) notes, "raising the minimum wage to $10.10 per hour would help a large, diverse group of workers, and indexing it to inflation would ensure that its real value does not deteriorate over time, as it has after past increases." Opponents of the minimum wage assert that wage levels should be determined by market conditions and supply and demand factors. They argue that forcing businesses to pay a higher minimum wage discourages the hiring of low-income workers. In addition, they believe that businesses would increase their prices to cover their higher expenses.

In February 2014 the Congressional Budget Office (CBO) published its analysis of the possible effects of raising the federal minimum wage. In *The Effects of a Minimum-Wage Increase on Employment and Family Income* (http://www.cbo.gov/sites/default/files/cbofiles/attachments/44995-MinimumWage.pdf), the CBO predicts that an increase to $10.10 per hour by the year 2016 would have mixed results. An estimated 900,000 people would be lifted above the federal poverty thresholds, which differ by family size and composition. (The CBO notes that the 2016 threshold for a three-person family would be $18,700 in annual income expressed in 2013 dollars.) On the down side, employers could eliminate as many as 500,000 jobs to cover their costs for the wage hike.

According to Juliana Goldman, in "Americans Split on Obama as 69% Back Minimum Wage Hike"

(Bloomberg.com, March 11, 2014), a poll conducted in March 2014 for Bloomberg News found that 69% of respondents favored raising the minimum wage to $10.10 per hour, and 28% oppose the idea. Another 3% were unsure on the matter. However, support for the wage hike declined after poll participants were told about the CBO's assessment that 500,000 jobs could be eliminated. More than half (57%) of the respondents said this trade-off would be "unacceptable," whereas 34% thought it would be "acceptable." The remaining 9% were unsure on the matter. As of December 2014, the U.S. Congress had not passed a bill raising the federal minimum wage. In June 2014 the city of Seattle, Washington, passed an ordinance to increase the local minimum wage to $15 per hour. Employers have up to seven years to implement the wage hike, depending on their size. During the November 2014 elections the voters in four states—Alaska, Arkansas, Nebraska, and South Dakota—approved minimum wage hikes. A listing of minimum wage laws by state is provided by the Department of Labor at http://www.dol.gov/whd/minwage/america.htm.

Employee Benefits

To attract and keep the best employees, many U.S. employers offer benefits and incentives with monetary value. The BLS publishes an annual report on employee benefits offered to civilian (nonmilitary), private industry, and state and local government workers. Federal workers, agricultural workers and private household workers are excluded. As of December 2014, the most recent report was "Employee Benefits in the United States—March 2014" (July 25, 2014, http://www.bls.gov/news.release/pdf/ebs2.pdf). The data are taken from the National Compensation Survey, which in 2014 included about 9,600 establishments in private industry and approximately 1,500 establishments in state and local governments. The BLS examines six types of benefits: retirement plans, medical care benefits, life insurance plans, paid sick leave, paid vacation, and paid holidays. Each benefit is analyzed in terms of the percentage of employees that had access to it. When known, the percentage of employees participating in a particular benefit type is provided.

For civilian employees the following results were obtained:

- Retirement plans—68% with access; 53% participating
- Medical care benefits—72% with access; 53% participating. (See Table 5.10.)
- Life insurance benefits—60% with access; 59% participating

TABLE 5.10

Employee access, participation, and take-up rates in percentage for medical care benefits, March 2014

[All workers = 100 percent]

Characteristics	Civilian[a]			Private industry			State and local government		
	Access	Participation	Take-up rate	Access	Participation	Take-up rate	Access	Participation	Take-up rate
All workers	72	53	74	69	50	72	87	73	83
Worker characteristics									
Full time	88	66	76	86	63	74	99	83	84
Part time	23	13	56	23	12	54	24	17	74
Union	94	79	83	94	78	83	95	80	84
Nonunion	68	49	72	67	47	71	81	67	83
Average wage within the following categories[b]									
Lowest 25 percent	38	22	59	34	20	57	68	55	80
Lowest 10 percent	22	12	54	20	10	51	53	42	78
Second 25 percent	77	56	73	74	52	70	92	78	86
Third 25 percent	88	69	79	86	66	77	94	78	84
Highest 25 perecent	94	74	79	93	71	77	97	81	83
Highest 10 percent	95	75	80	94	74	78	97	82	84
Establishment characteristics									
1 to 99 workers	58	41	72	57	41	71	74	63	85
1 to 49 workers	54	38	71	53	38	71	65	54	84
50 to 99 workers	70	51	72	69	49	71	87	75	86
100 workers or more	85	64	76	84	61	73	89	74	83
100 to 499 workers	80	58	72	80	56	71	86	71	83
500 workers or more	90	71	79	89	68	77	90	75	83

[a]Includes workers in the private nonfarm economy except those in private households, and workers in the public sector, except the federal government.
[b]Surveyed occupations are classified into wage categories based on the average wage for the occupation, which may include workers with earnings both above and below the threshold. The categories were formed using percentile estimates generated using employer costs for employee compensation (ECEC) data for March 2014.

SOURCE: Adapted from "Table 2. Medical Care Benefits: Access, Participation, and Take-up Rates, National Compensation Survey, March 2014," in *Employee Benefits in the United States—March 2014*, U.S. Department of Labor, Bureau of Labor Statistics, July 25, 2014, http://www.bls.gov/news.release/pdf/ebs2.pdf (accessed August 19, 2014)

- Paid sick leave—65% with access
- Paid vacations—74% with access
- Paid holidays—75% with access

The access rates for all six benefits were higher for full-time employees than for part-time employees. Union employees had higher access rates than nonunion employees. Across the board, the largest establishments (i.e., those with 500 or more workers) offered greater access to benefits than did smaller establishments. There were also significant differences based on wage level. For example, only 38% of civilian employees with average wages in the lowest 25% of the wage range had access to medical care benefits compared with 94% of the employees with average wages in the highest 25% of the wage range.

EMPLOYER-SUBSIDIZED HEALTH INSURANCE. As described in Chapter 1, during World War II (1939–1945) the government enforced wage freezes on private businesses. Companies struggling to attract and keep qualified workers greatly expanded their benefit programs, particularly employer-subsidized health insurance plans (ESIPs). In these plans, employers pay a portion (typically a large portion) of the monthly premiums on behalf of their employees. ESIPs have become very valuable to employees. Group health insurance plans are generally much more affordable and provide broader coverage than do individual plans. In addition, employees enrolled in ESIPs do not have to pay income taxes on the value of the premiums paid by their employers. The employer-paid portion is considered nonwage compensation to employees.

In "The Tax Exclusion for Employer-Sponsored Health Insurance" (February 2010, http://www.nber.org/papers/w15766.pdf), Jonathan Gruber of the National Bureau of Economic Research explains that in 1978 the Congress revised the federal tax code to make ESIPs even more attractive. For certain kinds of plans called "Section 125 cafeteria plans" employees are allowed to use pretax dollars to pay their share of the premiums. In other words, the money used to pay the employee portion is subtracted from the paycheck before income taxes are calculated. This feature also benefits employers because they pay certain "payroll" taxes (i.e., Social Security, Medicare, and unemployment insurance) based on the amounts that employees are paid. Imagine that an employee has gross pay of $1,000 per week, and is enrolled in a Section 125 cafeteria plan for which the employee's share of the premium is $75 per week. The employee's income tax and the employer's payroll taxes are calculated based on $925 per week, not $1,000 per week. Gruber indicates that as of 2010 approximately 80% of employees covered by employer-subsidized plans had access to Section 125 plans.

Health insurance policies fall into two categories: single coverage (i.e., for one person only) and family coverage. As shown in Table 5.11 and Table 5.12, the BLS's National Compensation Survey reveals that as of March 2014 civilian employers were paying, on average, 81% of the premium costs for single coverage and 69% of the premium costs for family coverage for ESIPs. There were differences for both plan types based on employee status and wages and establishment size. Overall, the employer-paid percentage was highest for full-time employees, union members, and employees with the highest wages. Larger establishments (i.e., those with 500 workers or more) paid a higher percentage than smaller firms.

According to the BLS, civilian employees enrolled in ESIPs in March 2014 paid a relatively small portion of their premiums: 19% for single coverage and 31% for family coverage. The Kaiser Family Foundation (KFF), a nonprofit organization, conducts an annual survey related to ESIPs. In "Employer Health Benefits: 2014 Annual Survey" (September 2014, http://files.kff.org/attachment/2014-employer-health-benefits-survey-full-report), the KFF notes that it found that employees in ESIPs in 2014 paid, on average, 18% of the premiums for single coverage and 29% for family coverage. These values are very close to those determined by the BLS.

As discussed in Chapter 3, the price of medical care, including health insurance premiums, has been increasing at a rapid rate. ESIPs have not been immune from rising costs. According to the KFF, the employee portion of premiums increased slightly from 27% in 2004 to 29% in 2014. In 2004 the average worker paid $2,661 in yearly premiums for family coverage; by 2014 that value had risen to $4,823, an increase of 81%. Employees enrolled in ESIPs have also seen their out-of-pocket expenses for deductibles, copayments, and coinsurance increase over time. For example, the average deductible more than doubled between 2006 and 2014 from $584 per year to $1,217 per year.

NONTRADITIONAL WORK ARRANGEMENTS
Working at Home

With technological advances such as Internet access, e-mail, and teleconferencing, working at home has become a viable option for many types of jobs. As employees conduct much of their daily work from home offices, employers are able to save on operating costs. Many employers will pay for computers, additional telephone lines, and even utilities to allow their employees to work from home offices. This allows them to reduce office space, which is one of the higher costs for an employer, especially in large metropolitan markets.

TABLE 5.11

Share of medical care plan premiums paid by employer and employee for single coverage, March 2014

[In percent]

Characteristics	Civilian[a]		Private industry		State and local government	
	Employer share of premium	Employee share of premium	Employer share of premium	Employee share of premium	Employer share of premium	Employee share of premium
All workers participating in single coverage medical plans	81	19	79	21	87	13
Worker characteristics						
Full time	81	19	79	21	88	12
Part time	74	26	72	28	82	18
Union	87	13	86	14	87	13
Nonunion	79	21	78	22	88	12
Average wage within the following categories[b]						
Lowest 25 percent	76	24	74	26	87	13
Lowest 10 percent	71	29	70	30	89	11
Second 25 percent	80	20	78	22	88	12
Third 25 percent	81	19	79	21	88	12
Highest 25 perecent	83	17	81	19	87	13
Highest 10 percent	83	17	81	19	88	12
Establishment characteristics						
1 to 99 workers	79	21	79	21	91	9
1 to 49 workers	80	20	79	21	92	8
50 to 99 workers	78	22	77	23	91	9
100 workers or more	81	19	79	21	87	13
100 to 499 workers	79	21	78	22	88	12
500 workers or more	83	17	80	20	87	13

[a]Includes workers in the private nonfarm economy except those in private households, and workers in the public sector, except the federal government.
[b]Surveyed occupations are classified into wage categories based on the average wage for the occupation, which may include workers with earnings both above and below the threshold. The categories were formed using percentile estimates generated using employer costs for employee compensation (ECEC) data for March 2014.

SOURCE: Adapted from "Table 3. Medical Plans: Share of Premiums Paid by Employer and Employee for Single Coverage, National Compensation Survey, March 2014," in *Employee Benefits in the United States—March 2014*, U.S. Department of Labor, Bureau of Labor Statistics, July 25, 2014, http://www.bls.gov/news.release/pdf/ebs2.pdf (accessed August 19, 2014)

Self-Employment

Self-employed workers are not on the payroll of a company. They may own or operate small businesses or work under contract arrangements with companies. Detailed information about self-employed business owners is provided in Chapter 6.

FOREIGN WORKERS IN THE UNITED STATES

Relatively high wages and favorable working conditions have attracted workers from around the world to the United States. There are two broad categories of foreign workers: those who have entered the country legally with the proper paperwork to pursue work and those who have entered illegally. Legal workers are tracked by the U.S. Citizenship and Immigration Service, formerly the U.S. Immigration and Naturalization Service.

Legal Foreign Workers

The Department of Labor explains in "Guestworker Programs" (2014, http://www.doleta.gov/Business/gw/guestwkr) that it issues a limited number of certifications to foreign workers to work in the United States on a temporary or permanent basis. According to the Department of Labor, "certification may be obtained in cases where it can be demonstrated that there are insufficient qualified U.S. workers available and willing to perform the work at wages that meet or exceed the prevailing wage paid for that occupation in the area of intended employment."

The U.S. Department of Homeland Security's Office of Immigration Statistics tracks the number of foreign workers entering the United States and publishes related data in annual reports. In *Nonimmigrant Admissions to the United States: 2013* (July 2014, http://www.dhs.gov/sites/default/files/publications/ois_ni_fr_2013.pdf), Katie Foreman and Randall Monger of the Department of Homeland Security note that just under 1.9 million temporary workers and trainees were admitted into the United States in fiscal year 2013 (October 1, 2012, to September 30, 2013). This value was down slightly from 1.9 million the previous fiscal year and from 2 million in fiscal year 2011.

Temporary foreign workers maintain the citizenship of their native country, and after fulfilling their contracts with U.S. employers they typically return to their country. Immigrant workers are people who have come to the United States through legal channels and intend to become citizens. They obtain jobs while waiting for their naturalization (the process of becoming a U.S. citizen) to be finalized.

TABLE 5.12

Share of medical care plan premiums paid by employer and employee for family coverage, March 2014

[In percent]

Characteristics	Civilian[a] Employer share of premium	Civilian[a] Employee share of premium	Private industry Employer share of premium	Private industry Employee share of premium	State and local government Employer share of premium	State and local government Employee share of premium
All workers participating in family coverage medical plans	69	31	68	32	71	29
Worker characteristics						
Full time	69	31	69	31	71	29
Part time	64	36	63	37	69	31
Union	81	19	84	16	77	23
Nonunion	66	34	66	34	64	36
Average wage within the following categories[b]						
Lowest 25 percent	58	42	58	42	63	37
Lowest 10 percent	57	43	57	43	56	44
Second 25 percent	67	33	66	34	73	27
Third 25 percent	70	30	70	30	71	29
Highest 25 perecent	73	27	72	28	74	26
Highest 10 percent	74	26	72	28	79	21
Establishment characteristics						
1 to 99 workers	63	37	62	38	71	29
1 to 49 workers	62	38	62	38	73	27
50 to 99 workers	64	36	63	37	69	31
100 workers or more	72	28	73	27	71	29
100 to 499 workers	70	30	70	30	69	31
500 workers or more	74	26	76	24	71	29

[a]Includes workers in the private nonfarm economy except those in private households, and workers in the public sector, except the federal government.
[b]Surveyed occupations are classified into wage categories based on the average wage for the occupation, which may include workers with earnings both above and below the threshold. The categories were formed using percentile estimates generated using employer costs for employee compensation (ECEC) data for March 2014.

SOURCE: Adapted from "Table 4. Medical Plans: Share of Premiums Paid by Employer and Employee for Family Coverage, National Compensation Survey, March 2014," in *Employee Benefits in the United States—March 2014*, U.S. Department of Labor, Bureau of Labor Statistics, July 25, 2014, http://www.bls.gov/news.release/pdf/ebs2.pdf (accessed August 19, 2014)

Illegal Foreign Workers

The issue of illegal immigration has become a heated topic. Much of the debate centers on the economic effect of undocumented workers (foreign workers who have entered the United States illegally). Some people claim that undocumented workers take jobs away from Americans and place a large burden on government-provided social programs. Others believe that undocumented workers are willing to take jobs that Americans do not want, such as low-paying, labor-intensive jobs with no benefits and little to no chance for advancement.

Employers are required by law to verify that new hires are U.S. citizens or foreigners with legal working status who are eligible to work in the United States. Job applicants have to show identification and documentation, including a Social Security card. However, the authenticity of these documents cannot be verified immediately. Thus, well-meaning businesses may unknowingly hire and train illegal workers who use fake documentation to obtain jobs.

There is little doubt that some businesses purposely hire illegal workers or at least ignore questionable paperwork to get inexpensive labor. Many critics maintain that the federal government's focus on terrorism and national security has diminished attention on issues that are related to undocumented workers. Others believe that businesses willing to hire the workers are to blame. Like many factors in the U.S. economy, the issue of undocumented workers is driven by supply and demand factors.

POLITICAL DEBATE AND PUBLIC PROTEST. The Department of Homeland Security publishes an annual estimate of the number of illegal immigrants residing in the United States. As of December 2014, the most recent report contained data compiled in January 2012. Bryan Baker and Nancy Rytina indicate in *Estimates of the Unauthorized Immigrant Population Residing in the United States: January 2012* (March 2013, http://www.dhs.gov/sites/default/files/publications/ois_ill_pe_2012_2.pdf) that there were about 11.4 million illegal immigrants in the United States in January 2012. This number was down from an estimated 11.5 million in January 2011. The Pew Research Center, a private organization, also compiles statistics on illegal immigrants. In "As Growth Stalls, Unauthorized Immigrant Population Becomes More Settled" (September 3, 2014, http://www.pewhispanic.org/2014/09/03/as-growth-stalls-unauthorized-immigrant-population-becomes-more-settled), Jeffrey S. Passel et al. estimate that in March 2013 there were

11.3 million unauthorized immigrants living in the United States. This value was down from its peak of 12.2 million in 2007. Many analysts believe job losses and other economic pressures of the Great Recession are responsible for the decrease.

Illegal immigration has become a politically charged and divisive issue. Some politicians advocate allowing many illegal immigrants already in the country the opportunity to obtain U.S. citizenship under certain conditions. This is called an amnesty provision by its critics and a conditional pathway to legal status by its supporters. The gist of the concept was advocated by President Bush.

During the summer of 2012 President Obama established the Deferred Action for Childhood Arrivals (DACA) program. It covers certain young illegal immigrants who were brought to the United States as children, have a clean criminal record, and meet other requirements. They can request temporary relief from deportation proceedings and apply for work authorization. They are called "DREAMers" because the provisions of the DACA program mirror those of the Development, Relief, and Education for Alien Minors (DREAM) Act, which was originally introduced in Congress in 2001, but failed to pass. In November 2014 President Obama furthered his immigration initiative by authorizing the Deferred Action for Parental Accountability (DAPA) program. It expands the pool of people eligible for the DACA program and provides opportunities for temporary deportation relief and work authorization to certain illegal immigrants who are the parents of either U.S. citizens or lawful permanent residents. As of December 2014, multiple lawsuits had been filed challenging the president's actions as unlawful; however, the lawsuits had not yet been heard in court.

The federal government has funded several provisions that are designed to improve security along the U.S.-Mexican border, including building a 700-mile (1,300-km) fence along the border. Although it is commonly referred to as a fence or wall, the structure includes steel fencing in some areas (primarily urban areas) and concrete barriers in other areas. The structure is not continuous. In addition, border security has been enhanced with larger numbers of border guards and new surveillance equipment. These measures are intended to stem the flow of illegal immigrants across the border and to cut down on the smuggling of illegal drugs and human trafficking.

Many state governments have grown frustrated by what they consider to be the failure of the federal government to stem the flow of illegal immigrants into the United States. According to the National Conference of State Legislatures (NCSL; http://www.ncsl.org/research/immigration/state-laws-related-to-immigration-and-immigrants.aspx), since the 1990s hundreds of laws addressing illegal immigration have been passed by state legislatures.

Some of the laws have not survived court challenges. The most controversial of these laws was Arizona's Senate Bill 1070, which was signed into law in April 2010. Among other provisions, the law gives Arizona law enforcement officials the power to detain anyone they suspect of being in the United States illegally. The constitutionality of the law was challenged in court by the federal government and by private organizations. Chau Lam and Ann Morse of the NCSL report in "U.S. Supreme Court Rules on Arizona's Immigration Enforcement Law" (June 25, 2012, http://www.ncsl.org/issues-research/immig/us-supreme-court-rules-on-arizona-immigration-laws.aspx) that in 2012 the U.S. Supreme Court struck down three sections of the law, but upheld the authority granted to law enforcement officers to determine the immigration status of lawfully stopped people. Since that time, other states have passed similar laws; however, some or all of the provisions of those laws have not passed court challenges.

U.S. JOBS GOING TO FOREIGN COUNTRIES

One consequence of the globalization of U.S. business has been offshoring (the transfer of jobs from the United States to other countries). This can occur when an entire business establishment, such as a factory or service center, is relocated to another nation or when certain jobs within a business are transferred to a foreign company. The latter is also an example of outsourcing (a business practice in which certain tasks within a company are contracted out to another firm; outsourcing may or may not involve sending work to another country). During the 1990s outsourcing became a popular means of reducing costs for some companies. Noncore functions, such as payroll management or housekeeping, are common examples in which outsourcing can be cost effective. However, offshoring is controversial because jobs move outside of the United States, primarily to developing countries, where labor costs are much cheaper.

Critics suggest that offshoring harms the U.S. economy by putting Americans out of work. Others claim that relocation of some operations to foreign countries has a limited effect on domestic employment. They argue that offshoring leads to lower prices for consumer and investment goods, with the ultimate effect of raising real wages (wages that are adjusted for changes in the price of consumer goods) and living standards in the United States.

LABOR UNIONS

Labor unions have had a significant impact on the U.S. workforce and labor policy. Unions are often able to secure higher wages and increased benefits for their members. The BLS reports in "Union Members—2013" (January 24, 2014, http://www.bls.gov/news.release/pdf/

FIGURE 5.7

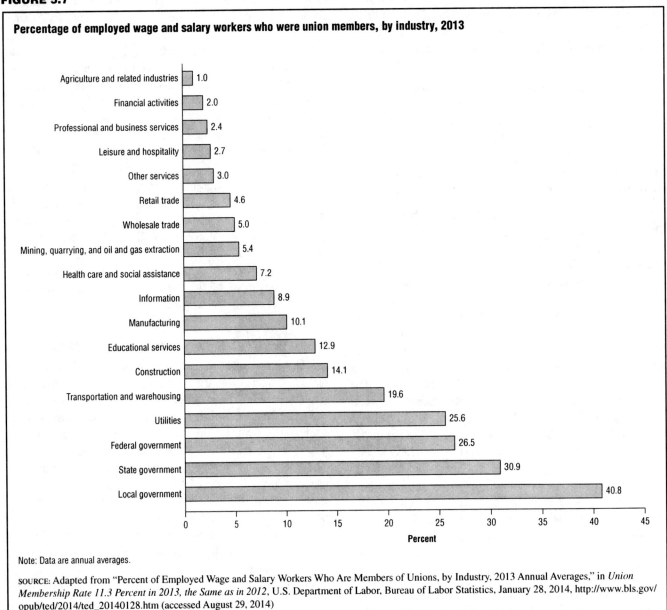

Percentage of employed wage and salary workers who were union members, by industry, 2013

Industry	Percent
Agriculture and related industries	1.0
Financial activities	2.0
Professional and business services	2.4
Leisure and hospitality	2.7
Other services	3.0
Retail trade	4.6
Wholesale trade	5.0
Mining, quarrying, and oil and gas extraction	5.4
Health care and social assistance	7.2
Information	8.9
Manufacturing	10.1
Educational services	12.9
Construction	14.1
Transportation and warehousing	19.6
Utilities	25.6
Federal government	26.5
State government	30.9
Local government	40.8

Note: Data are annual averages.

SOURCE: Adapted from "Percent of Employed Wage and Salary Workers Who Are Members of Unions, by Industry, 2013 Annual Averages," in *Union Membership Rate 11.3 Percent in 2013, the Same as in 2012*, U.S. Department of Labor, Bureau of Labor Statistics, January 28, 2014, http://www.bls.gov/opub/ted/2014/ted_20140128.htm (accessed August 29, 2014)

union2.pdf) that 14.5 million wage and salary workers were union members in 2013. This represented 11.3% of all wage and salary workers in the United States. The percentage is significantly lower than the rate of 20.1% reported in 1983 (the first year the BLS tracked union membership). According to the BLS, 35.3% of government workers were unionized in 2013, compared with only 6.7% of private industry workers. Figure 5.7 compares union membership by industry. Local governments had the most highly unionized employees at 40.8%. This group includes teachers, police officers, and firefighters. The highest rates of union membership in the private industries are found in the transportation and construction sectors.

Anecdotal evidence indicates that union membership is growing more popular in some industries. In "Reform, New Labor Issues Likely to Keep Unions Busy in 2014" (ModernHealthcare.com, January 15, 2014), Joe Carlson states that health care workers are increasingly interested in unionizing. According to Carlson, the workers are "feeling greater anxiety over wages and job security due to partly market and policy pressures to reduce healthcare costs." In "Not Your Grandpa's Labor Union" (BostonGlobe.com, April 6, 2014), Leon Neyfakh describes groups of music video dancers, jazz musicians, video game designers, and fast-food workers that have unionized or are considering doing so. Some college campuses have also seen a surge in unionization. According to Neyfakh, groups of graduate students, medical residents, and adjunct professors (part-time professors employed as needed, for example, a semester at a time) have joined labor unions. In March 2014 football players at Northwestern University made history when a regional representative of the National Labor Relations Board ruled that they are employees of the college and eligible to form a union. Northwestern appealed the ruling. As of December 2014, a final decision had not been reached in the case.

Despite their successes on behalf of American workers, contemporary labor unions continue to face opposition from employers, and often from employees, who question whether the benefits of being associated with a union are worth the cost. Union members are required to go on strike when the union has an unresolved grievance against an employer, and striking union members receive only a fraction of their income in strike pay. Workers who are part of a union may also find themselves facing fines for not abiding by the union bylaws.

For employers, unions can pose other problems. Business operations can be greatly interrupted by unresolved negotiations, whether or not they lead to a strike. Furthermore, because of the increased expenses associated with employing union members, a company's products or services might become less competitively priced in the marketplace. If sales are lost to foreign or nonunion competitors, companies may be forced to lay off employees or even go out of business.

The Gallup Organization has polled Americans regarding their opinions about labor unions since the 1930s. As of December 2014, the most recent polls were conducted in August 2014. Approval of labor unions reached its lowest point in 2009, when only 48% of respondents said they approve of labor unions. The percentage climbed to 53% in 2014. Historical support for unions was much higher, regularly 60% or greater through 2006. Support was highest during the 1950s, when 75% of those asked approved of unions.

PROTECTING AMERICAN WORKERS

The United States has enacted comprehensive labor laws to ensure that workplaces are operated safely and that workers are treated fairly. These include relatively strict laws to protect American workers from discrimination on the basis of gender, age, race, ethnicity, religion, sexual orientation, and other factors. In addition, as noted above, the FLSA offers protection for full- and part-time workers in private and government jobs and covers minimum wages, overtime pay, employer record keeping, and child labor.

Occupational Safety and Health Administration

The Occupational Safety and Health Administration (OSHA) was created in 1970 and operates within the Department of Labor. OSHA personnel set and enforce health and safety standards for workplaces and provide education, training, and assistance related to working conditions. Inspections are conducted randomly, in response to complaints, and after accidents that seriously injure or kill workers. The agency can impose civil fines on violators but lacks the authority to conduct criminal prosecutions.

OSHA also provides support for "whistle-blowers" who report unsafe working conditions or conduct certain activities related to workplace safety or health. In "OSHA Fact Sheet: Your Rights as a Whistleblower" (February 2013, https://www.osha.gov/OshDoc/data_General_Facts/whistleblower_rights.pdf), the agency notes that whistle-blowing employees who believe their employers have taken "unfavorable personnel action" against them can file a complaint with OSHA. The agency notes, "If the evidence supports the employee's allegation and a settlement cannot be reached, OSHA will generally issue an order, which the employer may contest, requiring the employer to reinstate the employee, pay back wages, restore benefits, and other possible remedies to make the employee whole."

According to OSHA, in "Commonly Used Statistics" (2014, https://www.osha.gov/oshstats/commonstats.html), the agency, in collaboration with its state partners, had about 2,200 inspectors in 2013 that conducted more than 39,000 inspections at workplaces. In total, OSHA reports that 4,405 workers were killed on the job in 2013, the lowest total since it began keeping fatality records in 1992. The agency states, "Since 1970, workplace fatalities have been reduced by more than 65 percent and occupational injury and illness rates have declined by 67 percent. At the same time, U.S. employment has almost doubled." Nevertheless, OSHA's regulations and enforcement policies evoke criticism, particularly from employers who believe that excessive government intervention hurts U.S. businesses.

U.S. Equal Employment Opportunity Commission

The U.S. Equal Employment Opportunity Commission (EEOC) enforces federal workplace discrimination laws. It consists of a general counsel and five commissioners who are appointed by the U.S. president and approved by the U.S. Senate. Besides its enforcement role, the EEOC has a training institute to educate employers on workplace discrimination and help them comply with the laws.

According to the EEOC, in "EEOC Charge Receipts by State" (2014, http://www1.eeoc.gov/eeoc/statistics/enforcement/state_13.cfm), in 2013 it received more than 94,000 discrimination complaints from throughout the country. Since 1997, the agency notes (2014, http://www.eeoc.gov/eeoc/statistics/enforcement/litigation.cfm), it has filed more than 5,000 lawsuits in the federal courts over violations of the laws that it enforces. These include the Civil Rights Act of 1964, the Americans with Disabilities Act of 1990, the Age Discrimination in Employment Act of 1967, and the Equal Pay Act of 1963.

CHAPTER 6
U.S. BUSINESSES

After all, the chief business of the American people is business.

—President Calvin Coolidge, January 17, 1925

Businesses are diverse in the United States. They range in size from the huge multinational corporation employing thousands of people to the self-employed individual. They include large and small businesses, home-based businesses, Internet-based businesses, and corporate and family farms. Businesses are a vital part of the American economic engine. They supply goods and services to the world. The consumption of business output is the primary driver behind the growth of the nation's gross domestic product (GDP; the total market value of final goods and services that are produced within an economy in a given year). Businesses also provide opportunities for employment, wealth-building, and investment.

Capitalism encourages business growth. However, businesses can become so large and powerful that they trigger concern about the lack of competition within an industry. Corporate fraud and accounting scandals have eroded the public's trust in the integrity of "big business." These negative feelings deepened during the latter half of the first decade of the 21st century because many Americans blamed irresponsible behavior by large corporations, particularly those in the financial industry, for causing the so-called Great Recession (which lasted from December 2007 to June 2009).

LEGAL STRUCTURES OF BUSINESSES

For legal and tax purposes, all U.S. businesses must be structured as one of several legally defined forms: sole proprietorships, business partnerships, corporations, or limited liability companies. Each offers both advantages and disadvantages to the business owner.

Sole Proprietorships

In a sole proprietorship one person owns and operates the whole business. Because the business and its owner are considered to be a single entity under the law, the owner assumes all the risk but also reaps all the benefits of the business. If the business fails, the sole proprietor may have to cover the losses from his or her personal assets, but if the business succeeds, he or she keeps all the profits. Sole proprietors typically pay lower taxes than those who head corporations or other forms of small businesses. Still, because almost all credit decisions are based on the owner's assets and credit history, it is often difficult for a business set up under this structure to borrow enough money to expand as rapidly as other kinds of businesses.

Business Partnerships

A business partnership has two or more co-owners. As in a sole proprietorship, members of a business partnership are legally recognized as one and the same with their company, meaning that they are personally responsible for the company's debts and other liabilities. Most partnerships start with the partners signing agreements that specify their duties in the business. Many states allow for silent partners, who invest start-up capital but have little role in the company's day-to-day affairs. (Start-up capital is the money used to start a new business.) A significant drawback to this form of business is that each of the partners is responsible for every other partner's actions. If a partner loses or steals money from the company, the other partners will have a legal responsibility to pay that debt.

There are three kinds of business partnerships: a general partnership is the simplest form, in which profits and liability are equally divided among partners or divided according to the terms of the signed agreement; a limited partnership allows partners to have limited liability for the company but also limited decision-making rights; and a joint venture, which is similar to a general partnership but is used only for single projects or for short periods.

Corporations

A corporation is an entity recognized by the state and federal government as entirely separate from its owner or owners. As such, a corporation can be taxed and sued and can enter into contractual agreements. Because it is an individual legal entity, a corporation allows its owners to have less personal liability for debts and lawsuits than a sole proprietorship or business partnership. Owners of corporations are considered to be shareholders, and they may elect a board of directors to oversee management of the company.

Although corporations are commonly thought of as large companies with hundreds or thousands of employees and publicly traded stock, this is not always the case. Owners of small businesses frequently incorporate as their business expands. All corporate owners must file "articles of incorporation" with their state government. For smaller businesses these forms are simple to fill out and file. One option is to file with the Internal Revenue Service as a subchapter S corporation. In an S corporation the owner must pay him- or herself wages like any other employee, but the structure also offers substantial tax flexibility. All corporations that are publicly traded have C corporation status. This means they have nearly unrestricted ownership and are subject to corporate taxes, paying at both the corporate and stockholder levels.

Limited Liability Company

The limited liability company is a combination of a corporation and a partnership in which the owners (or shareholders) have less personal liability for the company's debts and legal issues and have the benefit of simpler tax filings and more control over management issues.

BUSINESS STATISTICS

The U.S. Census Bureau collects data on U.S. businesses annually and conducts an economic census every five years in which it collects more comprehensive data about American businesses. As of December 2014, the most recent economic census data published were from the 2012 census. The Census Bureau divides businesses into two broad categories: employer firms and nonemployer firms.

- Employer firms—Businesses that have at least one paid employee at some time during the year. Employer firms vary greatly in size. For example, Wal-Mart (2014, http://news.walmart.com/walmart-facts/corporate-financial-fact-sheet) employed approximately 1.3 million people in the United States as of 2014, making it one of the largest employers in the nation. Under the Census Bureau's definition an employer firm could also include an individual business owner employing just one person, for example, a doctor who employs a receptionist.

- Nonemployer firms—In "Nonemployer Statistics: Definitions" (May 30, 2013, http://www.census.gov/econ/nonemployer/definitions.htm), the Census Bureau defines a nonemployer business as a business "that has no paid employees, has annual business receipts of $1,000 or more ($1 or more in the construction industry), and is subject to federal income taxes." Examples of nonemployers could include writers, artists, or other individuals who operate businesses without the help of paid employees.

Table 6.1 summarizes data from the 2012 Economic Census and related programs for employer and nonemployer firms. The government categorizes jobs using the North American Industry Classification System (NAICS). The NAICS is the standard classification system for businesses throughout North America. Note that Table 6.1 does not include data for firms engaged in construction or management of companies and enterprises. With these caveats, there were 6.4 million employer establishments operating in 2012 with 104.5 million employees and a total annual payroll of $4.6 trillion. The total value of sales, shipments, receipts, revenue or business done for the employer firms was almost $30 trillion. There were nearly 22.5 million nonemployer establishments operating in 2012 with a total value of sales, shipments, receipts, revenue or business done of $1 trillion.

Business Sizes—Number of Employees

Table 6.2 provides a breakdown by size of the 5.7 million employer firms operating in 2011. Note that the Census Bureau includes a range of 0–4 employees in Table 6.2. In "Statistics of U.S. Businesses (SUSB) Main" (October 16, 2014, https://www.census.gov/econ/susb/index.html), the agency explains that it collects business data during mid-March of each year. A business with no employees on the payroll at that time is still considered an employer firm if it had any paid employees at some time during the year.

As shown in Table 6.2, in 2011 nearly all employer firms (5,666,753 or 99.7%) had fewer than 500 employees each. They are considered "small businesses," using the most common definition of the term. The federal government uses more restrictive definitions for certain purposes. For example, the U.S. Small Business Administration (SBA) is a federal agency that was created in 1953 with the passage of the Small Business Act. The SBA's purpose is to support small businesses by offering financial and counseling assistance and ensuring that small businesses can compete against large companies in receiving government contracts. The size of what the SBA considers to be a small business varies by industry, and official size standards are determined by the agency's

TABLE 6.1

Business statistics, 2012

2012 NAICS code	Meaning of 2012 NAICS code	Number of establishments	Value of sales, shipments, receipts, revenue, or business done ($1,000)	Annual payroll ($1,000)	Number of employees	Number of nonemployer establishments	Nonemployer value of sales, shipments, receipts, revenue, or business done ($1,000)
21	Mining, quarrying, and oil and gas extraction	28,643	555,174,196	61,331,381	903,641	109,931	7,820,264
22	Utilities	17,804	522,282,839	58,635,047	655,358	18,452	786,823
23	Construction	S	S	S	S	2,346,798	127,049,119
31–33	Manufacturing	296,605	5,756,336,857	593,710,388	11,268,906	344,658	16,164,062
42	Wholesale trade	418,780	7,188,763,243	368,739,219	6,068,720	408,487	37,187,323
44–45	Retail trade	1,063,007	4,228,053,136	371,337,169	14,737,687	1,905,147	82,494,176
48–49	Transportation and warehousing	213,131	743,620,690	181,995,480	4,307,024	1,059,040	69,902,217
51	Information	134,652	1,231,918,569	263,505,162	3,206,226	327,795	11,784,901
52	Finance and insurance	471,754	3,532,178,296	522,739,524	6,217,104	720,598	52,045,924
53	Real estate and rental and leasing	348,565	490,644,280	88,122,302	1,980,320	2,389,906	227,427,897
54	Professional, scientific, and technical services	854,274	1,543,690,338	582,443,020	8,142,951	3,212,202	142,974,538
55	Management of companies and enterprises	S	S	S	S	N	N
56	Administrative and support and waste management and remediation services	385,314	724,942,308	345,629,713	10,217,859	2,006,177	42,443,032
61	Educational services	68,215	57,263,676	18,101,184	669,908	603,455	8,504,722
62	Health care and social assistance	830,813	2,051,106,989	804,364,242	18,587,467	1,943,028	59,887,058
71	Arts, entertainment, and recreation	124,347	201,437,811	63,634,983	2,092,370	1,236,539	30,281,203
72	Accommodation and food services	660,934	710,382,088	197,042,306	12,027,557	340,770	15,021,102
81	Other services (except public administration)	528,371	432,235,441	108,380,040	3,456,130	3,522,878	88,544,766
	Total	6,445,209	29,970,030,757	4,629,711,160	104,539,228	22,495,861	1,020,319,127

N = Not available or not comparable.
S = Withheld because estimate did not meet publication standards.
NAICS = North American Industry Classification System.
Note: Excludes NAICS Code 11—agriculture, forestry, fishing and hunting.

SOURCE: Adapted from "EC1200A1: All Sectors: Geographic Area Series: Economy-Wide Key Statistics: 2012," in *2012 Economic Census of the United States*, U.S. Department of Commerce, U.S. Census Bureau, August 8, 2014, http://factfinder2.census.gov/faces/tableservices/jsf/pages/productview.xhtml?src=bkmk (accessed August 21, 2014)

TABLE 6.2

Employer firm statistics, 2011

Enterprise employment size	Number of firms	Number of establishments	Employment	Annual payroll ($1,000)
Total	5,684,424	7,354,043	113,425,965	5,164,897,905
0–4	3,532,058	3,540,155	5,857,662	230,422,086
5–9	978,993	993,101	6,431,931	218,085,669
10–19	592,963	626,981	7,961,281	284,251,614
<20	5,104,014	5,160,237	20,250,874	732,759,369
20–99	481,496	651,624	18,880,001	746,085,051
100–499	81,243	350,197	15,867,437	690,509,553
<500	5,666,753	6,162,058	54,998,312	2,169,353,973
500+	17,671	1,191,985	58,427,653	2,995,543,932

Note: Firms with zero employees at the time of the survey were counted as employer firms if they had at least one employee during the previous year.

SOURCE: Adapted from "Number of Firms, Number of Establishments, Employment, and Annual Payroll by Enterprise Employment Size for the United States and States, Totals: 2011," in *Statistics of U.S. Businesses (SUSB) Main: Latest SUSB Annual Data*, U.S. Department of Commerce, U.S. Census Bureau, December 2013, http://www2.census.gov/econ/susb/data/2011/us_state_totals_2011.xls (accessed August 21, 2014)

Office of Size Standards, which issues standards according to a business's number of employees or its average annual receipts.

Many people perceive the U.S. economy as being dominated by large businesses, such as McDonald's and Microsoft. Although it is true that many of the world's largest companies are headquartered in the United States, small businesses collectively exert enormous influence on the U.S. economy. This includes nonemployer firms and employer firms. As shown in Table 6.2, in 2011 employer firms with fewer than 500 employees each employed nearly 55 million people total and had a combined annual payroll of almost $2.2 trillion. Only 17,671 firms had 500 or more employees each. These businesses employed more than 58 million people and had a combined annual payroll of almost $3 trillion. Thus, in 2011 small employer firms accounted for nearly half of total employment and payroll for all employer firms. Small businesses are found in numerous industries. The magazine *Inc.* publishes a list each year of the fastest growing private companies in the United States. In "The 2014 Inc. 5000" (http://www.inc.com/inc5000/list/2014), the editors put a small business called Fuhu at the top of the list. The company, which makes computer tablets and applications for children, had fewer than 300 employees worldwide in 2014.

Self-Employment

Self-employed people derive at least part of their income from sources other than employer payrolls. Thus, people who own and operate their own businesses are self-employed. So are people who sell their goods or services directly to customers and clients (i.e., outside of an employer-employee relationship). The self-employed are variously called independent contractors, entrepreneurs, proprietors, and freelancers.

There are various estimates of the number of self-employed people in the United States. As noted earlier, the Census Bureau estimates there were nearly 22.5 million nonemployer firms operating in 2012. (See Table 6.1.) In "Nonemployer Statistics" the agency states that "most nonemployers are self-employed individuals operating unincorporated businesses (known as sole proprietorships), which may or may not be the owner's principal source of income." However, the Census Bureau limits its definition of nonemployers to those with annual business receipts of $1,000 or more ($1 or more in the construction industry). Thus, its count of nonemployer firms does not include self-employed people making less than these amounts. In addition, of the nearly 6.5 million employer firms that were tabulated by the Census Bureau in 2012, there is a possibility that many of the smaller firms were owned and operated by self-employed people. (See Table 6.1; note that smaller firms are more likely than larger firms to be sole proprietorships, partnerships, or limited liability companies.)

BENEFITS AND DRAWBACKS. One of the main benefits of self-employment is the freedom of being one's own boss; that is, setting one's own hours, working conditions, pay rate, and so on. This level of independence is particularly appealing to self-motivated individuals who do not want to depend on an employer for their economic well-being. Of course, self-employed people miss out on coveted employer-supplied benefits, such as employer-subsidized medical insurance, employer contributions to employee retirement plans, and paid time off.

Taxation is another consideration. Employees typically pay most if not all of their income taxes through withdrawals from their paychecks. Likewise, employers deduct from employee paychecks Social Security and Medicare taxes. As is explained in Chapter 9, these taxes are assessed based on the amounts that employers pay their employees. Half of the amount is paid by employees through paycheck deductions and the other half is paid directly by employers to the government. Because self-employment income is not payroll-based, self-employed people are responsible for remitting directly to the government the income, Social Security, and Medicare taxes that they owe.

Some analysts complain that government programs and policies favor larger businesses, particularly those that are incorporated, over self-employed individuals. In "To Boost the Economy, Help the Self-Employed" (Businessweek.com, August 4, 2011), Richard Greenwald notes that many self-employed people are unable to take advantage of government programs, such as unemployment insurance and workers' compensation. As described in Chapter 5, employers pay money into state unemployment trust funds to provide a source of payments for employees who become unemployed and meet specific criteria. Likewise, the workers' compensation program is a form of insurance that employers purchase to provide payments to employees who become injured on the job.

BIG BUSINESS

As shown in Table 6.2, in 2011 there were 17,671 U.S. firms that had 500 or more employees each. Many large U.S. firms operate globally with offices, factories, or other establishments in numerous countries.

The business magazine *Forbes* publishes the Global 2000, an annual list of the world's 2,000 largest companies. In "The World's Biggest Public Companies" (May 7, 2014, http://www.forbes.com/global2000), *Forbes* ranks public companies using a composite score based on sales, assets, profits, and market value. The top-10 companies in the 2014 report and their primary business areas and main locations were:

- ICBC (banking; China)
- China Construction Bank (banking; China)
- Agricultural Bank of China (banking; China)
- JPMorgan Chase (banking; United States)

- Berkshire Hathaway (conglomerate; United States)
- Exxon Mobil (oil and gas; United States)
- General Electric (conglomerate; United States)
- Wells Fargo (banking; United States)
- Bank of China (banking; China)
- Petro China (oil and gas; China)

Five of the top-10 companies are based in the United States.

BUSINESS AND POLITICS

Large companies often have strong ties to the government. They have the resources to donate millions of dollars to political campaigns to elect sympathetic lawmakers and to otherwise encourage the passage of pro-business legislation. Likewise, lawmakers, eager to have companies locate facilities in their constituencies to boost local economies, may support policies that favor business interests to the detriment of other programs. Members of Congress may be more inclined to pass pro-business laws if their region has benefited from a large corporation's presence, or if they have received campaign contributions from such a company. This raises concerns that big businesses may be able to convince the government to favor their interests at the expense of the interests of other businesses, or even the population as a whole.

THE FINANCIAL INDUSTRY FALTERS

The U.S. financial industry consists of companies that are engaged primarily in banking, insurance, and investment. Large corporations, in particular, may be involved in a mixture of these enterprises. However, there is an important distinction between commercial banking and investment banking. John Waggoner and David J. Lynch explain in "Red Flags in Bear Stearns' Collapse" (USAToday.com, March 19, 2008) that commercial banks offer products, such as checking and savings accounts, and make loans. They are heavily regulated by the federal government and in most cases the deposits of the banks' customers are insured by the federal government through the Federal Deposit Insurance Corporation (FDIC). In contrast, investment banks facilitate and finance the buying and selling of investments, including stocks and bonds. Investment products are not FDIC insured. The clients of investment banks are mainly companies and government bodies, such as counties and cities. Waggoner and Lynch point out that investment banks receive much less government scrutiny and regulation than commercial banks.

As noted in Chapter 4, investment companies helped drive the housing boom during the first few years of the first decade of the 21st century by buying mortgage-backed securities (MBS) in the secondary mortgage market. The cash inflow allowed lenders to underwrite even more mortgages. At first, MBS were considered to be a wise investment, and financial institutions saw their stock values rise. However, many of the underlying mortgages were poorly underwritten, meaning that the applicants had not been properly screened by the mortgage companies. In many cases, applicants with poor credit histories were given complex mortgages with fluctuating interest rates and ballooning payments. The result was a recipe for disaster. The housing bubble burst and mortgage default rates skyrocketed. Ever since the bust, economists have argued about whether or not investment companies knew that the products they were buying contained these so-called toxic mortgages.

Credit Default Swaps

Investment always entails a certain amount of risk. The article "Credit Default Swaps" (NYTimes.com, March 9, 2012) explains that during the late 1990s the financial industry "invented" a type of insurance contract called the credit default swap (CDS) that was supposed to protect investors "against a default by a particular bond or security." However, CDS products are not technically insurance. The insurance industry is heavily regulated by the federal government. Insurance companies must show that they have the collateral (typically cash or other assets with immediately obtainable and verifiable worth) to cover losses that their clients might suffer.

The article notes that between 2000 and 2008 CDS sales skyrocketed from $900 billion to more than $30 trillion. Unbeknown to the investors, many of these products were "insuring" MBS that contained toxic mortgages. In addition, the swaps were not backed by enough cash collateral. Economists call this being undercapitalized. Some investment companies were even using swaps as collateral for the other swaps they were selling to investors.

In *The U.S. Financial Crisis: The Global Dimension with Implications for U.S. Policy* (January 30, 2009, http://graphics8.nytimes.com/packages/pdf/globaleconcrs.pdf), Dick K. Nanto of the Congressional Research Service examines in detail the role of CDSs in the Great Recession. Nanto points out that investors may have thought that CDSs were safe because the swaps were rated by credit rating firms. Credit rating firms are paid fees to assess the financial soundness of companies and investments. The firms use past historical data and computer models in their assessments. However, Nanto notes that CDSs were relatively new during the first decade of the 21st century. In addition, the credit rating firms used computer models that were supplied by the very companies that were issuing the CDSs. Nanto also suspects that the firms were advising clients "how to structure securities in order to receive higher ratings." Nanto concludes that "the large fees offered to credit rating firms for providing credit ratings were difficult for them to refuse in spite of doubts they might have had about the underlying quality of the securities."

Too Big to Fail?

By 2007 the toxic mortgages that had been initiated during the housing boom were going into default in record high numbers. These defaults devastated the value of the MBS. However, the supposed safety net (the CDSs) was too undercapitalized to cover the losses. The resultant effect was disastrous for the financial industry. Companies, many of them large corporations that had been in business for more than a century and that had weathered the Great Depression (1929–1939), were at the brink of failing. In early 2008 the Federal Reserve System, the national bank of the United States, brokered the sale of the investment bank Bear Stearns to JPMorgan Chase. Waggoner and Lynch note that the Federal Reserve agreed to cover $30 billion in bad assets held by Bear Stearns. The media called it the government's "too big to fail" approach, meaning that some banks are so vital to the overall success of the nation's financial industry that they cannot be allowed to fail. However, Waggoner and Lynch point out that it is more appropriate to say that some banks are "too interconnected to fail." They note that "Bear Stearns had a web of intertwined [investment] agreements with other banks, investment houses and corporations." Thus, the failure of one party in the web could bring down all the others.

Yalman Onaran notes in "Banks' Subprime Losses Top $500 Billion on Writedowns (Update1)" (Bloomberg.com, August 12, 2008) that by August 2008 banks and securities firms throughout the world had suffered losses totaling more than $500 billion due to declining MBS values. In September 2008 the investment firm Lehman Brothers declared bankruptcy after the government refused to bail out the company. However, later that month the government did intervene to save American International Group (AIG), a giant insurance corporation. Meanwhile, the administration of President George W. Bush (1946–) proposed a $700 billion bailout fund to rescue other struggling companies in the financial and automotive industries. It was called the Troubled Asset Relief Program (TARP) and was very unpopular politically. According to the article "Credit Crisis—The Essentials" (NYTimes.com, July 12, 2010), "Many Americans were angered by the idea of a proposal that provided billions of dollars in taxpayer money to Wall Street banks, which many believed had caused the crisis in the first place." Nevertheless, Congress approved the bailout measure. In February 2009 newly inaugurated President Barack Obama (1961–) proposed and Congress passed a $787 billion stimulus package that included funds that could be used to buy up toxic assets from troubled companies.

Bailout Update

The Congressional Budget Office (CBO) provides in *Report on the Troubled Asset Relief Program—April 2014* (2014, https://www.cbo.gov/sites/default/files/45260-TARP.pdf) an update on TARP funding as of March 12, 2014. The CBO divides TARP funding into four broad categories:

- Capital purchases and other support for financial institutions—$313 billion had been disbursed to assist over 700 businesses. Of the total, $293 billion had been repaid. Another $18 billion was written off by the government as unrecoverable (e.g., because a company went bankrupt or the stock purchased by the government had to be sold at a loss). The remaining $2 billion remained outstanding. The CBO anticipated no additional disbursements under this TARP category.

- Financial assistance to the automotive industry—$80 billion had been disbursed to assist General Motors and Chrysler and their associated financing businesses. Of the total, $59 billion had been repaid and $15 billion had been written off, leaving approximately $6 billion outstanding. The CBO anticipated no additional disbursements under this TARP category.

- Investment partnerships designed to increase liquidity in securitization markets—$19 billion had been disbursed through various public-private partnerships that were designed to encourage private investment in certain financial assets. Of the total, all had been repaid and no money had been written off, leaving zero outstanding. The CBO anticipated additional disbursements under this TARP category.

- Mortgage programs—$11 billion had been disbursed to programs such as the Home Affordable Modification Program, which is described in Chapter 4. No money had been repaid, and $11 billion had been written off. The CBO anticipated that $15 billion of additional disbursements would take place under this TARP category.

Financial Health of Banking Institutions

Commercial banks are heavily regulated by the government. They may be chartered (incorporated and authorized to conduct a certain business) at the state or federal level. The U.S. Department of the Treasury's Office of the Comptroller of the Currency (OCC) explains in "Answers & Solutions for Customers of National Banks" (2014, http://www.helpwithmybank.gov/national_banks/index.html) that two federal agencies (the OCC and the Federal Reserve) are responsible for supervising and regulating all federally chartered banks and credit unions and some state-chartered banks. The FDIC has regulatory authority for state-chartered banks that are not members of the Federal Reserve

System. All state-chartered banks are also supervised by state banking regulators.

Federal and state regulators closely monitor commercial financial institutions to ensure that they have enough assets to cover their obligations. The regulators close down banks that are in danger of failing and temporarily take over banks that fail. Following the Great Depression, a number of strict commercial bank regulations were implemented. The direct result was that bank failures became very rare. However, the housing industry bust and the crisis in the financial industry that set off the Great Recession in late 2007 triggered historically high numbers of bank failures. The FDIC reports in "Failed Bank List" (November 26, 2014, http://www.fdic.gov/bank/individual/failed/banklist.html) the following number of bank failures per year:

- 2000—2 failed banks
- 2001—4 failed banks
- 2002—11 failed banks
- 2003—3 failed banks
- 2004—4 failed banks
- 2005—0 failed banks
- 2006—0 failed banks
- 2007—3 failed banks
- 2008—25 failed banks
- 2009—140 failed banks
- 2010—157 failed banks
- 2011—92 failed banks
- 2012—51 failed banks
- 2013—24 failed banks
- 2014—17 failed banks through November 26

Each year *Forbes* magazine ranks the nation's 100 largest publicly traded banking institutions based on various measures of their financial health. In "America's Best and Worse Banks 2014" (Forbes.com, December 19, 2013), Kurt Baudenhausen discusses the rankings as of the third quarter of 2013 for the so-called Big Four, which are, by far, the largest U.S. banks based on their assets:

- JPMorgan Chase ($2.5 trillion in assets)—Ranked 54th, same as in 2012
- Bank of America ($2.1 trillion in assets)—Ranked 95th, down nine spots from 2012
- Citigroup ($1.9 trillion in assets)—Ranked 39th, up three spots from 2012
- Wells Fargo ($1.5 trillion in assets)—Ranked 73rd, up 11 spots from 2012

Overall Baudenhausen notes that many banks (including the Big Four) were still struggling to recover from the lingering effects of the housing crisis and the Great Recession, particularly poor repayment of loans by borrowers. In addition, some banks have been forced to spend large amounts of money on lawsuits brought against them for conduct the government believes precipitated the crisis. For example, in November 2013 JPMorgan Chase agreed to pay $13 billion to settle such litigation. In "Justice Department, Federal and State Partners Secure Record $13 Billion Global Settlement with JPMorgan for Misleading Investors about Securities Containing Toxic Mortgages" (November 19, 2013, http://www.justice.gov/opa/pr/justice-department-federal-and-state-partners-secure-record-13-billion-global-settlement), the U.S. Department of Justice calls the agreement "the largest settlement with a single entity in American history." Most of the money ($9 billion) will go to various federal and state agencies to settle claims related to the bank's involvement in packaging, marketing, and selling residential MBS including toxic mortgages. The remaining $4 billion is slated to provide "relief" to consumers through measures such as loan modifications.

BUSINESS ECONOMIC PERFORMANCE

Thousands of economic indicators are used to monitor the economic performance of American businesses. Two of the most prominent measures are gross private domestic investment (GPDI; a component of the GDP) and corporate profits—a component of the gross domestic income (GDI), which is national output based on the incomes earned and the costs incurred during production.

Gross Private Domestic Investment

As explained in Chapter 2, the nation's GDP includes four major components, the largest of which is personal consumption expenditures (PCE)—the amount spent by consumers on final goods and services. Businesses, of course, are the sources of the goods and services purchased by consumers. In addition, businesses spend money. Their spending is considered an investment because they are investing in the future of their companies.

The BEA captures business spending in the GPDI. Table 6.3 provides a breakdown of GPDI activities for 2010 through 2013 and estimated for 2014 based on data for the first two quarters of the year. Note that all the values shown are real, meaning they are inflation-adjusted. During 2013 the real GPDI totaled nearly $2.6 trillion. Table 2.2 in Chapter 2 indicates that the real GDP for 2013 was $15.7 trillion; thus, GPDI accounted for around 17% of the total GDP.

Table 6.4 shows the annual percent change in real GPDI for 1994 through 2013. In 2013 the real GPDI increased by 4.9%. Over the 20-year period shown in

TABLE 6.3

Real gross private domestic investment, 2010–13

[Billions of chained (2009) dollars]

	2010	2011	2012	2013	Seasonally adjusted at annual rates 2014	
					I	II
Gross private domestic investment	2,120.4	2,230.4	2,435.9	2,556.2	2,588.2	2,691.8
Fixed investment	2,056.2	2,186.7	2,368.0	2,479.2	2,536.1	2,572.7
Nonresidential	1,673.8	1,802.3	1,931.8	1,990.6	2,051.5	2,079.1
Structures	366.3	374.7	423.8	421.7	441.9	447.6
Equipment	746.7	847.9	905.6	947.2	974.8	991.4
Information processing equipment	281.4	285.9	295.0	304.0	298.1	312.9
Computers and peripheral equipment	—	—	—	—	—	—
Other	196.8	202.8	208.1	217.0	217.6	230.6
Industrial equipment	151.3	183.3	190.3	197.7	209.0	218.8
Transportation equipment	136.9	183.0	217.6	231.1	247.8	247.3
Other equipment	179.8	199.3	207.3	219.0	223.6	216.7
Intellectual property products	561.3	581.3	603.7	624.1	636.8	642.3
Software	254.2	271.8	287.0	295.9	300.0	302.8
Research and development	234.4	236.7	241.3	250.7	258.2	260.7
Entertainment, literary, and artistic originals	72.7	73.1	76.1	78.0	78.8	79.0
Residential	382.4	384.5	436.5	488.4	485.3	494.2
Change in private inventories	58.2	37.6	57.0	63.5	35.2	93.4
Farm	−7.0	1.4	−5.6	7.6	2.2	4.4
Nonfarm	65.9	36.6	65.9	55.2	33.3	90.0

Note: Users are cautioned that particularly for components that exhibit rapid change in prices relative to other prices in the economy, the chained-dollar estimates should not be used to measure the component's relative importance or its contribution to the growth rate of more aggregate series. Values are in billions of chained (2009) dollars. Quarterly data are seasonally adjusted at annual rates.

SOURCE: Adapted from "Table 3B. Real Gross Domestic Product and Related Measures," in *National Income and Product Accounts—Gross Domestic Product: Second Quarter 2014 (Advance Estimate) Annual Revision: 1999 through First Quarter 2014*, U.S. Department of Commerce, Bureau of Economic Analysis, July 30, 2014, http://www.bea.gov/newsreleases/national/gdp/2014/pdf/gdp2q14_adv.pdf (accessed August 7, 2014)

Table 6.4 real GPDI growth dropped to a low of −21.6% in 2009 (during the Great Recession) and reached a high of 12.9% the following year.

As shown in Table 6.3, the GPDI comprises two major components: fixed investment and change in private inventories. Fixed investment includes amounts invested by businesses in structures (e.g., nonresidential commercial buildings), equipment, and intellectual property products, such as software. During 2013 money invested in equipment totaled $947.2 billion, the most for any single GPDI component. Another $624.1 billion was invested in intellectual property products and $421.7 billion in structures. Note that residential fixed investment comprises a small portion of the GPDI and represents spending by landlords and owner-occupants on new houses and improvements to existing housing. Thus, residential fixed investment is not wholly an economic indicator for the business sector because it also captures real estate investment by consumers.

In *Concepts and Methods of the U.S. National Income and Product Accounts* (February 2014, http://www.bea.gov/national/pdf/chapter7.pdf), the BEA defines change in private inventories (or inventory investment) as "a measure of the value of the change in the physical volume of the inventories—additions less withdrawals—that businesses maintain to support their production and distribution activities." The agency indicates that this economic indicator is highly volatile, changing with great frequency, but provides valuable clues about business performance. The BEA states "inventory movement plays a key role in the timing, duration, and magnitude of business cycles, as unanticipated buildups in inventories may signal future cutbacks in production, and unanticipated shortages in inventories may signal future pickups in production." As shown in Table 6.3, the change in private inventories was $63.5 billion in 2013.

The BEA explains that the change in private inventories value can be positive or negative:

- Positive value—"indicates that total production (GDP) exceeded the sum of the final sales components of GDP in the current period and that the excess production was added to inventories."

- Negative value—"indicates that final sales exceeded production in the current period and that the excess sales were filled by drawing down inventories."

Corporate Profits

As explained in Chapter 2, the nation's GDI includes four major components: employee compensation, taxes on production minus subsidies, profits and surpluses, and consumption of fixed capital. One of the most closely monitored economic indicators for the business sector is called corporate profits and is tracked by the BEA. It is a measure of the income generated by corporations from

TABLE 6.4

Percentage change in real gross private domestic investment, 1994–2013

	1994	1995	1996	1997	1998	1999	2000	2001	2002	2003	2004	2005	2006	2007	2008	2009	2010	2011	2012	2013
Gross private domestic investment	**11.9**	**3.2**	**8.8**	**11.4**	**9.5**	**8.4**	**6.5**	**−6.1**	**−0.6**	**4.1**	**8.8**	**6.4**	**2.1**	**−3.1**	**−9.4**	**−21.6**	**12.9**	**5.2**	**9.2**	**4.9**
Fixed investment	8.2	6.1	8.9	8.6	10.2	8.8	6.9	−1.6	−3.5	4.0	6.7	6.8	2.0	−2.0	−6.8	−16.7	1.5	6.3	8.3	4.7
Nonresidential	7.9	9.7	9.1	10.8	10.8	9.7	9.1	−2.4	−6.9	1.9	5.2	7.0	7.1	5.9	−0.7	−15.6	2.5	7.7	7.2	3.0
Structures	1.8	6.4	5.7	7.3	5.1	0.1	7.8	−1.5	−17.7	−3.9	−0.4	1.7	7.2	12.7	6.1	−18.9	−16.4	2.3	13.1	−0.5
Equipment	12.3	12.1	9.5	11.1	13.1	12.5	9.7	−4.3	−5.4	3.2	7.7	9.6	8.6	3.2	−6.9	−22.9	15.9	13.6	6.8	4.6
Intellectual property products	4.0	7.3	11.3	13.0	10.8	12.4	8.9	0.5	−0.5	3.8	5.1	6.5	4.5	4.8	3.0	−1.4	1.9	3.6	3.9	3.4
Residential	9.0	−3.4	8.2	2.4	8.6	6.3	0.7	0.9	6.1	9.1	10.0	6.6	−7.6	−18.8	−24.0	−21.2	−2.5	0.5	13.5	11.9
Change in private inventories	—	—	—	—	—	—	—	—	—	—	—	—	—	—	—	—	—	—	—	—

SOURCE: Adapted from "Table 7. Real Gross Domestic Product: Percent Change from Preceding Year," in *National Income and Product Accounts—Gross Domestic Product: Second Quarter 2014 (Advance Estimate) Annual Revision: 1999 through First Quarter 2014*, U.S. Department of Commerce, Bureau of Economic Analysis, July 30, 2014, http://www.bea.gov/newsreleases/national/gdp/2014/pdf/gdp2q14_adv.pdf (accessed August 7, 2014)

TABLE 6.5

Corporate profits and their distribution, 2010–13 and first quarter of 2014

[Billions of dollars]

	2010	2011	2012	2013	2014 I
Corporate profits with inventory valuation and capital consumption adjustments	1,746.4	1,816.6	2,022.8	2,106.9	1,942.1
Less: taxes on corporate income	370.6	379.1	454.8	474.3	562.3
Equals: profits after tax with inventory valuation and capital consumption adjustments	1,375.9	1,437.5	1,568.0	1,632.6	1,379.8
Net dividends	564.0	703.7	857.1	959.6	902.8
Undistributed profits with inventory valuation and capital consumption adjustments	811.9	733.9	710.9	673.0	477.1
Addenda for corporate cash flow:					
Net cash flow with inventory valuation adjustment	2,094.9	2,071.5	2,066.5	2,080.8	1,919.5
Undistributed profits with inventory valuation and capital consumption adjustments	811.9	733.9	710.9	673.0	477.1
Consumption of fixed capital	1,262.5	1,298.8	1,348.5	1,402.1	1,435.1
Less: capital transfers paid (net)	−20.6	−38.8	−7.1	−5.7	−7.4
Addenda:					
Profits before tax (without inventory valuation and capital consumption adjustments)	1,840.7	1,806.8	2,136.1	2,235.3	2,297.2
Profits after tax (without inventory valuation and capital consumption adjustments)	1,470.1	1,427.7	1,681.3	1,761.1	1,734.9
Inventory valuation adjustment	−41.0	−68.3	−9.5	3.3	−24.6
Capital consumption adjustment	−53.3	78.1	−103.8	−131.8	−330.5

Note: Quarterly data are seasonally adjusted at annual rates.

SOURCE: Adapted from "Table 11A. Corporate Profits," in *National Income and Product Accounts—Gross Domestic Product: Second Quarter 2014 (Advance Estimate) Annual Revision: 1999 through First Quarter 2014*, U.S. Department of Commerce, Bureau of Economic Analysis, July 30, 2014, http://www.bea.gov/newsreleases/national/gdp/2014/pdf/gdp2q14_adv.pdf (accessed August 7, 2014)

---

the current production of goods and services. Corporate profits do not include capital gains or losses (i.e., changes in the value of capital assets, such as property, structures, and equipment). The BEA indicates in *Concepts and Methods of the U.S. National Income and Product Accounts* (February 2014, http://www.bea.gov/national/pdf/chapter13.pdf) that two mathematical adjustments are commonly made to corporate profits. An inventory valuation adjustment is made because "gains or losses that result from holding goods in inventory are not considered income from current production and thus are removed from business income." A capital consumption adjustment is made to adjust for the effects of depreciation (i.e., the depletion of an asset's value over time).

Table 6.5 lists corporate profits for 2010 through 2013 and annualized for 2014 based on data for the first quarter of the year. During 2013 corporate profits totaled $2.1 trillion. As shown in Table 6.5, the BEA also tracks three ways in which corporations use their profits:

- Corporate income taxes paid to the government ($474.3 billion in 2013)

- Net dividends paid to shareholders or corporate owners ($959.6 billion in 2013)

- Undistributed profits that are retained by corporations and used to finance capital investments, pay debt, hire new employees, or for other purposes ($673 billion in 2013)

Figure 6.1 shows corporate profits for the second quarter of 1994 through the first quarter of 2014. Corporations made huge gains during the first few years of the 21st century before suffering a dramatic decline during the Great Recession. Corporate profits then rebounded and peaked at $2.1 trillion during the fourth quarter of 2013. Preliminary data for the first quarter of 2014 predict that corporate profits for the year will be slightly lower at $1.9 trillion.

LABOR COSTS

The Bureau of Labor Statistics (BLS) within the U.S. Department of Labor tracks the costs incurred by the business sector to employ a workforce. In "Employer Costs for Employee Compensation—March 2014" (June 11, 2014, http://www.bls.gov/news.release/pdf/ecec.pdf), the BLS notes that it obtains data from the National Compensation Survey for nonfarm private and state and local government workers. As explained in Chapter 5, the 2014 survey included about 9,600 establishments in private industry and approximately 1,500 establishments in state and local governments. As shown in Table 6.6, the BLS estimated the employer cost for private workers during March 2014 at $29.99 per hour worked. This value included $20.96 in wages and salaries and $9.03 in total benefits. Overall, wages and salaries accounted for 69.9% of the total, and total benefits made up 30.1%. Insurance costs comprised the largest percentage (8.3%) of total benefits costs.

As of March 2014, private employers devoted $2.36 per hour toward health insurance for their employees. (See Table 6.6.) This amount comprised 7.9% of the total compensation cost for that month. Employer subsidizing of health insurance plans is explained in detail in Chapter 5. As noted in Chapter 1, the Centers for Medicare and Medicaid Services (CMS), an agency within the U.S.

FIGURE 6.1

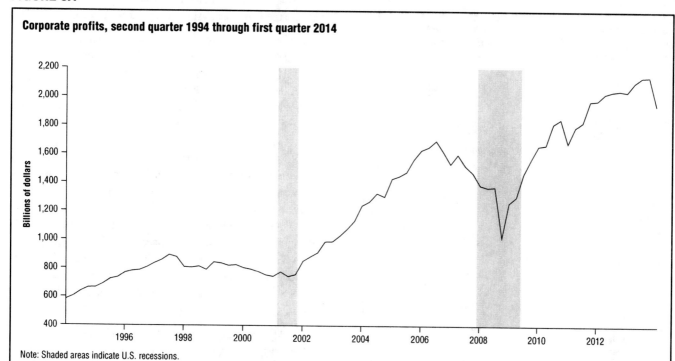

Corporate profits, second quarter 1994 through first quarter 2014

Note: Shaded areas indicate U.S. recessions.

SOURCE: "Corporate Profits with Inventory Valuation Adjustment (IVA) and Capital Consumption Adjustment (CCAdj)," in *FRED® Economic Data*, Federal Reserve Bank of St. Louis, July 30, 2014, http://research.stlouisfed.org/fredgraph.pdf?hires=1&type=application/pdf&chart_type=line&recession_bars=on&log_scales=&bgcolor=%23e1e9f0&graph_bgcolor=%23ffffff&fo=verdana&ts=12&tts=12&txtcolor=%23444444&show_legend=yes&show_axis_titles=yes&drp=0&cosd=1994-01-01&coed=2014-01-01&width=670&height=445&stacking=&range=Custom&mode=fred&id=CPROFIT&transformation=lin&nd=&ost=-99999&oet=99999&scale=left&line_color=%234572a7&line_style=solid&lw=2&mark_type=none&mw=1&mma=0&fml=a&fgst=lin&fq=Quarterly&fam=avg&vintage_date=&revision_date= (accessed August 21, 2014)

Department of Health and Human Services estimates total national spending on health care was $2.8 trillion in 2012. The CMS indicates that private businesses accounted for 21% of the total or $588 billion. Thus, health care spending represents a substantial labor cost for the business sector.

Another economic indicator related to labor costs is the employment cost index (ECI) which is published quarterly by the BLS. In *BLS Handbook of Methods* (July 11, 2013, http://www.bls.gov/opub/hom/pdf/homch8.pdf), the BLS states, "The ECI is a measure of the change in the cost of labor, independent of the influence of employment shifts among occupations and industry categories." The agency assesses changes in total compensation which includes wages, salaries, and employee benefits for the civilian (nonmilitary) workforce. The underlying data are obtained from the National Compensation Survey.

According to the BLS, in "Employment Cost Index—September 2014" (October 31, 2014, http://www.bls.gov/news.release/pdf/eci.pdf), the 12-month ECI in September 2014 was 2.2%, meaning that overall, compensation costs per hour worked increased by 2.2% from the previous year. Wages and salaries increased by 2.1%, and benefit costs increased by 2.4%. Figure 6.2 shows the percent change in ECI for total benefits and for health benefits only for 1982 through June 2014. In general, employer costs for health benefits increased by a greater percentage each year than did the costs for total benefits. This was true during the early years of the 21st century when political momentum built for comprehensive reform of the nation's health insurance system.

Table 1.4 in Chapter 1 lists the major provisions of the Patient Protection and Affordable Care Act (ACA), which was passed in 2010 and is commonly called Obamacare. The law requires certain employers to offer their employees health insurance coverage meeting specific standards or pay a penalty for failing to do so. This is known as the employer mandate or the employer penalty. The administration of President Barack Obama (1961—) delayed implementation of this provision until 2015 for employers with at least 100 full-time employees and until 2016 for employers with 50 to 99 full-time employees. In "Almost No One Wants the Obamacare Employer Mandate Now: Here's Why" (LATimes.com, July 28, 2014), Noam N. Levey indicates the employer mandate was added to the ACA because the law created government subsidies for lower-income Americans who buy health insurance plans via the government exchanges. Levey states, "Democrats who wrote the law worried that companies would be tempted to stop offering coverage [to their employees], shifting the cost to taxpayers." He

TABLE 6.6

Employer costs per hour worked for employee compensation and costs as a percentage of total compensation for private industry workers, March 2014

	Occupational group	
	All workers	
Compensation component	Cost	Percent
Total compensation	$29.99	100.0
Wages and salaries	20.96	69.9
Total benefits	9.03	30.1
Paid leave	2.09	7.0
Vacation	1.08	3.6
Holiday	0.63	2.1
Sick	0.26	0.9
Personal	0.11	0.4
Supplemental pay	0.85	2.8
Overtime and premium[a]	0.25	0.8
Shift differentials	0.06	0.2
Nonproduction bonuses	0.54	1.8
Insurance	2.50	8.3
Life	0.04	0.1
Health	2.36	7.9
Short-term disability	0.06	0.2
Long-term disability	0.05	0.2
Retirement and savings	1.15	3.8
Defined benefit	0.50	1.7
Defined contribution	0.66	2.2
Legally required benefits	2.44	8.1
Social Security and Medicare	1.75	5.8
Social Security[b]	1.40	4.7
Medicare	0.35	1.2
Federal unemployment insurance	0.04	0.1
State unemployment insurance	0.23	0.8
Workers' compensation	0.42	1.4

[a]Includes premium pay for work in addition to the regular work schedule (such as overtime, weekends, and holidays).
[b]Comprises the Old-Age, Survivors, and Disability Insurance (OASDI) program.
Note: The sum of individual items may not equal totals due to rounding.

SOURCE: Adapted from "Table 5. Employer Costs per Hour Worked for Employee Compensation and Costs as a Percent of Total Compensation: Private Industry Workers, By Major Occupational Group and Bargaining Unit Status, March 2014," in *Employer Costs for Employee Compensation—March 2014*, U.S. Department of Labor, Bureau of Labor Statistics, June 11, 2014, http://www.bls.gov/news.release/pdf/ecec.pdf (accessed August 20, 2014)

---

notes that this fear proved to be unfounded and that political support for the mandate was fading in 2014 because employers had not dropped their health insurance benefits in large numbers. The mandate does impose additional reporting requirements on employers, which Levey indicates "have made it deeply unpopular with business groups." In addition, some analysts worry that smaller employers subject to the mandate may transition some of their employees from full-time status to part-time status in order to avoid offering the required insurance or paying the penalty.

AGRICULTURE

The National Agricultural Statistics Service (NASS) within the U.S. Department of Agriculture (USDA) performs a census of agriculture every five years. As of December 2014, the most recent NASS results available were from the 2012 census. Table 6.7 summarizes key findings from the censuses dating back to 1982.

In *2012 Census of Agriculture: United States Summary and State Data, Volume 1, Geographic Area Series, Part 51* (May 2014, http://www.agcensus.usda.gov/Publications/2012/Full_Report/Volume_1,_Chapter_1_US/usv1.pdf), the NASS notes, "The census definition of a farm is any place from which $1,000 or more of agricultural products were produced and sold, or normally would have been sold, during the census year." As shown in Table 6.7, there were 2.1 million farms operating in 2012, down from 2.2 million farms in 2007. The national acreage devoted to farming also decreased from 922.1 million acres in 2007 to about 914.5 million acres in 2012. The market value of agricultural products sold increased from $297.2 billion in 2007 to $394.6 billion in 2012. Note that these values are expressed in current dollars rather than real dollars; thus, they do not account for the effects of deflation or inflation over time.

In "Ag and Food Sectors and the Economy" (April 8, 2014, http://ers.usda.gov/data-products/ag-and-food-statistics-charting-the-essentials/ag-and-food-sectors-and-the-economy.aspx), the USDA's Economic Research Service indicates that U.S. farms directly accounted for approximately 1% of the nation's GDP in 2012. The Economic Research Service notes, "The overall contribution of the agriculture sector to GDP is larger than this because sectors related to agriculture—forestry, fishing, and related activities; food, beverages, and tobacco products; textiles, apparel, and leather products; food services and drinking places—rely on agricultural inputs in order to contribute additional value to the economy." In addition, the Economic Research Service estimates that farms directly employed more than 2.6 million people in 2012. Nearly 14 million additional people were employed in jobs related to agriculture, for example, in food manufacturing and food services and drinking establishments.

FEDERAL REGULATION OF BUSINESS

Historically, U.S. economic philosophy has been to let the market operate with a minimum of government interference. This does not mean, however, that U.S. businesses go unregulated. Many local, state, and federal laws exist to protect the public and the economy from dangerous, unfair, or fraudulent activities by businesses. Major federal programs that oversee business activities are:

- U.S. Consumer Product Safety Commission (CPSC)—established in 1973 to protect the American public from unreasonable risks of serious injury or death from consumer products. Through a combination of voluntary and mandatory safety standards, the CPSC tries to prevent dangerous products from entering the market. If a product is found to be dangerous

FIGURE 6.2

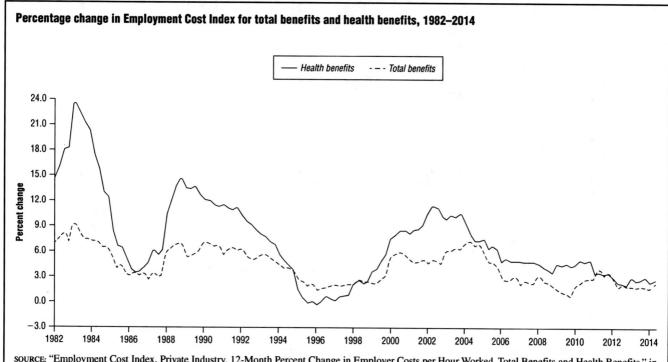

Percentage change in Employment Cost Index for total benefits and health benefits, 1982–2014

SOURCE: "Employment Cost Index, Private Industry, 12-Month Percent Change in Employer Costs per Hour Worked, Total Benefits and Health Benefits," in *Employment Cost Index: Health Benefits*, U.S. Department of Labor, Bureau of Labor Statistics, July 2014, http://www.bls.gov/web/eci/echealth.pdf (accessed August 21, 2014) IRAE157

after it has already been sold to consumers, the CPSC has the duty to inform the public and the power to force a recall of the product if it is deemed necessary.

- Employment Standards Administration (ESA)—one of the largest branches of the U.S. Department of Labor, the ESA is charged with enforcing a wide variety of labor laws dealing with minimum wage requirements, overtime pay standards, child labor protections, and unpaid leaves of absence. It also provides oversight of federal contractors concerning employment issues.

- U.S. Equal Employment Opportunity Commission (EEOC)—established in 1965, the EEOC is the primary federal agency responsible for preventing discrimination in the workplace. Its original purpose was to investigate violations of the Civil Rights Act of 1964, which prohibited discrimination in the workplace on the basis of race, color, national origin, sex, and religion. Over the years its powers have been expanded and it has been given responsibility to enforce other antidiscrimination laws.

- Federal Trade Commission (FTC)—created in 1914 with the passage of the Federal Trade Commission Act. Originally intended to combat the rise of business monopolies, the FTC grew to become the U.S. government's consumer protection agency, addressing consumer issues such as identity theft, false advertising, telemarketing and Internet scams, and anticompetition moves by businesses.

- Occupational Safety and Health Administration (OSHA)—establishes and enforces workplace safety standards. One or more OSHA standards covers almost every workplace in the United States.

- U.S. Environmental Protection Agency (EPA)—develops and enforces federal environmental regulations. The EPA keeps track of industrial pollutants and regularly updates its compliance codes for individual sectors and industries.

- U.S. Food and Drug Administration (FDA)—works to ensure that the food, drugs, and cosmetics sold in the United States are safe and effective. It establishes safety and sanitation standards for manufacturers of these goods, as well as quality standards that the goods themselves must meet. The FDA must prove that certain products, especially drugs, are safe and effective before they can be sold in the United States, and it can force products off the market if they are later discovered to be dangerous. In addition, the FDA ensures that the labeling of food, drugs, and cosmetics is complete and truthful.

Other Agencies

Besides the previously mentioned agencies, there are a number of other government agencies that regulate

TABLE 6.7

Statistics from the Census of Agriculture, selected years, 1982–2012

All farms	2012	2007	2002	1997	Not adjusted for coverage			
					1997	1992	1987	1982
Farms (number)	2,109,303	2,204,792	2,128,982	2,215,876	1,911,859	1,925,300	2,087,759	2,240,976
Land in farms (acres)	914,527,657	922,095,840	938,279,056	954,752,502	931,795,255	945,531,506	964,470,625	986,796,579
Average size of farm (acres)	434	418	441	431	487	491	462	440
Farms by size								
1 to 9 acres	223,634	232,849	179,346	205,390	153,515	166,496	183,257	187,665
10 to 49 acres	589,549	620,283	563,772	530,902	410,833	387,711	412,437	449,252
50 to 179 acres	634,047	660,530	658,705	694,489	592,972	584,146	644,849	711,652
180 to 499 acres	346,038	368,368	388,617	428,215	402,769	427,648	478,294	526,510
500 to 999 acres	142,555	149,713	161,552	179,447	175,690	186,387	200,058	203,925
1,000 to 1,999 acres	91,273	92,656	99,020	103,007	101,468	101,923	102,078	97,395
2,000 acres or more	82,207	80,393	77,970	74,426	74,612	70,989	66,786	64,577
Total cropland (farms)	**1,551,654**	**1,685,339**	**1,751,450**	**1,857,239**	**1,661,395**	**1,697,137**	**1,848,574**	**2,010,609**
Market value of agricultural products sold ($1,000)	394,644,481	297,220,491	200,646,355	201,379,812	196,864,649	162,608,334	136,048,516	131,900,223
Average per farm (dollars)	187,097	134,807	94,245	90,880	102,970	84,459	65,165	58,858
Crops, including nursery and greenhouse crops ($1,000)	212,397,074	143,657,928	95,151,954	100,668,794	98,055,656	75,228,256	58,931,085	62,256,087
Livestock, poultry, and their products ($1,000)	182,247,407	153,562,563	105,494,401	100,711,018	98,808,993	87,380,078	77,117,431	69,644,136
Farms by value of sales[a]:								
Less than $2,500	788,310	900,327	826,558	693,026	496,514	422,767	490,296	536,327
$2,500 to $4,999	191,422	200,302	213,326	265,667	228,477	231,867	262,918	278,208
$5,000 to $9,999	214,245	218,531	223,168	267,575	237,975	251,883	274,972	281,802
$10,000 to $24,999	244,954	248,285	256,157	293,639	274,040	301,804	326,166	340,254
$25,000 to $49,999	152,873	154,732	157,906	179,629	170,705	195,354	219,636	248,828
$50,000 to $99,999	129,366	125,456	140,479	163,510	158,160	187,760	218,050	251,501
$100,000 to $499,999	232,955	240,873	240,746	282,422	277,194	286,951	263,698	274,580
$500,000 or more	155,178	116,286	70,642	70,408	68,794	46,914	32,023	27,800
Farms by legal status for tax purposes:								
Family or individual	1,828,946	1,906,335	1,909,598	1,922,590	1,643,424	1,653,491	1,809,324	1,945,639
Partnership	137,987	174,247	129,593	185,607	169,462	186,806	199,559	223,274
Corporation	106,716	96,074	73,752	90,432	84,002	72,567	66,969	59,792
Other-cooperative, estate or trust, institutional, etc	35,654	28,136	16,039	17,247	14,971	12,436	11,907	12,271
Average age of principal operator (years)	58.3	57.1	55.3	54.0	54.3	53.3	52.0	50.5
Total farm production expenses[b] ($1,000)	328,939,354	241,113,666	173,199,216	157,752,357	150,590,993	130,779,261	108,138,053	(NA)

[a]Data for 1982 exclude abnormal farms.
[b]Data for 2002 and prior years are based on a sample of farms.
(NA) = Not applicable.

SOURCE: Adapted from "Table 1. Historical Highlights: 2012 and Earlier Census Years," in *2012 Census of Agriculture: United States Summary and State Data, Volume 1, Geographic Area Series, Part 51*, U.S. Department of Agriculture, May 2014, http://www.agcensus.usda.gov/Publications/2012/Full_Report/Volume_1,_Chapter_1_US/usv1.pdf (accessed August 21, 2014)

specific industries or aspects of the economy. Some of them are well known, whereas many others may be virtually unknown to people outside the fields they regulate. A few examples are:

- Federal Communications Commission—regulates the telecommunications industry, including all television, radio, satellite, cable, and wire services in the United States and its territories
- Federal Energy Regulatory Commission—regulates the national transmission network for oil, natural gas, and electricity
- Federal Maritime Commission—regulates the waterborne foreign commerce of the United States
- National Highway Traffic Safety Administration—regulates automobile design and safety
- Office of Surface Mining—regulates surface coal mining
- U.S. Securities and Exchange Commission—regulates the stock market

Government Regulation and Deregulation

Since the late 1970s the federal and many state governments have lessened their restrictions on certain industries. Called deregulation, this process allows industries to set their own standards and control their own systems of pricing and other business functions. For example, beginning in 1938 the airline industry was regulated by a federal body called the Civil Aeronautics Board, which controlled airlines' schedules, flying routes, and prices. To stimulate competition in the industry, Congress passed the Airline Deregulation Act of 1978. The industry experienced a flood of new airlines that offered low fares to compete with the established airlines. Although deregulation has actually caused some problems with larger airlines having too much control (or monopolizing) of the industry and with overly crowded flight routes, most economists agree that the result has been a safer, but cheaper, air transportation system. Other industries that have experienced some degree of deregulation include electric utilities, telephone services, trucking, railroads, and banking.

MARKET POWER

One of the foundations of a capitalistic economy is competition. Competition for customers among sellers theoretically ensures that buyers receive the lowest price. Likewise, competition among buyers drives up demand, helping to ensure that sellers have profitable markets for their products. However, sometimes market power becomes concentrated within an industry, for example, with only one company wielding great control over supply and/or demand factors.

Monopolies

If an industry becomes dominated by one seller, the lack of competition allows that entity to set prices in the marketplace—a situation known as monopolization. The federal government has long fought against monopolization in most private U.S. industries. In 1890 Congress passed the Sherman Antitrust Act to strengthen competitive forces in the economy. Section 2 of the law states: "Every person who shall monopolize, or attempt to monopolize, or combine or conspire with any other person or persons, to monopolize any part of the trade or commerce among the several States, or with foreign nations, shall be deemed guilty of a felony."

The law, however, proved difficult to enforce as Congress and the courts argued over its interpretation. In 1911 the U.S. Supreme Court established a historic legal standard in *Standard Oil Co. of New Jersey v. United States* (221 U.S. 1) that the law "prohibits all contracts and combination which amount to an unreasonable or undue restraint of trade in interstate commerce." Mathew Ingram explains in "A Google Monopoly Isn't the Point" (Businessweek.com, September 23, 2011) that "for the purposes of U.S. antitrust law, at least—being a monopoly isn't illegal. What is illegal is either acquiring that monopoly by nefarious or anticompetitive means or using that dominant position in a way that harms the market for those services." Thus, the legal standard focuses on any negative effects on consumers.

It should be noted that the government has allowed monopolies to form in certain industries that are deemed vital to the public interest. Two examples are telephone services and providers of electric power. For nearly 100 years the private company AT&T was allowed to monopolize the telephone services industry in the United States. In 1974 the U.S. Department of Justice filed suit against AT&T, accusing it of using unfair business practices. The suit was finally settled nearly a decade later and resulted in the breakup of the giant company into several smaller business units. Likewise, private utility companies have been permitted by the government to monopolize electric power service in certain geographic areas.

Companies that become hugely successful and powerful in their industry face an increased risk of being accused of using monopolistic practices by their competitors. Dominance also brings more intense scrutiny by federal regulators. This was particularly true for companies that dominated markets in the information technology sector, which came under fire by their competitors and U.S. and foreign governments beginning in the 1990s. For example, the software giant Microsoft paid millions of dollars in fines after losing antitrust cases brought against it.

CORPORATE BEHAVIOR AND RESPONSIBILITY

Businesses play a vital role in the economic well-being of the United States. Besides economic performance, Americans also expect businesses to behave in a legally and socially responsible manner. There is no public or political consensus on the exact social responsibilities of businesses. Nevertheless, it is recognized that the decisions and practices of company officials, particularly of large corporations, affect not only employees and investors but also the communities in which businesses are located. Fraud and corruption at the corporate level can adversely affect large numbers of people. Likewise, poor performance by businesses in meeting environmental, health, or consumer-protection standards has detrimental effects on society at large.

Big Tobacco: An Industry under Attack

During the 1990s several state governments brought lawsuits against the nation's major tobacco firms to recoup taxpayer money that was spent treating sick smokers under state Medicaid programs. In 1998 a settlement was reached in which the companies agreed to pay a total of $246 billion spread among the governments of all 50 states. The payments are to be made over a 25-year period. The settlement also required the tobacco companies to change their advertising methods and to reduce their political lobbying efforts.

In 1999 the federal government filed its own lawsuit, *United States v. Philip Morris Inc.* (116 F. Supp. 2d 131), against the tobacco companies, alleging that the defendants had engaged in a decades-long scheme to "defraud the American public" regarding the safety of cigarette smoking. The case centered on internal documents obtained from tobacco companies that seemed to demonstrate that the companies were well aware that nicotine was addictive and that cigarette smoking caused lung cancer. The litigation dragged on for seven years. In 2006 Judge Gladys Kessler (1938–) of the U.S. District Court for the District of Columbia ruled that the cigarette companies had committed civil violations of the Racketeer Influenced and Corrupt Organizations Act. However, rulings by other courts in 2005 meant that the government could not receive billions of dollars in penalty fines that it had sought from the tobacco companies. Kessler did issue an injunction ordering the companies to remove terms such as *light* and *ultra light* from cigarette packaging. Both sides filed appeals in the case that proved to be unsuccessful.

These lawsuits represent an unusual occurrence in U.S. history because the state and federal governments brought financial pressure on an entire industry. In 2009 Congress passed the Family Smoking Prevention and Tobacco Control Act, which gave the FDA regulatory control over cigarettes and other forms of tobacco. According to Duff Wilson, in "Senate Approves Tight Regulation over Cigarettes" (NYTimes.com, June 11, 2009), the act does not allow the FDA to ban smoking or nicotine, but only to "set standards that could reduce nicotine content and regulate chemicals in cigarette smoke." However, Wilson notes that industry analysts believe the tobacco companies will continue to prosper because "as long as they have a market of addicted customers, even if that clientele is dwindling, they can raise prices to remain profitable."

Public Perception of Big Business

The Gallup Organization regularly conducts polls that ask respondents about their opinions of various institutions in American society. The results from a June 2014 poll are shown in Table 6.8. Only about one-fifth (21%) of those asked expressed a "great deal" (9%) or "quite a lot" (12%) of confidence in big business. In fact, big business rated 14th out of 17 societal institutions that were listed by Gallup for the public's level of confidence. The military garnered the highest rating, with 74% of those asked expressing a "great deal" or "quite a lot" of confidence. Small business was the second highest, with

TABLE 6.8

Poll respondents rate their level of confidence in various institutions in American society, June 2014

NOW I AM GOING TO READ YOU A LIST OF INSTITUTIONS IN AMERICAN SOCIETY. PLEASE TELL ME HOW MUCH CONFIDENCE YOU HAVE IN EACH ONE—A GREAT DEAL, QUITE A LOT, SOME, OR VERY LITTLE?

[Sorted by most to least confidence in 2014]

	% A "great deal" and "quite a lot" of confidence
The military	74
Small business	62
The police	53
The church or organized religion	45
The medical system[a]	34
The U.S. Supreme Court	30
The presidency	29
The public schools	26
Banks	26
The healthcare system[b]	23
The criminal justice system	23
Newspapers	22
Organized labor	22
Big business	21
News on the Internet	19
Television news	18
Congress	7

[a]Based on 510 respondents.
[b]Based on 517 respondents.

SOURCE: Rebecca Riffkin, "Now I am going to read you a list of institutions in American society. Please tell me how much confidence you have in each one—a great deal, quite a lot, some, or very little?" in *Public Faith in Congress Falls Again, Hits Historic Low*, The Gallup Organization, June 19, 2014, http://www.gallup.com/poll/171710/public-faith-congress-falls-again-hits-historic-low.aspx (accessed August 22, 2014). Copyright © 2014 Gallup, Inc. All rights reserved. The content is used with permission; however, Gallup retains all rights of republication

FIGURE 6.3

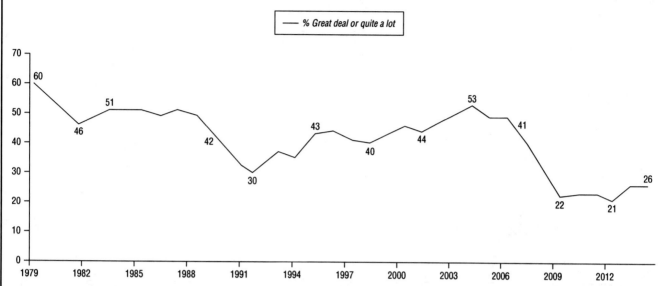

Poll respondents with a great deal or quite a lot of confidence in banks, 1979–2014

NOW I AM GOING TO READ YOU A LIST OF INSTITUTIONS IN AMERICAN SOCIETY. PLEASE TELL ME HOW MUCH CONFIDENCE YOU, YOURSELF, HAVE IN EACH ONE—A GREAT DEAL, QUITE A LOT, SOME, OR VERY LITTLE?

SOURCE: Rebecca Riffkin, "Americans' Confidence in Banks, 1979–2014 Trend," in *In U.S., Confidence in Banks Remains Low*, The Gallup Organization, June 26, 2014, http://www.gallup.com/poll/171995/confidence-banks-remains-low.aspx (accessed August 22, 2014). Copyright © 2014 Gallup, Inc. All rights reserved. The content is used with permission; however, Gallup retains all rights of republication.

62% of respondents providing a favorable opinion. Banks (presumably both large and small) received a much lower rating. Only 26% of respondents had a "great deal" or "quite a lot" of confidence in banks. As shown in Figure 6.3, this value was up from its low point of 21% in 2012 but is far below the more favorable ratings given to banks before the Great Recession occurred.

According to Gallup, the public perception of big business has been relatively poor dating back to the 1970s. (See Figure 6.4.) Americans soured even more on big business during the first decade of the 21st century for a variety of reasons. One factor was a string of high-profile corporate scandals. Most of the misdeeds involved deceptive accounting practices that enriched a handful of top executives but hurt thousands of employees and investors. Large corporations enveloped in scandal have included Enron (an international broker of commodities such as natural gas, water, coal, and steel), the telecommunications corporations WorldCom and Quest, and manufacturer Tyco International. All four companies had senior executives sent to prison and seriously eroded public confidence in big business. As explained earlier in this chapter, in 2008 the federal government decided to bail out several large financial and insurance corporations that were in danger of failing. This decision was widely unpopular with the American public because it believed the Great Recession had been caused, in large part, by these corporations. Public dissatisfaction increased when many of the rescued corporations subsequently paid out large bonuses to their executives.

FIGURE 6.4

Poll respondents with a great deal or quite a lot of confidence in small business and big business, 1973–2014

NOW I AM GOING TO READ YOU A LIST OF INSTITUTIONS IN AMERICAN SOCIETY. PLEASE TELL ME HOW MUCH CONFIDENCE YOU, YOURSELF, HAVE IN EACH ONE—A GREAT DEAL, QUITE A LOT, SOME, OR VERY LITTLE?

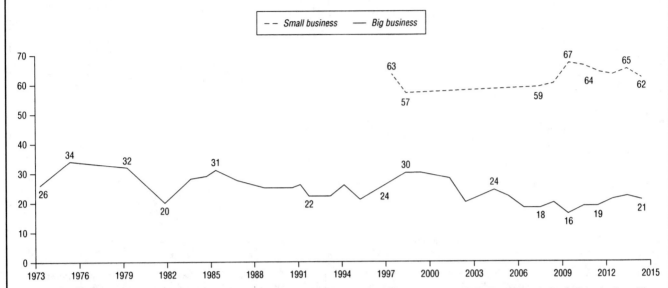

SOURCE: Rebecca Riffkin, "Americans' Confidence in Small Business vs. Confidence in Big Business," in *In U.S., Confidence in Banks Remains Low*, The Gallup Organization, June 26, 2014, http://www.gallup.com/poll/171995/confidence-banks-remains-low.aspx (accessed August 22, 2014). Copyright © 2014 Gallup, Inc. All rights reserved. The content is used with permission; however, Gallup retains all rights of republication.

CHAPTER 7
SAVING AND INVESTING

If you would be wealthy, think of saving as well as getting.

—Benjamin Franklin, *The Way to Wealth* (1758)

Saving and investing are two sides of the same coin. The purpose of saving is to put aside money for use in the future. Saved money can actually make money if it is put into a bank account that earns interest. This is basically a low-risk investment with a low rate of return, but it does preserve the money for the future. Investing is another matter. It means exchanging money for assets that may or may not go up in value over time. Investments that go up in value reap profits for the investor, and those profits can be modest or extravagant. Investments that go down in value are another story. Some or even all of the original money invested is lost. Thus, investing entails risk, particularly in a market-driven economy where fluctuations in supply and demand determine the profitability of investments. At a macroeconomic level, the U.S. economy thrives on investing; it provides money for business growth and government expenses. From a microeconomic standpoint, Americans are urged to save and/or gainfully invest some of their earnings to ensure that they have a safety net in the event of a personal financial crisis and to sustain them after they retire.

PERSONAL SAVING RATE

The personal saving rate is a government-measured rate that tracks how much money Americans have available for saving and investing. It is calculated by the U.S. Department of Commerce's Bureau of Economic Analysis (BEA) using data from many sources on income, taxes, government revenues and expenses, and personal expenses. The rate is actually a ratio of two BEA measures: disposable personal income (DPI) and personal saving.

Chapter 3 describes how DPI is derived from personal income (e.g., wages and salaries) minus tax and nontax payments made to the government. Table 7.1 shows DPI amounts for 2012 and 2013 and estimates for 2014 based on quarterly data. In 2012 and 2013 the DPI totaled $12.4 trillion and $12.5 trillion, respectively. In the second quarter of 2014 the DPI totaled nearly $13 trillion when seasonally adjusted at annual rates. Personal saving is determined by subtracting personal outlays (spending) from the DPI. Thus, personal saving is the money that people have left over for saving and investing. As shown in Table 7.1, personal saving amounts in 2012 and 2013 were $896.2 billion and $608.1 billion, respectively. Based on data from the second quarter of 2014, the seasonally adjusted and annualized amount for 2014 was forecast at $682.9 billion.

The personal saving amount is divided by the DPI to show what percentage of the DPI is available for saving and investing. As shown in Table 7.1, in 2012 and 2013 the personal saving rate was 7.2% and 4.9%, respectively. In the second quarter of 2014 the personal saving rate was 5.3% when seasonally adjusted at annual rates.

Because the personal saving rate is based on so many other calculated variables, any small errors in the dependent variables will be exaggerated in the rate itself. In addition, the BEA excludes from its definition of income certain wealth components such as capital gains (which is an increase in the value of an asset). As a result, the BEA admits that the personal saving rate gives an incomplete picture of household savings behavior. However, it is useful for tracking changes over time.

Personal Saving Rate Trends

Figure 7.1 shows the personal saving rate from June 2004 through June 2014. From the first quarter of 2005 to the third quarter of 2007 the rates were regularly less than 3% per quarter. In late 2007 the rate began a steep upward climb. Throughout 2008 and early 2009 it was typically greater than 5%. Analysts believe this rise was driven by fear. The housing market bust in the latter half of the first decade of the 21st century and the Great Recession gutted home values and greatly reduced the value of other investments into which Americans had put their money (and faith). After the recession ended in 2009 the saving rate decline somewhat and hovered between around 5% and 7% through 2012. After a brief spike above 10% in 2012 it declined to the range of 4% to 5.5%.

INVESTING BASICS

As noted earlier, personal saving is the amount of money that people have left over for saving and investing. Most people wish not simply to preserve the value of the money, but to use the money to make more money. This is the incentive for investing.

Three of the most important considerations to investors are liquidity, risk, and reward. Liquidity is the ease with which an investment can be transferred into cash. Some people keep cash under their mattresses or in other household hiding places. These savers greatly value liquidity; that is, their cash is quickly and easily accessible whenever they need it. However, the cash is at risk of being stolen or destroyed (e.g., by fire), and it earns no interest. Even worse, the cash actually loses its spending power over time because of inflation.

One of the fundamentals of investing is that reward is usually related to risk; that is, low-risk investments reap relatively low rewards, whereas high-risk investments (if successful) reap higher rewards. Banks, credit unions, and other financial institutions offer accounts in which cash can be held for safekeeping and perhaps earn interest. The level of liquidity depends on the type of account that is

TABLE 7.1

Disposable personal income and its disposition, 2012–13 and first and second quarters 2014

[Billions of dollars]

	2012	2013	Seasonally adjusted at annual rates 2012 I	II
Disposable personal income	12,384.0	12,505.1	12,775.8	12,968.5
Less: personal outlays	11,487.9	11,897.1	12,146.9	12,285.7
Equals: personal saving	896.2	608.1	629.0	682.9
Personal saving as a percentage of disposable personal income	7.2	4.9	4.9	5.3

Note: Quarterly values are seasonally adjusted at annual rates.

SOURCE: Adapted from "Table 2. Personal Income and Its Disposition (Years and Quarters)," in *Personal Income and Outlays: June 2014, Revised Estimates: 1999 through May 2014*, U.S. Department of Commerce, Bureau of Economic Analysis, August 1, 2014, http://www.bea.gov/newsreleases/national/pi/2014/pdf/pi0614.pdf (accessed August 19, 2014)

FIGURE 7.1

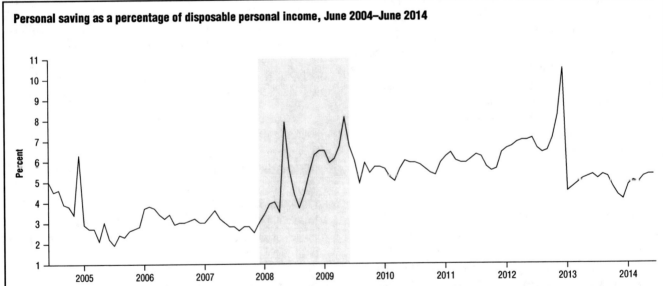

Personal saving as a percentage of disposable personal income, June 2004–June 2014

Note: Shaded areas indicate U.S. recessions.

SOURCE: "Personal Saving Rate," in *FRED® Economic Data*, Federal Reserve Bank of St. Louis, August 1, 2014, http://research.stlouisfed.org/fredgraph.pdf?hires=1&type=application/pdf&chart_type=line&recession_bars=on&log_scales=&bgcolor=%23e1e9f0&graph_bgcolor=%23ffffff&fo=verdana&ts=12&tts=12&txtcolor=%23444444&show_legend=yes&show_axis_titles=yes&drp=0&cosd=2004-06-01&coed=2014-06-01&width=670&height=445&stacking=&range=10yrs&mode=fred&id=PSAVERT&transformation=lin&nd=&ost=-99999&oet=99999&scale=left&line_color=%234572a7&line_style=solid&lw=2&mark_type=none&mw=1&mma=0&fml=a&fgst=lin&fq=Monthly&fam=avg&vintage_date=&revision_date= (accessed August 22, 2014)

chosen. This is generally a very low-risk investment option. Government entities offer investment options, such as bonds, that are considered to be low risk, because the investments are backed by government money.

The riskiest types of investments, and those with the highest returns, are typically offered by companies. Securities markets (stocks, bonds, and mutual funds) and commodities markets (raw materials and foreign currencies and securities) in the United States are used by corporations to raise money for their business operations and by banks and individuals to build wealth and, in some cases, to pay for retirement. These markets have fueled periods of astounding economic growth (called bull markets), but they have also been at the center of downturns (called bear markets) and disastrous economic crashes, creating the need for an extensive regulatory system. Despite regulations, however, the markets occasionally see high-profile scandals involving major figures in the business world.

Securities are financial assets that give holders ownership or creditor rights in a particular organization. The word usually refers to stocks (which are sometimes called equities). Historically, stocks have appreciated faster than inflation has increased, allowing people to build greater wealth than if they attempted to save money in traditional accounts. Investments can also serve as collateral for certain loans. Besides stocks, there are other types of securities that can be bought and sold on the open market, including bonds and mutual funds. Commodities are typically tangible products (usually raw materials) that are bought and sold in bulk. Financial instruments such as foreign currencies and securities of the U.S. and foreign governments are also called commodities. Commodities can refer either to the material itself or to a contract to buy the item in the future.

In the broadest sense, any money expenditure that returns a profit is considered to be an investment. Thus, the cost of a college education can be considered to be an investment because it will likely increase earnings potential in the future. However, in this chapter, investments are limited to things that have easily measured financial worth.

FINANCIAL ASSETS

The government does not track the amount of money that people invest, but it does track the value of assets in which households have invested. The Federal Reserve System, the national bank of the United States, in cooperation with the Internal Revenue Service, collects detailed financial information on American families every three years through the Survey of Consumer Finances (SCF). The primary focus of the SCF is to determine Americans' net worth, which is discussed in detail in Chapter 8. As part of its SCF calculations, the Federal Reserve gathers information about household assets, which it divides into two categories: financial assets and nonfinancial assets. Financial assets include bank accounts, stocks, bonds, investment funds, and other similar assets. Nonfinancial assets include real estate, vehicles, business equity, and other similar assets. This chapter will focus on financial assets, while Chapter 8 focuses on nonfinancial assets.

According to Jesse Bricker et al. of the Federal Reserve, in *Changes in U.S. Family Finances from 2010 to 2013: Evidence from the Survey of Consumer Finances* (September 2014, http://www.federalreserve.gov/pubs/bulletin/2014/pdf/scf14.pdf), just over 6,000 interviews were conducted as part of the 2013 survey. The resulting data are believed to be representative of nearly 123 million U.S. families.

Table 7.2 provides a listing of the percentage of families holding different types of financial assets. Demographic and economic information is shown for the families and the heads of households. In 2013 nearly all (94.5%) of surveyed families owned some kind of financial asset. According to Bricker et al., for families that held assets the mean (average) value of held financial assets was $270,100. The median value was $21,200, meaning that half of the asset holders had assets worth less than that value and the other half had assets worth more than that value.

Transaction Accounts and Certificates of Deposit

Bricker et al. note that transaction accounts include the following:

- Checking, savings, and money market deposit accounts
- Money market mutual funds
- Call or cash accounts at brokerages

Checking, saving, and money market deposit accounts are accounts held at financial institutions, such as banks or credit unions, in which customers can deposit money. In most cases, deposits up to $250,000 are insured by funds backed by the federal government. This ensures depositors that their money will be repaid in case the financial institution goes out of business. The Federal Deposit Insurance Corporation, an independent government agency, insures funds at banks; the National Credit Union Administration administers a fund for federally insured credit unions.

According to the U.S. Securities and Exchange Commission, in "Money Market Funds" (2014, http://www.sec.gov/answers/mfmmkt.htm), money market mutual funds are mutual funds that are "required by law to invest in low-risk securities." As a result, they have relatively low

TABLE 7.2

Family holdings of financial assets, 2013

Family characteristic	Transaction accounts	Certificates of deposit	Savings bonds	Bonds	Stocks	Pooled investment funds	Retirement accounts	Cash value life insurance	Other managed assets	Other	Any financial asset
					Percentage of families holding asset						
All families	93.2	7.8	10.0	1.4	13.8	8.2	49.2	19.2	5.2	7.7	94.5
Percentile of income											
Less than 20	78.9	4.8	4.0	*	4.2	2.2	8.8	10.9	1.5	7.4	81.8
20–39.9	90.8	6.3	5.1	1.0	5.5	2.7	27.9	16.0	3.2	8.2	92.9
40–59.9	97.2	7.6	8.0	1.2	9.3	6.0	50.8	19.3	4.4	8.8	98.2
60–79.9	99.1	9.2	14.7	0.7	14.4	8.5	70.4	23.8	6.3	6.5	99.8
80–89.9	99.8	9.6	18.5	*	25.4	12.3	83.7	23.1	8.6	6.4	99.8
90–100	100.0	11.9	18.2	6.9	45.3	30.8	92.7	28.7	12.9	8.4	100.0
Age of head (years)											
Less than 35	90.2	5.2	9.0	*	7.2	4.2	39.3	9.2	1.3	9.2	92.6
35–44	91.8	4.4	12.5	0.6	14.3	6.3	55.4	13.3	2.0	5.4	93.1
45–54	91.8	6.7	10.4	0.6	14.7	8.2	56.5	17.1	3.9	9.0	93.3
55–64	94.6	5.8	10.8	1.6	15.5	10.6	59.3	24.4	7.1	7.5	95.7
65–74	97.1	11.7	9.4	4.0	18.4	11.8	48.0	29.4	11.9	8.2	97.4
75 or more	96.7	18.8	6.8	2.7	15.3	10.3	29.0	30.4	9.2	5.5	97.2
Family structure											
Single with child(ren)	83.7	3.4	6.7	*	2.4	1.5	29.0	16.1	2.9	9.6	86.6
Single, no child, age less than 55	90.4	3.9	5.1	*	10.0	4.7	41.5	8.1	1.4	10.0	91.9
Single, no child, age 55 or more	93.7	11.8	7.0	1.7	12.6	8.3	35.6	21.4	8.4	6.9	95.0
Couple with child(ren)	94.4	6.7	14.8	0.9	15.6	9.0	60.1	17.5	4.4	5.4	95.8
Couple, no child	97.3	10.5	10.6	2.9	19.5	12.0	58.6	27.0	7.3	8.5	97.6
Education of head											
No high school diploma	78.8	4.5	2.8	*	2.8	*	14.1	12.0	*	4.9	81.9
High school diploma	90.4	5.1	6.6	0.6	6.4	3.0	37.6	18.3	3.3	6.7	92.2
Some college	94.1	8.1	10.3	0.9	9.4	4.7	43.8	18.7	4.6	9.6	95.5
College degree	99.1	10.7	14.7	2.7	24.9	16.3	71.2	22.2	8.2	8.3	99.4
Race or ethnicity of respondent											
White non-Hispanic	96.4	9.6	12.8	2.0	17.4	10.7	56.7	21.5	6.9	7.6	97.3
Nonwhite or Hispanic	86.5	4.1	4.4	0.2	6.3	3.0	34.0	14.5	1.8	7.7	88.9
Current work status of head											
Working for someone else	94.5	6.7	11.5	0.6	13.7	7.6	59.5	17.0	3.7	6.9	95.9
Self-employed	95.9	8.3	10.7	3.3	23.0	14.8	53.2	21.1	7.8	11.7	96.4
Retired	93.0	11.1	7.8	2.8	12.2	8.3	32.2	25.4	8.2	7.1	94.1
Other not working	78.1	*	5.0	*	6.6	3.0	24.1	8.8	*	10.5	81.2
Housing status											
Owner	97.5	10.0	13.0	2.0	18.2	11.2	60.5	23.8	7.2	6.8	98.1
Renter or other	85.2	3.6	4.5	0.3	5.4	2.6	28.2	10.6	1.6	9.2	87.8
Percentile of net worth											
Less than 25	80.0	1.3	3.4	*	1.6	1.1	17.3	7.5	*	6.9	83.2
25–49.9	95.0	4.1	6.7	*	5.2	1.9	40.0	16.7	1.0	6.7	96.6
50–74.9	98.2	9.1	12.4	*	11.4	6.0	57.8	21.6	4.6	7.8	98.5
75–89.9	99.5	17.0	17.4	1.5	28.0	15.1	76.9	28.6	12.2	8.7	99.6
90–100	99.6	15.8	18.1	10.1	50.0	36.7	89.3	34.4	19.2	10.4	99.8

*Ten or fewer observations.

SOURCE: "6. Family Holdings of Financial Assets, By Selected Characteristics of Families and Type of Asset, 1989–2013 Surveys," in *Survey of Consumer Finances: Historic Tables and Charts*, Board of Governors of the Federal Reserve System, September 4, 2014, http://www.federalreserve.gov/econresdata/scf/scfindex.htm (accessed September 13, 2014)

---

risks compared with other mutual funds. However, unlike deposit accounts at banks and credit unions, money market mutual funds are not federally insured. Neither are call or cash accounts at brokerages, which are accounts in which depositors keep cash to buy securities.

Compared with other types of investments, transaction accounts offer high liquidity and very low risk. The trade-off is that transaction accounts typically offer lower rates of return, such as interest rates, than other investments.

Certificates of deposit (CDs) are savings accounts in which money is placed for a predetermined amount of time, commonly one to five years, in exchange for payment of a set interest rate throughout that period. There are penalties for early withdrawal of the money. CDs are a low-risk investment; however, because they offer limited liquidity, their rates of return are generally better than those for transaction accounts.

As shown in Table 7.2, 93.2% of U.S. families had transaction accounts in 2013. Bricker et al. indicate that mean holdings were $36,300 while the median amount was $4,100. In addition, 7.8% of families held one or more CDs in 2013. According to Bricker et al., the mean value of CD holdings was $64,500, and the median value was $16,000.

Bonds

A bond is basically an IOU (which is short for "I Owe yoU") from an issuing entity that promises to pay back the borrowed amount plus interest at a specified future date (the maturity date). Bonds have a face value, called par, and that amount defines the amount of the debt. A bond offers returns to holders in two ways. The issuing organization will either make regular interest payments on the bond or initially sell the bond at a much lower price than the face value. After a certain amount of time (often many years), the holder can redeem the bond for face value. Bonds are sold to investors by government entities, companies, and nonprofit organizations.

The price of a bond depends on a number of factors, including the organization's creditworthiness and the interest rate. Generally, the better the organization's credit rating, the higher the price of the bond. If the organization begins to have financial problems that will affect its ability to repay the bonds, the price of those bonds will go down. One of the best-known rating companies for bonds is Standard & Poor's, which rates issuing organizations on a scale ranging from AAA to D. Bonds may be short term or long term. Long-term bonds are riskier than short-term bonds and therefore tend to pay higher interest rates.

The government issues a variety of securities, including bonds, for the purpose of earning revenue. Government entities sell bonds to raise funds for public projects, such as road improvement or school construction. Some of the best known bond types sold by the federal government are savings bonds, Treasury bills (T-bills), and Treasury notes (T-notes). Savings bonds are not marketable securities. They can be sold or redeemed only by the U.S. Department of the Treasury. As shown in Table 7.2, 10% of U.S. families owned savings bonds in 2013. Bricker et al. indicate the mean value was $6,400, and the median value was $1,000. T-bills are short-term government securities sold by the Treasury Department that mature within a few days or up to 26 weeks. The customer purchases a T-bill for less than its face value and then receives face value at maturity. For example, a customer might pay $90 up-front for a $100 T-bill. When the T-bill matures, the customer will receive the $100. T-bills can be bought and sold in other markets. T-notes have maturity periods lasting two, three, five, and 10 years. They earn a fixed rate of interest every six months. T-notes can be sold by the customer before the maturity date.

Companies also sell bonds in order to raise money. In essence, they are borrowing money from individuals, companies, and government entities willing to invest in businesses.

Only 1.4% of families surveyed in the 2013 SCF owned bonds other than savings bonds in 2013. (See Table 7.2.) Ownership of this asset was concentrated in households at the higher income levels. According to Bricker et al., the mean value of family holdings of bonds other than savings bonds was $581,700; the median value was $94,500.

Stocks

To raise money to operate and expand a company, its owners will often sell part of the company. A company that wants to raise money this way must first organize itself as a legal corporation. At that time, it creates shares of stock, which are small units of ownership in the company. These shares of stock can then be sold to raise funds for the company. Those who own them are called shareholders in the company. They have the right to attend shareholder meetings, inspect corporate documents, and vote on certain matters that affect the company. Shareholders may also have preemptive rights, which means they are able to buy new shares before they are offered to the public so that existing shareholders can maintain their percentage of ownership in a company.

Not all corporations offer their shares for sale to the public. When a company chooses to do so, its first sale of shares is called an initial public offering (IPO). IPO stock is purchased by investors at a price set by the company. The money paid for each share of stock is then available to the company for its business operations. In return, shareholders can receive benefits in two forms: dividends and appreciation. Dividends are a portion of the company's profits that are distributed to shareholders. Not all companies that issue stock pay dividends. Those that do usually pay them every quarter (a quarter is three consecutive months of the year; there are four quarters in a fiscal year), and, even though each share of stock might earn only a few pennies in dividends, the total amounts to individual or institutional investors who hold large amounts of shares can be enormous. Appreciation is a gradual increase in the value of a share over time. If a corporation prospers, a shareholder can sell his or her share to someone else for a higher price than he or she originally paid for it. There is no guarantee that a stock will appreciate; however, it is quite possible that it will depreciate over time.

PRICING SHARES. When a corporation creates shares, it determines the price per share for the IPO. From then on the price of each share depends on the public's perception of how well the corporation is doing. If a corporation reports higher profits, then the price per share will likely increase. The challenge for investors is that they cannot predict the future, and stock prices tend to fluctuate up and down over time. A variety of events can influence a stock's price, from the release of a popular new product to news that a company's chief executive officer is being investigated for fraud.

The market for stocks sold by shareholders to other shareholders is called the secondary stock market. It would be almost impossible for all shareholders to find buyers for their shares on their own when they choose to sell. To make it easier for shareholders to buy and sell shares, companies affiliate with a particular stock exchange that handles share transactions. The two most prominent exchanges in the United States are the New York Stock Exchange (NYSE) and the National Association of Securities Dealers Automated Quotations (NASDAQ). The NYSE and NASDAQ are themselves publicly traded companies. The United States also hosts the American Stock Exchange in New York City, the Boston Stock Exchange, the Philadelphia Stock Exchange, the Chicago Stock Exchange, and the Pacific Exchange in San Francisco, California. Additionally, there are stock exchanges in most countries throughout the world.

As shown in Table 7.2, 13.8% of the surveyed families owned publicly traded stocks in 2013. According to Bricker et al., the mean value of stock holdings was $294,300, and the median value was $27,000.

DIFFERENCES BETWEEN BONDS AND STOCK. Bonds differ from shares of stock in several important respects. First, any organization can issue bonds, whereas only corporations can issue stock. For that reason, unincorporated businesses and federal, state, and local governments use bonds to raise money. Unlike shares of stock, bonds provide no ownership interest in the company. The organization's only obligation to the bondholder is to pay the debt and interest. Third, bonds are usually less risky for the purchaser than stocks because the organization is legally obligated to pay the debt, whereas if a corporation has financial difficulties, it is not permitted to pay anything to shareholders until it has paid off its creditors. However, the rate of return on investment for stocks is generally higher than on bonds to compensate for the higher risk factor. Like stocks, though, bonds are traded by investors for prices that may be different from the par value. Investors who buy bonds are buying the right to receive the interest payments and to redeem the bond.

Pooled Investment Funds

Pooled investment funds are funds that combine ownership of various kinds of stocks and bonds. Bricker et al. note that the funds include mutual funds, real estate investment trusts, and hedge funds. Mutual funds are the most common type.

In a mutual fund, the money of multiple investors is pooled and then invested in stocks, bonds, or both. The managers of the mutual fund then buy and sell the stocks and bonds on behalf of the investors. Most investors try to diversify investments; that is, they put money into a number of different types of investments, rather than just one or two (a person's total investments are called his or her portfolio). That way, if one investment loses money, another investment may make enough profit to compensate for the loss.

However, for small investors it can be difficult to diversify the holdings in their portfolios. It takes time to evaluate different investments, and small investors may only be able to afford to buy one or two shares of each stock. Most brokers have a minimum purchase requirement that is higher than what the average investor can afford. Mutual funds were developed to solve such problems; by combining their money, small investors are able to diversify.

Unlike the prices of stocks and bonds, the price of a mutual fund is determined by the fund manager, rather than by the open market. This price, called the net asset value, is based on the fund manager's estimation of the fund's value at a particular time. Mutual funds may be purchased either directly from the fund manager or through a broker or other intermediary. The latter is more expensive because the investor will be required to pay fees. Mutual funds provide income to investors in two ways. First, if the mutual fund sells stocks or bonds at a profit or receives dividends or interest payments on bonds, these gains can be paid to investors as distributions. Second, the price of the mutual fund itself may go up, in which case investors can sell their mutual funds for more than they paid.

As shown in Table 7.2, 8.2% of surveyed families owned pooled investment funds in 2013. Bricker et al. note that the average value of the funds held by families was $462,900; the median value was $80,000.

Retirement Accounts

Retirement accounts are specially designed to save and invest money deposited by investors while they are young to provide a source of funds when they retire. Retirement fund managers invest the money in various ways, including stocks, bonds, mutual funds, and real estate. Given that Social Security retirement benefits are relatively low compared with a person's career income and that the long-term solvency (the state of having enough money to pay all debts) of Social Security continues to be in question, retirement funds are essential to the baby boom and later generations as they approach retirement age. (Baby boomers are people who were born between 1946 and 1964.)

Retirement accounts were the most commonly held type of financial asset in the 2013 SCF. Nearly half (49.2%) of all families surveyed in 2013 had a retirement account. (See Table 7.2.) According to Bricker et al., the average value of the accounts was $201,300, and the median value was $59,000.

Retirement accounts include both individually purchased and employer-sponsored accounts. The latter are particularly appealing to employees because some employers contribute money or company stock to their employees' retirement accounts. Some retirement accounts feature appealing tax advantages. For example, 401(k) plans are funds that workers can contribute to on a before-tax basis and that grow tax-free until the money is withdrawn. This is beneficial because retirees (by virtue of their lower incomes) are typically in lower income tax brackets than they were during their working years.

In *2014 Investment Company Fact Book* (2014, http://www.ici.org/pdf/2014_factbook.pdf), the Investment Company Institute indicates that U.S. retirement assets totaled $17.1 trillion in 2014. Most of the money ($15 trillion, or 88% of the total) was invested in mutual funds.

Cash Value Life Insurance

Cash value life insurance combines the chief benefit of a life insurance policy (i.e., a payout upon the policyholder's death) with an investment vehicle. The policyholder's premiums are invested, for example, in stocks, bonds, or mutual funds, by the fund administrator.

As shown in Table 7.2, 19.2% of surveyed families in 2013 owned cash value life insurance. According to Bricker et al., the average value of this asset was $35,300; the median value was $8,000. These values represent the current cash value of the life insurance policies, not the death penalties that will be paid out at some point in the future.

Other Financial Assets

The 2013 SCF revealed that 5.2% of the surveyed families had a managed asset in 2013 other than the ones described so far. (See Table 7.2.) These include personal annuities and trusts with an equity interest and managed investment accounts. Bricker et al. note the average value of these assets was $370,600, and the median value was $100,000.

As shown in Table 7.2, 7.7% of the surveyed families owned some other type of financial asset in 2013. According to Bricker et al., the average value of these assets was $55,100; the median value was $4,000. These assets include oil and gas leases, royalties, proceeds from lawsuits or estates in settlement, loans made by households to other people, and futures contracts. Futures contracts typically involve commodities.

COMMODITIES. In a broad sense, anything that can be bought or sold is a commodity. The term *commodity*, in the narrow sense used in this chapter, means a contract to buy or sell something that will be available in the future. These sorts of agreements are traded in commodities exchanges. Two important exchanges in the United States are the Chicago Board of Trade and the Kansas City Board of Trade.

There are two basic types of commodities. Futures are standardized contracts in which the seller promises to deliver a particular good to the buyer at a specified time in the future, at which point the buyer will pay the seller the price called for in the contract. Options on futures (which are usually simply called options) are more complicated. Depending on their exact terms, they establish the right of the buyer of the option to either buy or sell a futures contract for a specified price. Options that establish the right to buy a futures contract are *call* options. Those that establish the right to sell a futures contract are *put* options. In either case, the buyer of the option only has a limited time in which he or she can exercise his or her right, but the buyer is also free not to exercise the right at all.

The meaning of a commodities contract has changed over the years. Raw materials and agricultural commodities have been traded through commodities exchanges since the mid-19th century. More recently, commodities markets have expanded to include trading in foreign currencies, U.S. and foreign government securities, and U.S. and foreign stock indexes.

Because contracts are made before the goods are actually available, commodities are by their very nature speculative. Buyers purchase commodities because they think their value may increase over time, whereas the sellers think their value may decrease. For example, the seller of a grain futures contract may believe there will be a surplus of grain that will drive down prices, whereas the buyer thinks a shortage of grain will drive up prices. It is the speculative nature of commodities that makes them interesting to investors. Even if they have no need for the goods that the commodities contracts represent, speculative investors can make a profit by buying the commodities contracts at low prices and then selling them to others when prices rise. Commodities respond differently than stocks and bonds to market forces such as inflation; therefore, they can be a valuable part of a diversified portfolio. However, they are riskier and more difficult to understand.

HOMEOWNERSHIP AS AN INVESTMENT

One of the largest investments made by most Americans is the purchase of a home. Because real estate tends to appreciate (increase in value), buying a home is considered to be a relatively low-risk investment. However, this conventional wisdom proved faulty when the U.S. housing market crashed in the latter half of the first decade of the 21st century. As noted in Chapter 4, home prices actually depreciated (decreased in value) beginning in 2008. Likewise, high foreclosure rates meant that many Americans lost both their home and a substantial investment in their future. As shown in Table 7.3, the nation's homeownership rate peaked at 69.4% in 2004 and then declined to 64.8% by the second quarter of 2014.

The Federal Reserve compiles home mortgage data on a quarterly and annual basis. Table 7.4 lists the amount of home mortgage debt held by households for 1979 through 2013. The value peaked in 2007 at $10.6 trillion and then declined to $9.4 trillion in 2013. Data for the first quarter of 2014 indicate that the annualized value for 2014 would also be about $9.4 trillion. Chapter 4 includes a detailed discussion on mortgages and trends in home pricing.

TABLE 7.3

U.S. home ownership rate, by quarter, first quarter 1995–second quarter 2014

[In percent]

Year	Homeownership rates (seasonally adjusted)			
	First quarter	Second quarter	Third quarter	Fourth quarter
2014	65.0	64.8		
2013	65.2	65.1	65.1	65.1
2012	65.6	65.6	65.3	65.3
2011	66.5	66.0	66.1	65.9
2010	67.2	67.0	66.7	66.5
2009	67.4	67.4	67.4	67.2
2008	67.9	68.1	67.7	67.5
2007	68.5	68.3	68.0	67.8
2006	68.6	68.8	68.9	68.8
2005	69.2	68.8	68.7	68.9
2004	68.7	69.4	68.9	69.0
2003	68.1	68.2	68.3	68.5
2002*	67.9	67.8	67.9	68.2
2001	67.6	67.9	67.9	67.9
2000	67.1	67.3	67.5	67.5
1999	66.7	66.7	66.8	66.9
1998	66.0	66.1	66.6	66.5
1997	65.5	65.7	65.8	65.8
1996	65.3	65.4	65.4	65.4
1995	64.4	64.7	64.8	65.1

*Revised in 2002 to incorporate information collected in Census 2000.
Note: As new quarterly data are input, previous quarters' seasonally adjusted homeownership rates may change.

SOURCE: Robert R. Callis and Melissa Kresin, "Table 4SA. Homeownership Rates for the United States: 1995 to 2014, Seasonally Adjusted (in Percent)," in *Residential Vacancies and Homeownership in the Second Quarter 2014*, U.S. Department of Commerce, U.S. Census Bureau, July 29, 2014, http://www.census.gov/housing/hvs/files/qtr214/q214press.pdf (accessed August 14, 2014)

GOVERNMENT REGULATION OF THE STOCK MARKET SYSTEM

The prices of stocks, bonds, and commodities fluctuate with supply and demand factors. When fluctuations are created as a result of greed or corruption, or by the creation of artificial and unsustainable conditions, the results can be disastrous. Such was the case in 1929, when the stock market crashed and ushered in the period known as the Great Depression (1929–1939). The exact causes of the 1929 crash and the ensuing depression are complex and reach far beyond U.S. borders. However, certain conditions related to the U.S. stock market were significant contributors to the economic disaster. The federal government under President Franklin D. Roosevelt (1882–1945), who served from 1933 to 1945, passed

TABLE 7.4

Household home mortgage debt outstanding, 1979 through first quarter 2014

[Billions of dollars; quarterly figures are seasonally adjusted. Data shown are on an end-of-period basis.]

	Domestic nonfinancial sectors		
	Total	Households home mortgage	Consumer credit
1979	1,272.8	826.7	354.6
1980	1,389.1	926.5	358.0
1981	1,498.2	998.3	377.9
1982	1,567.5	1,031.2	396.7
1983	1,719.8	1,116.4	444.9
1984	1,939.8	1,243.3	526.6
1985	2,272.0	1,450.2	610.6
1986	2,529.5	1,649.0	666.4
1987	2,747.1	1,828.6	698.6
1988	3,036.3	2,054.8	745.2
1989	3,305.1	2,260.1	809.3
1990	3,567.8	2,489.3	824.4
1991	3,755.0	2,667.4	815.6
1992	3,959.4	2,840.4	824.8
1993	4,201.3	2,999.0	886.2
1994	4,525.0	3,165.5	1,021.2
1995	4,844.5	3,319.2	1,168.2
1996	5,182.3	3,537.3	1,273.9
1997	5,489.2	3,753.2	1,344.2
1998	5,901.4	4,054.7	1,441.3
1999	6,375.3	4,431.6	1,553.6
2000	6,959.5	4,813.9	1,741.3
2001	7,625.7	5,322.0	1,891.8
2002	8,437.9	6,028.2	1,997.0
2003	9,472.5	6,909.9	2,102.9
2004	10,544.3	7,859.4	2,220.1
2005	11,721.3	8,912.7	2,320.6
2006	12,946.4	9,910.0	2,461.9
2007	13,830.1	10,611.4	2,615.7
2008	13,849.6	10,579.0	2,650.6
2009	13,546.3	10,417.6	2,552.8
2010	13,214.8	9,912.7	2,647.4
2011	13,052.9	9,697.5	2,755.9
2012	13,044.2	9,481.7	2,923.6
2013	13,146.1	9,386.2	3,097.4
2014—first quarter	13,199.1	9,351.4	3,148.2

SOURCE: Adapted from "D.3. Credit Market Debt Outstanding by Sector," in *Federal Reserve Statistical Release: Z.1 Financial Accounts of the United States: Flow of Funds, Balance Sheets, and Integrated Macroeconomic Accounts—First Quarter 2014*, Board of Governors of the Federal Reserve System, June 5, 2014, http://www.federalreserve.gov/releases/z1/current/z1.pdf (accessed August 14, 2014)

a number of laws that were designed to prevent the sort of abuses of the market that led to the Great Depression, laws that form the basis for the modern market regulatory system.

An Unregulated System

At the time of the 1929 crash, the stock market was largely unregulated. In the months before the crash, there were signs that the system was beginning to collapse under its own weight, but the industrialists who owned most of the real wealth fed millions of dollars into the market to stabilize it. They were successful for a time, but the artificial conditions created through margin buying (the buying of many market shares at a deflated value) and wild speculation eventually brought the whole system down. Most people lost all, or nearly all, of the money they had invested in the stock market.

Regulation of Securities

Beginning in 1933 Congress enacted a series of laws that were designed to regulate the securities markets. The Securities Act of 1933 (sometimes referred to as the Truth in Securities law) was a reaction against the events that led up to the stock market crash of 1929. Its purpose was relatively simple: to protect investors by ensuring that they receive full information on the securities offered for sale to the public and to prohibit fraud in such sales. The Securities Act set up a system whereby most corporations that wanted to offer shares for sale to the public had to register their securities; the registration information was then made available to the public for review. The required registration forms (which are still being used in the 21st century) contained information on the company's management and business structure, its financial statements, and the securities it was offering for sale. This system was intended to protect investors by making available any information they needed to make informed decisions about their investments, although the truth or accuracy of the information was not guaranteed.

With the passage of the Securities Exchange Act of 1934, Congress established the U.S. Securities and Exchange Commission (SEC), which regulates the entire U.S. securities industry. The SEC expanded the Securities Act of 1933 to require more stringent reporting of publicly traded companies and all other entities involved in securities transactions, including stockbrokers, dealers, transfer agents, and exchanges. Additionally, large companies with more than $10 million in assets and 500 shareholders are required to file regular reports with the SEC detailing their finances and business dealings. The Securities Exchange Act of 1934 also explicitly outlawed illegal insider trading.

Since the 1930s, additional legislation has been passed to increase oversight of the securities and commodities futures trading industries, to prevent fraud, and to reassure investors.

INVESTMENT FRAUD

Even with careful oversight, fraudulent activities can occur in the securities industry. Securities fraud and the ensuing scandals are devastating to investors and to the markets as a whole. There were many high-profile instances of accounting scandals and securities fraud during the first decade of the 21st century.

Illegal Insider Trading

Insider trading is the buying or selling of stock by someone who has information about the company that other stockholders do not have. Most often, it refers to directors, officers, or employees buying or selling their own company's stock. Insider trading alone is not illegal, but insiders must report their stock transactions to the SEC. Insider trading becomes illegal when it is not reported to the SEC and breaches a fiduciary duty to the corporation; that is, when it violates a duty to act in the corporation's best interests. Most often, this happens when someone within the company has confidential information and uses it as the basis for a stock transaction. For example, if a company officer knows that the company is going to file for bankruptcy the next day and sells his or her shares of the company's stock because the stock price will likely plummet, the trading is illegal. The same goes for someone who gives the information to an outside stockholder so he or she can act on it.

The best-known case of illegal insider trading during the first decade of the 21st century involved the company ImClone Systems, the media mogul Martha Stewart (1941–), and the stockbroker Peter Bacanovic (1962–). The SEC alleged that Bacanovic passed confidential information to Stewart and that she sold her stock in ImClone because of that information. The SEC also accused Stewart and Bacanovic of trying to cover up the matter afterward by lying to federal investigators. The insider trading charge against Stewart was dropped, but she was convicted of lying to investigators and obstruction of justice. Bacanovic was convicted of most of the charges against him. Stewart and Bacanovic were both sentenced to five months in prison.

Overvaluing and Accounting Scandals

Overvaluing is the overstatement of income by companies with the assistance of their accountants to create an inflated impression of financial success among investors, thereby increasing the value of stock. During the first decade of the 21st century Wall Street experienced many scandals concerning such accounting practices at major corporations. The most egregious and notorious breaches of regulations occurred at three companies: Enron (an international broker of commodities such as natural gas, water, coal, and steel), the telecommunications corporation WorldCom, and the manufacturer Tyco International. All three became targets of SEC investigations, with company executives brought up on criminal fraud charges and investors losing billions of dollars.

Ponzi Schemes

A Ponzi scheme is a fraudulent investment deal in which the person perpetrating the scheme convinces potential investors that he or she has special investment know-how and can gain them large returns on their investment. The fraudster collects money from an initial group of investors. He or she then convinces additional people to invest and uses part of their money to pay the dividends that the first group is expecting. The initial investors are so pleased that they invest even more money, and so the scheme continues. The fraudster must continually recruit new investors and/or collect more money from existing investors to keep the scam going.

In "In Ponzi We Trust" (SmithsonianMag.com, December 1998), Mary Darby notes that the scheme is named after Charles Ponzi (1883–1949), an investment broker who conducted such a scam during the late 1910s and early 1920s in Boston, Massachusetts. Ponzi convinced investors he could earn them huge returns on their investments through a scheme that involved international postal reply coupons. His suave and confident manner won over many investors who made Ponzi a very rich and celebrated man for a short time. Eventually the scam was exposed, and the public learned that Ponzi owed millions to duped investors. He went to prison before being deported back to his native Italy. According to Darby, "Half-a-dozen banks crashed in the aftermath of Ponzi's fall. His note holders received less than 30 cents on the dollar."

There are many variations of the classic Ponzi scheme. In 2008 the investment broker Bernard Madoff (1938–) confessed to his sons that he had been running a Ponzi-type investment scam for more than a decade. Madoff eventually pleaded guilty and in June 2009 was sentenced to 150 years in prison. In "The Madoff Scam: Meet the Liquidator" (CBSNews.com, June 20, 2010), the CBS program *60 Minutes* reports that Madoff had collected nearly $36 billion from thousands of investors. Approximately half of that money went missing (i.e., could not be found in Madoff's company accounts) before the scam collapsed. As of December 2014, Madoff remained in federal prison. His scam is considered to be the largest Ponzi scheme in U.S. history.

SURVEY OF HOUSEHOLD ECONOMICS AND DECISIONMAKING

Previous chapters have described the Survey of Household Economics and Decisionmaking, which is conducted by the Federal Reserve. As of December 2014, the most recent results were reported in *Report on the Economic Well-Being of U.S. Households in 2013*

(July 2014, http://www.federalreserve.gov/econresdata/ 2013-report-economic-well-being-us-households-201407 .pdf). Data for the sample population of 4,134 people are considered representative of the U.S. population as a whole.

Table 7.5 provides a breakdown of the specific savings purposes noted by 2,270 survey respondents in 2013. Overall, more than half of those asked were saving for retirement (58%) or for "unexpected expenses" (52.6%). Nearly half (49.1%) said they were saving "just to save." Other popular purposes named by respondents included saving to pay off debts (26.5%), saving for their children's benefit (20.5%), and saving for purchases such as cars or appliances (20.4%).

The Federal Reserve asked 3,163 nonretirees about their saving habits for retirement purposes. As shown in Table 7.6, nearly a third (30.9%) of those asked had no retirement savings or pension. The value was much higher for people aged 18 to 29 years (50.5%) than for people aged 60 years and older (15.4%). The largest portion (43.7%) of respondents indicated they were saving for retirement through defined contribution pension plans offered by their employers. These are plans to which employees (and some employers) make regular fixed contributions. One common plan is called a 401(k) plan, which is named after the section of the tax code that created it. The money in such a plan is not taxed until it is withdrawn. Likewise, 403(b) plans, or tax-sheltered annuities, offer retirement savings options for certain employees of public schools and nonprofit organizations.

TABLE 7.5

Percentage of survey respondents saving for specific purposes, 2013

[Percent, except as noted]

Response	Rate
Education (yours or someone else's)	17.9
Retirement	58.0
Your children	20.5
Major appliance, car, or other big purchase (excluding a home)	20.4
Home purchase	13.0
Pay off debts	26.5
Unexpected expenses	52.6
Just to save	49.1
Taxes	15.0
To leave behind some inheritance or charitable donation	10.6
Other	4.9
Refused	0.7
Number of respondents	2,270

SOURCE: "Table C.80. Which of the Following Categories Are You Saving Money For?" in *Report on the Economic Well-Being of U.S. Households in 2013*, Board of Governors of the Federal Reserve System, July 2014, http://www.federalreserve.gov/econresdata/2013-report-economic-well-being-us-households-201407.pdf (accessed August 9, 2014)

TABLE 7.6

Percentage of survey respondents with and without retirement savings or pensions, by age group, September 2013

[Percent, except as noted]

	18–29	30–44	45–59	60+	Overall
No retirement savings or pension	50.5	27.8	23.0	15.4	30.9
Social Security Old-Age benefits	17.5	31.5	46.4	67.6	36.3
401(k), 403(b), thrift or other defined contribution pension plan through an employer	30.3	52.8	47.9	37.1	43.7
Defined benefit pension through an employer (i.e., pension based on a formula, your earnings, and years of service)	7.0	16.0	27.0	25.9	18.2
Individual Retirement Account (IRA)	11.2	23.5	29.2	31.9	23.0
Savings outside a retirement account (e.g., a brokerage account, savings account)	15.4	19.3	28.6	33.3	22.7
Real estate or land	4.4	8.9	16.2	20.5	11.3
Other	1.7	3.4	4.1	4.1	3.2
Total number of respondents					3,163

Note: Among those who are not currently retired.

SOURCE: "Table 18. What Type(s) of Retirement Savings or Pension do You (or Your Spouse/Partner) Have? (by Age)?" in *Report on the Economic Well-Being of U.S. Households in 2013*, Board of Governors of the Federal Reserve System, July 2014, http://www.feder-alreserve.gov/econresdata/2013-report-economic-well-being-us-households-201407.pdf (accessed August 9, 2014)

CHAPTER 8
WEALTH IN THE UNITED STATES

All communities divide themselves into the few and the many. The first are the rich and the well-born, the other the mass of the people.

—Alexander Hamilton, 1787

In the United States a small percentage of people have enormous wealth and the rest of the population has far less wealth. Is this a natural and acceptable result of capitalism or an economic injustice that must be righted for the good of society? This is a debate that has raged since the nation was founded. Some people believe the accumulation of great wealth is possible for anyone in the United States as the reward for hard work, ingenuity, and wise decision making. However, other people believe U.S. political, business, and social systems are unfairly structured so as to limit wealth building by certain segments of the population.

The federal government does not measure the overall wealth of individual Americans or the population as a whole. However, it does compile data on related economic indicators, specifically income and net worth (assets minus liabilities). The distribution of these indicators provides valuable information about the distribution of wealth in the United States.

U.S. INCOME AND ITS DISTRIBUTION

The U.S. Department of Commerce's Bureau of Economic Analysis (BEA) provides estimates of the nation's total personal income. As shown in Table 3.1 in Chapter 3, in 2013 personal income totaled $14.2 trillion. Personal income includes employee compensation (wages, salaries, and supplements), proprietors' income (income earned by the owners of unincorporated businesses), rental income, income from interest and dividends, and transfer receipts (e.g., unemployment benefits or Social Security benefits).

Historically, employee compensation has accounted for the largest component of personal income. In 2013 employee compensation accounted for $8.8 trillion, or 62% of the $14.2 trillion in personal income. (See Table 3.1 in Chapter 3.) Thus, employment factors are extremely important to an analysis of income distribution. Chapter 5 provides data on employment and unemployment for workers in various industries and for specific population segments. For example, Table 5.4 in Chapter 5 shows unemployment rates by age, class, race, and ethnicity in July 2014. Unemployment rates differed substantially by demographic categories. The overall unemployment rate for people aged 16 years and older was 6.2%. The highest unemployment rate for nonteenagers was 11.4% for African Americans. The rates for whites (5.3%) and Asian Americans (4.5%) were much lower.

Weekly Earnings

The U.S. Bureau of Labor Statistics (BLS) relies on income data collected by the U.S. Census Bureau as part of its Current Population Survey (CPS), which includes approximately 60,000 households, as described in Chapter 5.

In *Usual Weekly Earnings of Wage and Salary Workers: Third Quarter 2014* (October 24, 2014, http://www.bls.gov/news.release/pdf/wkyeng.pdf), the BLS notes that CPS participants who work full time are asked about their "usual" pretax earnings. The earnings may include tips, overtime pay, or commissions. The BLS calculates the median values (half the values are lower and half the values are higher) for all workers and for specific demographic and occupational groups. For the third quarter of 2014 the BLS reports median seasonally adjusted earnings for full-time wage and salary workers aged 16 years and older. (Note that this data set excludes self-employed workers.) The agency provides the following values for selected demographic groups:

- Total men and women—$790 per week
- Men—$870 per week

- Women—$715 per week
- White men—$896 per week
- African American men—$679 per week
- Asian American men—$1,087 per week
- Hispanic men—$617 per week
- White women—$733 per week
- African American women—$608 per week
- Asian American women—$795 per week
- Hispanic women—$553 per week

The data show substantial differences in median weekly earnings based on sex, race, and ethnic origin.

CPS Annual Social and Economic Supplement

The Census Bureau also collects income data as part of the CPS Annual Social and Economic Supplement (CPS ASEC). Data from 2013 are presented and analyzed by Carmen DeNavas-Walt and Bernadette D. Proctor of the Census Bureau in *Income and Poverty in the United States: 2013* (September 2014, http://www.census.gov/content/dam/Census/library/publications/2014/demo/p60-249.pdf). The 2013 data were collected from approximately 68,000 households. DeNavas-Walt and Proctor define income as "money income before taxes," and include the following:

- Earnings
- Unemployment and workers' compensation
- Social Security and Supplemental Security Income
- Public assistance
- Veterans' payments
- Survivor and disability benefits
- Pension or retirement income
- Interest and dividends
- Rents, royalties, estates, and trusts
- Educational assistance
- Alimony and child support
- Financial assistance from outside of the household
- Other income

This definition excludes capital gains (profits realized from the sale of assets) and the value of noncash benefits, such as food stamps, Medicare (a federal health insurance program for people aged 65 years and older and people with disabilities), Medicaid (a state and federal health insurance program for low-income people), public housing, and employer-provided fringe benefits.

DeNavas-Walt and Proctor report that the Census Bureau uses six mathematical and statistical methods to measure income inequality. Three of these methods—mean logarithmic deviation of income, the Theil index, and the Atkinson measure—are complex and not widely used. The remaining three methods and their results are explained in the following sections.

QUINTILE SHARES. The quintile shares method ranks all the households from the lowest to the highest income. The households are divided into five groups of equal population size (i.e., quintiles). For example, the lowest quintile includes households making less than the 20th percentile of income, and the top quintile includes households making the 80th to 100th percentile of income. The income of each group is then divided by the total income for all groups. If income equality exists, each group will account for 20% of the total income.

Table 8.1 shows quintile shares of income based on CPS ASEC data from 1973, 1983, 1993, 2003, and 2013. In 2013 the lowest quintile of U.S. households accounted for 3.2% of total income, whereas the highest quintile accounted for 51% of total income. The lowest quintile lost income share over time, from 4.2% in 1973 to 3.2% in 2013. The second, third, and fourth quintiles also lost income share over the decades. The second quintile fell from 10.4% income share in 1973 to 8.4% income share in 2013. The third quintile decreased from 17% income share in 1973 to 14.4% income share in 2013. The fourth quintile fell from 24.5% in 1973 to 23% in 2013. The highest quintile gained income share over the decades, from 43.9% in 1973 to 51% in 2013. As noted in Table 8.1, the mean (average) household income of the highest quintile in 2013 was $185,206. The mean household income of the lowest quintile in 2013 was $11,651.

GINI INDEX. The Gini index (or Gini coefficient) is a mathematically derived value that is used to describe the inequality in a data distribution. It was developed during the early 1900s by the Italian statistician Corrado Gini (1884–1965). When used to compute income inequality, a value of zero indicates perfect equality, whereas a value of one indicates that all income is made by only one family.

As shown in Table 8.1, in 2013 the Gini index was 0.476. In 1973 it was 0.400. Thus, the Gini index indicates that income inequality increased over the past four decades.

RATIO OF INCOME PERCENTILES. The top rows in Table 8.1 list household incomes at selected percentiles. Below the rows are listed household income ratios. Each household income ratio is a percentile income limit divided by another percentile income limit; for example, the 95th/50th ratio in 2013 was $196,000 divided by $51,939, which equals 3.78. In other words, the income

TABLE 8.1

Selected measures of household income dispersion, selected years 1973–2013

[Income in 2013 CPI-U-RS adjusted dollars]

Measures of income dispersion	2013*	2003	1993	1983	1973
Measure					
Household income at selected percentiles					
10th percentile limit	12,401	13,345	12,201	11,686	12,302
20th percentile limit	20,900	22,778	20,585	19,892	20,606
40th percentile limit	40,187	43,063	39,177	36,989	39,623
50th (median)	51,939	54,865	49,594	46,425	49,262
60th percentile limit	65,501	68,968	61,583	56,770	58,626
80th percentile limit	105,910	110,023	95,725	85,794	84,410
90th percentile limit	150,000	149,708	129,761	112,288	108,957
95th percentile limit	196,000	195,203	166,111	141,152	135,668
Household income ratios of selected percentiles					
90th/10th	12.10	11.22	10.64	9.61	8.86
95th/20th	9.38	8.57	8.07	7.10	6.58
95th/50th	3.78	3.56	3.35	3.04	2.75
80th/50th	2.04	2.01	1.93	1.85	1.71
80th/20th	5.07	4.83	4.65	4.31	4.10
20th/50th	0.40	0.42	0.42	0.43	0.42
Mean household income of quintiles					
Lowest quintile	11,651	12,661	11,682	11,232	11,899
Second quintile	30,509	32,522	29,616	28,216	29,917
Third quintile	52,322	55,207	49,644	46,649	49,072
Fourth quintile	83,519	87,386	77,149	70,002	70,592
Highest quintile	185,206	186,284	160,736	128,525	126,309
Top 5 percent	322,343	320,744	275,877	194,187	194,555
Shares of household income of quintiles					
Lowest quintile	3.2		3.6	4.0	4.2
Second quintile	8.4		9.0	9.9	10.4
Third quintile	14.4	14.8	15.1	16.4	17.0
Fourth quintile	23.0	23.4	23.5	24.6	24.5
Highest quintile	51.0	49.8	48.9	45.1	43.9
Top 5 percent	22.2	21.4	21.0	17.0	16.9
Summary measures					
Gini index of income inequality	0.476	0.464	0.454	0.414	0.400
Mean logarithmic deviation of income	0.578	0.530	0.467	0.397	0.355
Theil	0.415	0.397	0.385	0.288	0.270
Atkinson:					
e = 0.25	0.100	0.095	0.092	0.072	0.068
e = 0.50	0.196	0.187	0.178	0.147	0.136
e = 0.75	0.298	0.283	0.266	0.226	0.210

CPS ASEC = Current Population Survey Annual Social and Economic Supplement.
CPI-U-RS = Consumer Price Index for all Urban Consumers, Research Series.
*Data are based on the CPS ASEC sample of 68,000 addresses. The 2014 CPS ASEC included redesigned questions for income and health insurance coverage. All of the approximately 98,000 addresses were eligible to receive the redesigned set of health insurance coverage questions. The redesigned income questions were implemented to a subsample of these 98,000 addresses using a probability split panel design. Approximately 68,000 addresses were eligible to receive a set of income questions similar to those used in the 2013 CPS ASEC and the remaining 30,000 addresses were eligible to receive the redesigned income questions. The source of the 2013 data for this table is the portion of the CPS ASEC sample which received the income questions consistent with the 2013 CPS ASEC, approximately 68,000 addresses.

SOURCE: Adapted from Carmen DeNavas-Walt and Bernadette D. Proctor, "Table A-2. Selected Measures of Household Income Dispersion: 1967 to 2013," in *Income and Poverty in the United States: 2013*, U.S. Department of Commerce, U.S. Census Bureau, September 2014, http://www.census.gov/content/dam/Census/library/publications/2014/demo/p60-249.pdf (accessed September 17, 2014).

of households in the 95th percentile was 3.78 times the income of households in the 50th percentile. These ratios are useful for seeing how income inequality changes over time. In 1973 the 95th/50th ratio was 2.75. Thus, the income gap between high-income households and medium-income households widened between 1973 and 2013. The 95th/50th ratio is commonly called the "top half" inequality measure.

Survey of Consumer Finances

As explained in Chapter 7, every three years the Federal Reserve and the Internal Revenue Service (IRS) conduct a Survey of Consumer Finances (SCF) to collect detailed financial information on American families. More than 6,000 interviews were conducted as part of the 2013 survey. The resulting data are believed representative of nearly 123 million U.S. families. Summary SCF data are discussed by Jesse Bricker et al. in *Changes in U.S. Family Finances from 2010 to 2013: Evidence from the Survey of Consumer Finances* (September 2014, http://www.federalreserve.gov/pubs/bulletin/2014/pdf/scf14.pdf). More detailed data are presented in tables and charts provided by the Federal Reserve (http://www.federalreserve.gov/econresdata/scf/scfindex.htm).

The definition of income used in the SCF includes:

- Wages
- Self-employment and business income
- Taxable and tax-exempt interest
- Dividends
- Realized capital gains
- Food stamps and other related support programs provided by the government
- Pensions and withdrawals from retirement accounts
- Social Security
- Alimony and other support payments
- Miscellaneous sources of income for all members of the primary economic unit in the household

Bricker et al. divide U.S. families into income percentiles (based on pretax income) for 2013 and compare the mean values to those obtained in 2010 as follows:

- Less than 20 percentile of income—$15,200 mean income (down 8% from 2010)
- 20 to 39.9 percentile of income—$30,500 mean income (down 6% from 2010)
- 40 to 59.9 percentile of income—$48,700 mean income (down 1% from 2010)
- 60 to 79.9 percentile of income—$77,900 mean income (up 1% from 2010)
- 80 to 89.9 percentile of income—$121,700 mean income (down 1% from 2010)
- 90 to 100 percentile of income—$223,200 mean income (up 10% from 2010)

It should be noted that Bricker et al. call these values "usual" income, rather than actual income. Usual income does not include unexpected and temporary deviations from regular income, such as short bouts of unemployment or unexpected bonuses. Between 2010 and 2013 the mean income declined or was relatively flat for all but the highest income percentile. Overall, the bottom 90th percentile accounted for 52.7% of income in 2013, compared with 47.3% for the 90th to 100th percentile. In other words, U.S. families in the top income percentile received nearly half of all income in 2013.

Bricker et al. track trends in income inequality dating back to the 1989 SCF. They note that the income share of the bottom 90th percentile of the distribution fell from around 58% in 1989 to 52.7% in 2013. The income share of the top third percentile increased from 25% in 1989 to 30.5% in 2013. The income share for the 90th to the 97th percentiles was flat, at just under 17% between 1989 and 2013.

Congressional Budget Office

In *The Distribution of Household Income and Federal Taxes, 2010* (December 2013, http://www.cbo.gov/sites/default/files/cbofiles/attachments/44604-AverageTaxRates.pdf), the Congressional Budget Office (CBO) uses data from the IRS's Statistics of Income and the Census Bureau's CPS to assess the distribution of household income and federal taxes. The CBO considers household income to include market income and government transfers. Market income consists of the following components:

- Labor income
- Business income
- Capital gains and capital income (e.g., dividends)
- Income received in retirement for past services
- Other sources of income

Government transfers include cash payments and in-kind benefits from social insurance and other government assistance programs, such as Social Security and Medicare.

The CBO divides income into quintiles (five groups) of equal population. Table 8.2 shows the income quintiles for 2010 and lists the market income, government transfers, federal taxes, and after-tax income for each quintile. Overall, the lowest quintile accounted for just 2.3% of total market income, compared with 57.9% for the highest quintile. However, the lowest quintile was allocated a much greater share of government transfers than higher quintiles. The lowest quintile received more than one-third (36.2%) of Social Security and Medicare transfers and nearly half (47%) of other government transfers. By contrast, the highest quintile received 11.4% of Social Security and Medicare transfers and 6.2% of other government transfers. The portion of federal taxes paid by the lowest quintile in 2010 was nearly negligible, at less than 0.05%. The highest quintile paid a much greater share of federal taxes, at 69.3%.

Table 8.3 compares before-tax and after-tax income by income quintile for 2010. The table shows how government transfers and federal tax policy help alleviate a small amount of the before-tax income inequality between quintiles. The lowest quintile accounted for 5.1% of before-tax income, but 6.2% of after-tax income. The highest quintile accounted for 51.9% of before-tax income, but 48.1% of after-tax income.

The CBO has collected income data since 1979. It notes that real (inflation-adjusted) after-tax income for the lowest quintile grew 49% between 1979 and 2010, or an average of 1.3% per year. Real after-tax income for the middle three quintiles (the 21st to 80th percentiles) increased 40% between 1979 and 2010, for an annual

TABLE 8.2

Distribution of market income, government transfers, and federal taxes, by market income group, 2010

Market income group	Market income	Government transfers		Federal taxes	After-tax income
		Social Security and Medicare	Other transfers		
		Average amount (dollars)[a]			
Lowest quintile	8,100	14,200	8,500	[b]	30,800
Second quintile	30,700	10,300	4,900	2,500	43,400
Middle quintile	54,800	7,900	2,900	8,100	57,400
Fourth quintile	87,700	5,500	1,900	16,100	78,900
Highest quintile	234,400	5,200	1,300	58,900	181,900
All quintiles	**79,300**	**8,900**	**4,100**	**16,600**	**75,500**
		Share (percent)			
Lowest quintile	2.3	36.2	47.0	[c]	9.3
Second quintile	7.4	22.1	22.8	2.8	11.0
Middle quintile	13.0	16.7	13.3	9.2	14.3
Fourth quintile	21.0	11.7	8.8	18.4	19.8
Highest quintile	57.9	11.4	6.2	69.3	47.2
All quintiles	**100.0**	**100.0**	**100.0**	**100.0**	**100.0**

[a]Income amounts have been rounded to the nearest $100.
[b]Between zero and $50.
[c]Between zero and 0.05 percent.

Notes: Market income is composed of labor income, business income, capital gains, capital income (excluding capital gains), income received in retirement for past services, and other sources of income. Government transfers are cash payments and in-kind benefits from social insurance and other government assistance programs. Federal taxes include individual and corporate income taxes, social insurance (or payroll) taxes, and excise taxes. After-tax income is the sum of market income and government transfers, minus federal tax liabilities. Income groups are created by ranking households by before-tax income. Quintiles (fifths) contain equal numbers of people.

SOURCE: "Distribution of Market Income, Transfers, and Federal Taxes, by Market Income Group, 2010," in *The Distribution of Household Income and Federal Taxes, 2010*, Congressional Budget Office, December 2013, http://www.cbo.gov/sites/default/files/cbofiles/attachments/44604-AverageTaxRates.pdf (accessed August 26, 2014)

---

average growth rate of 1.1%. The top quintile experienced the largest gain of 64%, or 1.6% per year between 1979 and 2010. Within the latter quintile the top first percentile experienced phenomenal income growth of 201% over the same period. This equates to an average increase of 3.6% per year.

AGGREGATE U.S. NET WORTH

As noted earlier, net worth is defined as assets minus liabilities. The Federal Reserve computes the aggregate net worth of U.S. households and nonprofit organizations (NPOs) on a quarterly and annual basis. These values are published in tabular form in "B.100 Balance Sheet of Households and Nonprofit Organizations" as part of the *Federal Reserve Statistical Release: Z.1 Financial Accounts of the United States*. Table 8.4 shows aggregate net worth data between 2008 and 2013 and the first quarter of 2014. Overall, the total net worth of U.S. households and NPOs was $81.8 trillion at the end of the first quarter of 2014. This value was based on total assets of $95.5 trillion, including $67.2 trillion in financial assets and $28.3 trillion in nonfinancial assets.

Financial Assets

Financial assets are financial devices that have worth because they have perceived value. The most obvious examples are stocks and bonds. Other devices often counted as financial assets include pensions (which are promises of future income), life insurance policies with cash value, and insurance policies on tangible assets because they protect valuable resources.

Table 8.4 and Figure 8.1 provide data collected by the Federal Reserve regarding the aggregate financial assets of households and NPOs. At the end of the first quarter of 2014, the assets totaled $67.2 trillion. The largest component was pension entitlements ($19.8 trillion, or 29% of the total). This category includes defined value pension plans, but excludes Social Security. Other large financial assets were corporate equities ($13.5 trillion, or 20% of the total), and deposits and currency ($9.8 trillion, or 15% of the total). Deposits include monies in checking and savings accounts and in money market funds.

Table 7.2 in Chapter 7 includes detailed SCF data on financial assets. In the 2013 survey 94.5% of all families had at least one financial asset. The most widely held financial assets were transaction accounts (93.2%). These include checking accounts, savings accounts, money market deposit accounts, money market mutual funds, and call accounts at brokerages. Nearly half (49.2%) of the families surveyed had retirement accounts. This category does not include Social Security benefits and certain employer-sponsored defined benefit plans.

Overall, Table 7.2 in Chapter 7 shows that families that were the most likely to hold financial assets were

TABLE 8.3

Distribution of before-tax and after-tax income, by income group, 2010

Income group	Average income (2010 dollars)* 2010	Share of income (percent) 2010
Before-tax income		
Lowest quintile	24,100	5.1
Second quintile	44,200	9.6
Middle quintile	65,400	14.2
Fourth quintile	95,500	20.4
Highest quintile	239,100	51.9
All quintiles	92,200	100.0
81st to 90th percentiles	134,600	14.6
91st to 95th percentiles	181,600	9.9
96th to 99th percentiles	286,400	12.5
Top 1 percent	1,434,900	14.9
After-tax income		
Lowest quintile	23,700	6.2
Second quintile	41,000	10.9
Middle quintile	57,900	15.4
Fourth quintile	80,600	21.0
Highest quintile	181,800	48.1
All quintiles	75,500	100.0
81st to 90th percentiles	108,700	14.3
91st to 95th percentiles	142,400	9.5
96th to 99th percentiles	215,200	11.5
Top 1 percent	1,013,100	12.8

*Income amounts have been rounded to the nearest $100.
Notes: Average federal tax rates are calculated by dividing federal tax liabilities by before-tax income. Before-tax income is the sum of market income and government transfers. Market income is composed of labor income, business income, capital gains, capital income (excluding capital gains), income received in retirement for past services, and other sources of income. Government transfers are cash payments and in-kind benefits from social insurance and other government assistance programs. After-tax income is the sum of market income and government transfers, minus federal tax liabilities. Federal taxes include individual and corporate income taxes, social insurance (or payroll) taxes, and excise taxes. Income groups are created by ranking households by before-tax income. Quintiles (fifths) contain equal numbers of people; percentiles (hundredths) contain equal numbers of people as well.

SOURCE: Adapted from "Table 1. Distribution of Before- and After-Tax Income, by Income Group, 2009 and 2010," in *The Distribution of Household Income and Federal Taxes, 2010*, Congressional Budget Office, December 2013, http://www.cbo.gov/sites/default/files/cbofiles/attachments/44604-AverageTaxRates.pdf (accessed August 26, 2014)

---

those in the highest net worth categories. For example, 83.2% of families in the lowest category (less than the 25th percentile) had any financial asset in 2013, compared with 99.8% of families in the top net worth category (90th to 100th percentile).

Nonfinancial Assets

Nonfinancial assets are tangible things that have value in and of themselves. Examples include gold, land, houses, cars, boats, artwork, jewelry, and all other durable consumer goods with recognizable value in the marketplace.

In its calculation of aggregate net worth, the Federal Reserve includes only five components in nonfinancial assets. (See Table 8.4.) At the end of the first quarter of 2014, the assets were valued at:

- Real estate held by households—$22.8 trillion
- Consumer durable goods—$5 trillion
- Real estate held by NPOs—$2.7 trillion
- NPO equipment—$326.2 billion
- NPO intellectual property products—$141.6 billion

Figure 8.2 shows the breakdown of the assets by percentage value. Real estate held by households accounted for the vast majority (71%) of nonfinancial assets, followed by consumer durable goods (18%), real estate held by NPOs (9%), and NPO-owned equipment and software (2%).

Table 8.5 provides a detailed breakdown of the types of nonfinancial assets that were held by families in 2013 based on data from the SCF. Nearly all (91%) of the families had at least one nonfinancial asset. The most widely held nonfinancial assets were vehicles (86.3%) and primary residences (65.2%). Housing wealth has historically been a large component of total family wealth. As shown in Table 8.5, there were substantial differences between net worth percentiles regarding ownership of primary residences. Only 18.5% of people in the less than 25th percentile of net worth owned a primary residence in 2013, compared with 96.6% of people in the 90th to 100th percentile.

Liabilities

Assets alone do not provide an indication of the nation's wealth status. Debts and other obligations, known as liabilities, must be subtracted. These liabilities totaled almost $13.8 trillion at the end of the first quarter of 2014. (See Table 8.4.) Credit market instruments accounted for the vast majority of this total, at $13.1 trillion. Home mortgages ($9.3 trillion) were the single largest credit market instrument, followed by consumer credit ($3.1 trillion). Both of these liability types are described in detail in Chapter 4.

Trends in Aggregate Net Worth

As noted earlier, the total net worth of U.S. households and NPOs totaled $81.8 trillion at the end of the first quarter of 2014. (See Table 8.4.) Table 8.6 compares this value to net worth values calculated by the Federal Reserve dating back to 2004. Net worth was approximately $56.6 trillion in 2004 and climbed to $67.8 trillion in 2007, the year in which the Great Recession began. Net worth fell dramatically in 2008 primarily due to a huge drop in the value of real estate. This drop is illustrated in Figure 8.3, which shows owners' equity in real estate for households. Household equity (market value minus mortgage liability) nosedived during the recession before beginning to climb again. Between 2012 and 2013 net worth grew from $70.8 trillion to $80.3 trillion, an increase of 13%.

TABLE 8.4

Balance sheet of households and nonprofit organizations, 2008–13 and first quarter 2014

[Billions of dollars; amounts outstanding end of period, not seasonally adjusted. Sector includes domestic hedge funds, and personal trusts.]

	2008	2009	2010	2011	2012	2013	2014 I
Assets	71,476.2	73,030.3	77,130.1	78,258.0	84,441.4	94,042.3	95,548.6
Nonfinancial assets	24,813.9	23,671.0	23,323.3	23,265.8	25,007.3	27,544.4	28,329.6
Real estate	19,861.4	18,693.4	18,330.9	18,111.2	19,711.8	22,069.7	22,820.4
Households[a,b]	17,444.1	16,913.9	16,347.4	15,939.7	17,394.5	19,407.5	20,165.4
Nonprofit organizations	2,417.3	1,779.5	1,983.6	2,171.5	2,317.2	2,662.2	2,654.9
Equipment (nonprofits)[c]	268.5	279.5	290.6	304.6	315.1	323.7	326.2
Intellectual property products (nonprofits)[c]	105.4	110.0	115.0	123.6	132.4	140.0	141.6
Consumer durable goods[c]	4,578.6	4,588.1	4,586.7	4,726.4	4,848.0	5,011.0	5,041.4
Financial assets	46,662.4	49,359.3	53,806.9	54,992.2	59,434.1	66,497.9	67,219.0
Deposits	8,172.0	8,090.0	8,059.4	8,736.8	9,241.5	9,572.3	9,783.0
Foreign deposits	56.9	50.2	49.7	46.9	45.1	48.4	43.6
Checkable deposits and currency	294.9	396.0	423.6	752.0	897.8	1,004.7	1,096.7
Time and savings deposits	6,239.3	6,330.8	6,455.9	6,827.7	7,191.2	7,388.7	7,551.6
Money market fund shares	1,580.9	1,313.0	1,130.2	1,110.2	1,107.4	1,130.4	1,091.1
Credit market instruments	5,156.0	5,619.9	5,834.0	5,425.5	5,422.2	5,446.0	5,262.8
Open market paper	6.0	22.7	21.1	19.4	18.8	15.0	14.3
Treasury securities	188.7	859.7	1,134.4	715.6	941.0	935.4	841.2
Agency- and GSE-backed securities	1,057.7	358.3	353.7	304.6	154.2	125.9	14.1
Municipal securities	1,720.9	1,828.1	1,871.8	1,808.3	1,665.8	1,626.3	1,603.7
Corporate and foreign bonds	1,946.5	2,324.6	2,248.3	2,379.0	2,468.8	2,578.0	2,626.2
Other loans and advances[d]	29.7	26.8	26.2	23.4	20.9	25.9	27.2
Mortgages	111.8	110.9	100.1	100.8	86.9	80.4	78.4
Consumer credit (student loans)	94.6	88.8	78.4	74.5	65.6	59.1	57.8
Corporate equities[a]	5,931.0	7,496.1	8,995.3	9,025.4	10,412.8	13,309.6	13,502.0
Mutual fund shares[e]	3,306.7	4,115.8	4,600.2	4,502.9	5,408.7	6,890.1	7,058.9
Security credit	737.4	701.7	725.2	726.1	757.0	815.5	818.7
Life insurance reserves	1,049.8	1,109.2	1,137.2	1,203.6	1,186.1	1,242.2	1,256.7
Pension entitlements[f]	14,071.9	15,209.5	16,751.6	17,126.1	18,093.8	19,563.8	19,766.4
Equity in noncorporate business[g]	7,471.3	6,216.2	6,895.6	7,366.9	8,038.4	8,760.8	8,868.9
Miscellaneous assets	766.3	800.9	808.2	878.8	873.6	897.6	901.5
Liabilities	14,278.2	14,049.5	13,766.5	13,566.0	13,626.8	13,768.2	13,784.8
Credit market instruments	13,849.6	13,546.3	13,214.8	13,052.9	13,044.2	13,146.1	13,147.5
Home mortgages[h]	10,579.0	10,417.6	9,912.7	9,697.5	9,481.7	9,386.2	9,349.1
Consumer credit	2,650.6	2,552.8	2,647.4	2,755.9	2,923.6	3,097.4	3,103.6
Municipal securities[i]	259.5	265.4	263.2	255.5	241.0	227.8	227.6
Depository institution loans n.e.c.	26.4	−15.9	61.0	11.5	62.6	92.7	124.4
Other loans and advances	133.2	133.7	136.1	138.1	139.3	141.3	141.3
Commercial mortgages[j]	200.9	192.6	194.3	194.3	195.9	200.8	201.6
Security credit	164.8	203.0	278.2	238.9	303.7	339.2	351.7
Trade payables[i]	236.7	278.2	248.8	250.0	254.0	255.0	256.0
Deferred and unpaid life insurance premiums	27.0	22.1	24.7	24.3	24.9	27.9	29.6
Net worth	57,198.0	58,980.9	63,363.7	64,692.0	70,814.6	80,274.1	81,763.8
Memo:							
Replacement-cost value of structures:							
Residential	12,985.7	12,680.7	12,730.1	12,846.1	13,218.9	14,154.5	14,469.3
Households	12,780.0	12,479.0	12,526.6	12,640.3	13,010.0	13,930.8	14,240.6
Nonprofit organizations	205.6	201.7	203.5	205.8	208.9	223.7	228.7
Nonresidential (nonprofits)	1,537.8	1,487.7	1,500.8	1,554.1	1,600.3	1,682.8	1,695.9
Disposable personal income (SAAR)	10,995.4	10,937.2	11,243.7	11,787.4	12,245.8	12,476.2	12,719.0
Household net worth as percentage of disposable personal income (SAAR)	520.2	539.3	563.5	548.8	578.3	643.4	642.8
Owners' equity in household real estate[j]	6,865.1	6,496.3	6,434.7	6,242.1	7,912.8	10,021.3	10,816.3
Owners' equity as percentage of household real estate[k]	39.4	38.4	39.4	39.2	45.5	51.6	53.6

U.S. NET WORTH DISTRIBUTION

Bricker et al. use SCF data to rank families by their net worth (assets minus liabilities). In 2013 the mean net worth for each income percentile was:

- Less than 20 percentile of usual income—$64,600 mean net worth
- 20 to 39.9 percentile of usual income—$113,100 mean net worth
- 40 to 59.9 percentile of usual income—$164,800 mean net worth
- 60 to 79.9 percentile of usual income—$350,900 mean net worth
- 80 to 89.9 percentile of usual income—$631,400 mean net worth
- 90 to 100 percentile of usual income—$3.3 million mean net worth

TABLE 8.4

Balance sheet of households and nonprofit organizations, 2008–13 and first quarter 2014 [CONTINUED]

[Billions of dollars; amounts outstanding end of period, not seasonally adjusted. Sector includes domestic hedge funds, and personal trusts.]

[a]At market value.
[b]All types of owner-occupied housing including farm houses and mobile homes, as well as second homes that are not rented, vacant homes for sale, and vacant land.
[c]At replacement (current) cost.
[d]Syndicated loans to nonfinancial corporate business by nonprofits and domestic hedge funds.
[e]Value based on the market values of equities held and the book value of other assets held by mutual funds.
[f]Includes public and private defined benefit and defined contribution pension plans and annuities, including those in IRAs, at life insurance companies. Excludes Social Security.
[g]Net worth of nonfinancial noncorporate business and owners' equity in unincorporated security brokers and dealers.
[h]Includes loans made under home equity lines of credit and home equity loans secured by junior liens.
[i]Liabilities of nonprofit organizations.
[j]Real estate (households) minus liabilities (home mortgages).
[k]Owners' equity in household real estate divided by real estate (households).

SOURCE: Adapted from "B.100. Balance Sheet of Households and Nonprofit Organizations," in *Federal Reserve Statistical Release: Z.1 Financial Accounts of the United States: Flow of Funds, Balance Sheets, and Integrated Macroeconomic Accounts—First Quarter 2014*, Board of Governors of the Federal Reserve System, June 5, 2014, http://www.federalreserve.gov/releases/z1/current/z1.pdf (accessed August 14, 2014)

FIGURE 8.1

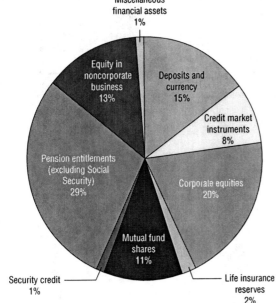

Financial assets of households and nonprofit organizations at end of first quarter 2014

SOURCE: Adapted from "B.100. Balance Sheet of Households and Nonprofit Organizations," in *Federal Reserve Statistical Release: Z.1 Financial Accounts of the United States: Flow of Funds, Balance Sheets, and Integrated Macroeconomic Accounts—First Quarter 2014*, Board of Governors of the Federal Reserve System, June 5, 2014, http://www.federalreserve.gov/releases/z1/current/z1.pdf (accessed August 14, 2014)

Table 8.7 shows the breakdown by income sources for various net worth percentiles in 2013. The data indicate that people at the bottom of the net worth profile (those in the less than 25th percentile) obtained the vast majority (73.1%) of their income from wages. By contrast, people at the top end of the spectrum (those in the 90th to 100th percentile) obtained less than half (47%) of their income from wages. A business, farm, or self-employment provided around a quarter (24.7%) of the income for families in the top percentile.

Bricker et al. provide mean and median net worth values by net worth percentiles, as follows:

- Less than 25 percentile of net worth—mean, −$13,400; and median, less than $50
- 25 to 49.9 percentile of net worth—mean, $35,900; and median, $31,300
- 50 to 74.9 percentile of net worth—mean, $177,700; and median, $168,100
- 75 to 89.9 percentile of net worth—mean, $546,200; and median, $505,800
- 90 to 100 percentile of net worth—mean, $4,024,800; and median, $1,871,800

It should be noted that the mean net worth for people in the less than 25th percentile was negative. This means that, on average, their debts were greater than their assets.

Bricker et al. examine changes in the wealth (net worth) distribution over time by comparing the portion of wealth held by various percentile groups. The wealth share of the 97th to 100th percentile increased from 44.8% in 1989 to 54.5% in 2013. The portion held by the 90th to 97th percentile was relatively flat at 19% to 22% over the entire period. It was 20.9% in 2013. The wealth share of the bottom 90th percentile of the distribution declined from 33.2% in 1989 to 24.7% in 2013. (It should be noted that the individual percentages do not sum to 100% due to rounding.) Overall, the top percentile group (90th to 100th percentile) held just over 75% of total wealth in 2013, whereas the remaining population held around 25% of total wealth.

FIGURE 8.2

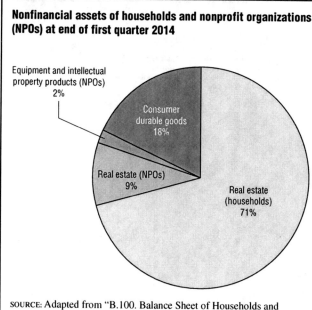

Nonfinancial assets of households and nonprofit organizations (NPOs) at end of first quarter 2014

SOURCE: Adapted from "B.100. Balance Sheet of Households and Nonprofit Organizations," in *Federal Reserve Statistical Release: Z.1 Financial Accounts of the United States: Flow of Funds, Balance Sheets, and Integrated Macroeconomic Accounts—First Quarter 2014*, Board of Governors of the Federal Reserve System, June 5, 2014, http://www.federalreserve.gov/releases/z1/current/z1.pdf (accessed August 14, 2014)

Family Characteristics

Table 8.8 is a compilation of SCF data that provides a detailed breakdown of net worth in 2013 by family characteristics. It should be noted that the data are based on actual income rather than on usual income and have not been adjusted for inflation. Thus, the net worth distributions differ from the values given by Bricker et al. Table 8.8, however, provides useful information about net worth differences between different types of households in 2013. For example, net worth varied considerably based on family structure. Couples with no children had a much higher mean net worth ($941,400) than did other family types. Likewise, homeowners ($773,400) had a much higher mean net worth than nonhomeowners ($70,400). The age, education, occupation, and work status of the heads of households were also important factors. Households with heads aged 65 to 74 years and those with college degrees had higher net worths than their counterparts. The same was true for household heads that were self-employed and those that worked in managerial or professional occupations. Lastly, net worth values differed based on the race and ethnicity of the respondents (i.e., the household members who responded to the survey questions). Households with white non-Hispanic respondents had a much higher mean net worth ($696,500) than those with nonwhite or Hispanic respondents ($184,200).

POVERTY IN THE UNITED STATES

In the United States poverty is officially defined and measured in different ways by different government entities. At the federal level it is measured using two methods: poverty thresholds and poverty guidelines. Poverty thresholds are set by the Census Bureau, which also tracks poverty populations in the United States. The thresholds specify minimum income levels for different sizes and categories of families. For example, the Census Bureau (2014, https://www.census.gov/hhes/www/poverty/data/threshld) notes that in 2013 the poverty threshold for a family of four including two related children under the age of 18 years in the household was $23,624 per year. Thus, a family matching this description and making less than that amount in 2013 was considered to be in poverty.

Poverty guidelines (or federal poverty levels) are published by the U.S. Department of Health and Human Services (HHS) for the 50 states and the District of Columbia. According to the HHS, the poverty guidelines are a simplification of the poverty thresholds and are used for administrative purposes, such as determining eligibility for specific federal programs. As of December 2014, the most recent guidelines were published in "Annual Update of the HHS Poverty Guidelines" (*Federal Register*, vol. 79, no. 14, January 22, 2014).

The Poverty Rate

The Census Bureau calculates the number of people in poverty and the poverty rate using income-based poverty thresholds. As of December 2014, the most recent data were collected in 2013 as part of the CPS ASEC and are analyzed by DeNavas-Walt and Proctor. As noted earlier, the 2013 CPS ASEC survey included around 68,000 households. The data were used to estimate annual poverty levels for the nation as a whole.

According to DeNavas-Walt and Proctor, 45.3 million people in the United States lived at or below the federal poverty level in 2013. (See Figure 8.4.) This represented 14.5% of the total U.S. population and was up from around 12% in 2000. The number of people in poverty has varied widely over the past few decades. In 1960 approximately 40 million people in the United States lived in poverty. This value dropped to fewer than 25 million people during the 1970s and then began increasing, reaching around 39 million during the early 1990s. The number declined to around 31 million in 2000 and then gradually increased over the ensuing years.

As shown in Figure 8.4, the poverty rate was around 22% during the early 1960s, and then dropped to around 11% during the early 1970s. It topped 15% during the early 1990s and then decreased to around 12% in 2000, before rising again.

There were dramatic demographic differences in the U.S. poverty rate in 2013. For example, DeNavas-Walt and Proctor note that men had a rate of 13.1%, compared with 15.8% for women. Age was also a distinguishing factor. The poverty rate for people under the age of 18 years was 19.9%; for people aged 18 to 64 years, 13.6%;

TABLE 8.5

Family holdings of nonfinancial assets, 2013

Family characteristic	Vehicles	Primary residence	Other residential property	Equity in nonresidential property	Business equity	Other	Any nonfinancial asset
				Percentage of families holding asset			
All families	86.3	65.2	13.3	7.1	11.7	6.6	91.0
Percentile of income							
Less than 20	61.9	37.5	2.1	2.5	4.1	4.3	71.9
20–39.9	85.5	52.9	5.8	3.4	5.4	4.3	90.7
40–59.9	92.5	63.7	10.1	6.2	8.7	5.4	95.0
60–79.9	95.6	80.6	16.1	9.2	13.6	7.3	98.2
80–89.9	96.4	88.6	25.2	11.0	17.9	9.6	99.0
90–100	95.4	93.7	39.7	17.8	35.5	13.7	99.8
Age of head (years)							
Less than 35	82.7	35.6	4.7	1.8	6.5	5.1	84.9
35–44	89.9	61.7	9.5	5.6	15.6	4.4	92.8
45–54	87.7	69.1	15.9	6.9	14.6	7.0	91.8
55–64	89.2	74.2	18.6	8.7	15.5	7.4	92.9
65–74	89.4	85.8	21.5	13.9	11.0	10.4	95.9
75 or more	76.0	80.2	12.3	9.3	4.4	6.0	89.6
Family structure							
Single with child(ren)	79.8	48.9	5.5	2.7	5.8	3.5	86.2
Single, no child, age less than 55	74.9	35.2	4.7	2.8	7.2	4.3	78.5
Single, no child, age 55 or more	73.9	66.8	10.2	4.5	5.7	8.1	85.6
Couple with child(ren)	94.4	72.9	14.9	6.6	16.7	6.1	97.0
Couple, no child	93.6	78.5	21.5	13.7	14.8	8.7	96.4
Education of head							
No high school diploma	75.4	51.8	5.9	3.4	3.2	1.5	82.3
High school diploma	84.4	64.0	9.0	5.6	8.2	4.1	89.1
Some college	86.6	56.4	12.0	6.5	11.0	7.7	90.4
College degree	90.7	74.2	19.5	9.7	17.3	9.5	95.4
Race or ethnicity of respondent							
White non-Hispanic	90.3	73.9	16.0	8.6	13.9	7.7	94.9
Nonwhite or Hispanic	78.1	47.4	7.8	4.0	7.2	4.4	83.2
Current work status of head							
Working for someone else	90.8	62.6	12.2	5.2	6.7	5.3	93.3
Self-employed	88.6	77.7	25.5	16.1	70.0	11.7	95.5
Retired	79.1	72.9	13.4	8.6	3.3	7.9	87.8
Other not working	72.7	35.8	4.6	4.0	3.3	5.1	78.1
Housing status							
Owner	93.6	100.0	17.9	9.9	15.4	7.6	100.0
Renter or other	72.6	*	4.7	1.9	4.7	4.7	74.3
Percentile of net worth							
Less than 25	66.3	18.5	2.4	*	3.4	2.8	68.8
25–49.9	91.0	57.7	4.2	2.2	5.1	4.3	96.6
50–74.9	92.9	89.8	13.2	7.7	10.8	6.6	99.3
75–89.9	95.4	93.2	25.6	13.9	17.8	9.8	99.4
90–100	94.3	96.6	45.1	24.2	41.7	16.8	99.5

*Ten or fewer observations.

SOURCE: "9. Family Holdings of Nonfinancial Assets and of Any Asset, By Selected Characteristics of Families and Type of Asset, 1989–2013 Surveys," in *Survey of Consumer Finances. Historic Tables and Charts*, Board of Governors of the Federal Reserve System, September 4, 2014, http://www.federalreserve.gov/econresdata/scf/scfindex.htm (accessed September 13, 2014)

and for people aged 65 years and older, 9.5%. Poverty rates in 2013 by race and ethnicity were:

- African Americans—27.2%
- Hispanics—23.5%
- Asian Americans—10.5%
- Whites—12.3%
- Non-Hispanic whites—9.6%

DeNavas-Walt and Proctor note that the 2013 CPS ASEC included native-born Americans, foreign-born naturalized U.S. citizens, and noncitizens. Poverty rate estimates for these groups on a national level were:

- Noncitizens—22.8%
- Native-born Americans—13.9%
- Foreign-born naturalized U.S. citizens—12.7%

According to DeNavas-Walt and Proctor, among family types, the highest poverty rate (30.6%) was with families headed by unmarried females. By contrast, the poverty rate for families headed by unmarried males was 15.9%, and for families headed by married couples it was only 5.8%. Poverty rates also differed by work experience. The rate for people who did not work at least one week in 2013 was 32.3%, compared with 17.5% for people who worked less than full time throughout the year. The rate was only 2.7% for those who worked full time year-round.

Poverty and Economic Well-Being

The poverty rates calculated by the Census Bureau are based on cash income only. There is widespread agreement that these rates do not adequately characterize the standard of living of the nation's poorest people. DeNavas-Walt and Proctor acknowledge that the poverty estimates do not consider the value of noncash benefits. These include government assistance programs, such as Medicare and Medicaid, government subsidies for food and housing, and employer-provided fringe benefits.

SURVEY OF INCOME AND PROGRAM PARTICIPATION. The federal government uses other methods to measure what it calls the "economic well-being" of the nation's households. The Census Bureau conducts the Survey of Income and Program Participation (SIPP). SIPP surveys are conducted monthly and follow the same participants for a multiyear period, called a panel. Each panel typically lasts two to four years. The Census Bureau reports

TABLE 8.6

Net worth of households and nonprofit organizations at end of period, 2004–13 and first quarter 2014

Year	Household net worth*
2004	56,581
2005	62,604
2006	67,392
2007	67,832
2008	57,198
2009	58,981
2010	63,364
2011	64,692
2012	70,815
2013	80,274
2014: first quarter	81,764

*Includes nonprofit organizations. Billions of dollars; amounts outstanding end of period, not seasonally adjusted.

SOURCE: Adapted from "Household Net Worth and Growth of Domestic Nonfinancial Debt," in *Federal Reserve Statistical Release: Z.1 Financial Accounts of the United States: Flow of Funds, Balance Sheets, and Integrated Macroeconomic Accounts—First Quarter 2014*, Board of Governors of the Federal Reserve System, June 5, 2014, http://www.federalreserve.gov/releases/z1/current/z1.pdf (accessed August 14, 2014)

FIGURE 8.3

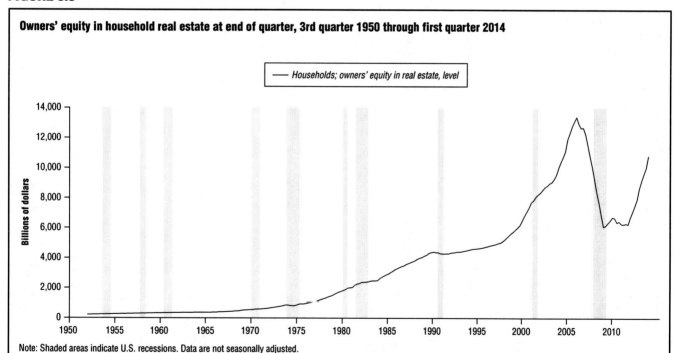

Owners' equity in household real estate at end of quarter, 3rd quarter 1950 through first quarter 2014

Note: Shaded areas indicate U.S. recessions. Data are not seasonally adjusted.

SOURCE: "Households; Owners' Equity in Real Estate, Level," in *FRED® Economic Data*, Federal Reserve Bank of St. Louis, June 5, 2014, http://research.stlouisfed.org/fredgraph.pdf?hires=1&type=application/pdf&chart_type=line&recession_bars=on&log_scales=&bgcolor=%23e1e9f0&graph_bgcolor=%23ffffff&fo=verdana&ts=12&tts=12&txtcolor=%23444444&show_legend=yes&show_axis_titles=yes&drp=0&cosd=1949-10-01&coed=2014-01-01&width=670&height=445&stacking=&range=Custom&mode=fred&id=OEHRENWBSHNO&transformation=lin&nd=&ost=-99999&oet=99999&scale=left&line_color=%234572a7&line_style=solid&lw=2&mark_type=none&mw=1&mma=0&fml=a&fgst=lin&fq=Quarterly%2C+End+of+Period&fam=avg&vintage_date=&revision_date= (accessed August 24, 2014)

TABLE 8.7

Before-tax family income, by income source and percentile of net worth, selected years 1989–2013

Income source	Percentile of net worth					All families
	Less than 25	25–49.9	50–74.9	75–89.9	90–100	
1989 Survey of Consumer Finances						
Wages	78.5	82.2	76.1	72.4	43.3	65.4
Interest or dividends	0.1	1.5	2.1	4.8	14.1	6.4
Business, farm, self-employment	1.6	3.5	3.7	9.1	22.9	11.0
Capital gains	a	0.4	1.9	2.6	13.1	5.5
Social Security or retirement	7.8	8.9	11.1	9.3	5.2	8.1
Transfers or other	11.8	3.5	5.1	1.8	1.3	3.6
Total	**100**	**100**	**100**	**100**	**100**	**100**
1992 Survey of Consumer Finances						
Wages	74.3	81.1	72.6	63.9	43.3	62.9
Interest or dividends	0.3	0.5	1.8	4.4	11.9	5.2
Business, farm, self-employment	1.9	3.4	6.2	8.8	29.1	13.2
Capital gains	a	0.1	0.4	1.5	4.5	1.9
Social Security or retirement	7.9	7.2	9.7	9.8	5.4	7.7
Transfers or other	15.5	7.7	9.4	11.6	5.9	9.0
Total	**100**	**100**	**100**	**100**	**100**	**100**
1995 Survey of Consumer Finances						
Wages	80.1	85.5	80.3	72.4	50.2	69.7
Interest or dividends	0.2	0.4	1.3	3.3	13.2	5.4
Business, farm, self-employment	1.9	2.7	5.5	8.2	24.1	11.4
Capital gains	0.7	0.5	1.0	1.2	5.9	2.5
Social Security or retirement	7.9	7.9	10.0	14.3	5.3	8.6
Transfers or other	9.2	3.1	1.9	0.7	1.3	2.4
Total	**100**	**100**	**100**	**100**	**100**	**100**
1998 Survey of Consumer Finances						
Wages	85.1	86.8	79.9	71.9	46.9	68.5
Interest or dividends	0.1	0.4	1.3	3.6	9.1	4.2
Business, farm, self-employment	1.6	3.5	4.5	9.6	27.0	12.8
Capital gains	0.1	0.5	0.6	2.4	11.1	4.5
Social Security or retirement	7.3	7.0	12.1	11.3	4.7	8.1
Transfers or other	5.8	1.8	1.6	1.2	1.3	1.8
Total	**100**	**100**	**100**	**100**	**100**	**100**
2001 Survey of Consumer Finances						
Wages	84.2	86.9	82.3	74.5	48.5	68.5
Interest or dividends	0.1	0.4	1.1	3.6	8.2	4.2
Business, farm, self-employment	1.9	1.3	4.4	7.2	22.7	11.4
Capital gains	b	0.3	1.1	2.6	15.3	6.7
Social Security or retirement	8.9	8.9	10.1	10.7	4.2	7.6
Transfers or other	4.8	2.2	1.0	1.3	1.0	1.6
Total	**100**	**100**	**100**	**100**	**100**	**100**
2004 Survey of Consumer Finances						
Wages	82.0	85.5	79.3	72.4	52.8	69.7
Interest or dividends	b	0.3	0.7	1.9	8.2	3.5
Business, farm, self-employment	1.1	2.6	5.1	8.6	22.0	11.1
Capital gains	b	b	b	1.2	8.1	3.1
Social Security or retirement	9.6	9.2	13.2	15.3	8.2	10.9
Transfers or other	7.3	2.5	1.7	0.7	0.8	1.8
Total	**100**	**100**	**100**	**100**	**100**	**100**
2007 Survey of Consumer Finances						
Wages	79.7	80.0	77.5	72.2	46.0	64.3
Interest or dividends	0.1	0.3	0.7	1.9	7.7	3.7
Business, farm, self-employment	1.8	5.3	7.2	8.0	25.2	13.9
Capital gains	0.1	0.4	1.3	2.9	14.2	6.6
Social Security or retirement	9.6	10.8	11.8	14.1	6.1	9.5
Transfers or other	8.7	3.2	1.6	0.8	0.7	1.9
Total	**100**	**100**	**100**	**100**	**100**	**100**

in "SIPP Introduction and History" (December 23, 2013, http://www.census.gov/programs-surveys/sipp/about/sipp-introduction-history.html) that the 2004 SIPP panel lasted from February 2004 to January 2008. The 2008 panel began in September 2008 and ended in December 2013; the 2014 panel began in February 2014.

In "Well-Being Main" (September 6, 2013, http://www.census.gov/hhes/well-being/index.html), the Census Bureau notes that "extended measures of well-being gauge how people are faring at the household level using such factors as possession of consumer durables, housing and neighborhood conditions, and the meeting of basic

TABLE 8.7

Before-tax family income, by income source and percentile of net worth, selected years 1989–2013 [CONTINUED]

Income source	Percentile of net worth					All families
	Less than 25	25–49.9	50–74.9	75–89.9	90–100	
2010 Survey of Consumer Finances						
Wages	75.7	80.8	75.0	69.3	54.5	67.2
Interest or dividends	b	0.1	0.4	1.6	8.4	3.5
Business, farm, self-employment	3.7	4.6	6.3	7.7	24.7	13.0
Capital gains	b	0.2	0.2	−0.2	3.2	1.3
Social Security or retirement	9.4	9.5	15.7	19.9	7.6	11.8
Transfers or other	11.1	4.7	2.5	1.7	1.6	3.2
Total	100	100	100	100	100	100
2013 Survey of Consumer Finances						
Wages	73.1	79.4	72.6	69.1	47.0	62.5
Interest or dividends	b	0.1	0.4	1.2	7.7	3.4
Business, farm, self-employment	3.1	3.1	4.2	7.7	24.7	13.0
Capital gains	b	0.1	0.7	1.4	10.2	4.6
Social Security or retirement	11.2	12.5	18.8	18.6	8.5	13.0
Transfers or other	12.6	4.8	3.3	2.0	1.9	3.6
Total	100	100	100	100	100	100

aLess than 0.05 percent.
bTen or fewer observations in any of the types of income.
Note: Nominal values.

SOURCE: Adapted from "2. Amount of Before-Tax Family Income, Distributed By Income Sources, By Percentile of Net Worth, 1989–2013 Surveys," in *Survey of Consumer Finances: Historic Tables and Charts*, Board of Governors of the Federal Reserve System, September 4, 2014, http://www.federalreserve.gov/econresdata/scf/scfindex.htm (accessed September 13, 2014)

---

needs." The SIPP program surveys people about their consumer possessions and seeks to determine if their basic needs are being met for housing and food.

As of December 2014, the most recently published SIPP data relevant to "economic well-being" were from 2011. They are summarized by Julie Siebens of the Census Bureau in *Extended Measures of Well-Being: Living Conditions in the United States: 2011* (September 2013, http://www.census.gov/prod/2013pubs/p70-136.pdf). In addition, the Census Bureau (http://www.census.gov/hhes/well-being/publications/extended-11.html) provides links to detailed tables of related data. For example, in "Table 1. Percent of Households Reporting Consumer Durables, for Householders 15 Years and Older, by Selected Characteristics: 2011" (September 5, 2013, http://www.census.gov/hhes/well-being/files/p70-136/tab1.xls), the Census Bureau lists the percentages of households classified as having "income below poverty" during 2011 that owned the following consumer goods:

- Refrigerator—97.8%
- Stove—96.6%
- Television—96.1%
- Microwave—93.2%
- Air conditioner—83.4%
- Videocassette recorder—83.2%
- Cellular phone—80.9%
- Clothes washer—68.7%
- Clothes dryer—65.3%
- Computer—58.2%
- Telephone—54.9%
- Dishwasher—44.9%
- Food freezer—26.2%

MEDICAL INSURANCE COVERAGE. As described in Chapter 1, the Medicaid program is jointly funded by the federal and state governments. Each state sets its own eligibility criteria and benefits that are within broad federal guidelines. In "Eligibility" (2014, http://www.medicaid.gov/medicaid-chip-program-information/by-topics/eligibility/eligibility.html), the Centers for Medicare and Medicaid Services indicates that eligibility can hinge on family income compared with the federal poverty levels, other income standards, or existing participation in other government programs. According to the Kaiser Family Foundation (KFF), in "Medicaid Expansion under the Affordable Care Act" (*Journal of the American Medical Association*, vol. 309, no. 12, March 27, 2013), in 2013 Medicaid eligibility was limited to "specific low-income groups," including the disabled, the elderly, children, pregnant woman, and some parents. Thus, many low-income childless adults were not eligible.

The Patient Protection and Affordable Care Act, which is commonly called the Affordable Care Act (ACA) or Obamacare, was passed in 2012 and mandated that all states expand their Medicaid programs to cover more people. Specifically, it required that eligibility be

TABLE 8.8

Family net worth, by family characteristics and housing status, 2013

[Thousands of dollars]

Family characteristic	2013 Median	2013 Mean
All families	81.4	528.4
Percentile of income		
Less than 20	6.2	87.5
20–39.9	21.5	111.4
40–59.9	61.8	170.1
60–79.9	158.7	333.8
80–89.9	298.4	629.9
90–100	1,134.5	3,248.0
Age of head (years)		
Less than 35	10.5	75.4
35–44	47.1	347.5
45–54	105.4	526.0
55–64	165.7	795.4
65–74	232.1	1,047.3
75 or more	195.0	611.4
Family structure		
Single with child(ren)	14.2	129.1
Single, no child, age less than 55	14.1	148.0
Single, no child, age 55 or more	107.9	372.8
Couple with child(ren)	93.0	587.2
Couple, no child	213.7	941.4
Education of head		
No high school diploma	17.3	107.7
High school diploma	52.4	199.7
Some college	46.8	318.2
College degree	218.7	1,015.5
Race or ethnicity of respondent		
White non-Hispanic	141.9	696.5
Nonwhite or Hispanic	18.1	184.2
Current work status of head		
Working for someone else	62.1	314.8
Self-employed	359.5	2,121.1
Retired	128.5	501.1
Other not working	9.1	135.2
Current occupation of head		
Managerial or professional	192.6	1,047.8
Technical, sales, or services	31.7	267.6
Other occupation	49.2	172.3
Retired or other not working	90.8	431.4
Housing status		
Owner	195.5	773.4
Renter or other	5.4	70.4
Percentile of net worth		
Less than 25	*	−13.0
25–49.9	31.4	35.9
50–74.9	100.2	177.7
75–89.9	505.1	546.3
90–100	1,871.6	3,962.4

*Less than 0.05 ($50).
Note: Nominal values.

SOURCE: Adapted from "4. Family Net Worth, by Selected Characteristics of Families, 1989–2013 Surveys," in *Survey of Consumer Finances: Historic Tables and Charts*, Board of Governors of the Federal Reserve System, September 4, 2014, http://www.federalreserve.gov/econresdata/scf/scfindex.htm (accessed September 13, 2014)

---

extended to adults with an income below 138% of the federal poverty levels. However, as noted in Chapter 1, the U.S. Supreme Court ruled in 2012 that Congress could not force the states to expand their Medicaid programs. As a result, the expansion became optional. In "Implementing the ACA: Medicaid Spending and Enrollment Growth for FY 2014 and FY 2015" (October 14, 2014, http://kff.org/report-section/implementing-the-aca-medicaid-spending-enrollment-growth-issue-brief), Robin Rudowitz et al. of the KFF indicate that 28 states and the District of Columbia expanded their Medicaid programs as of late 2014. Wisconsin did a partial expansion to cover adults with an income up to 100% of the federal poverty levels. In addition, Pennsylvania was expected to expand its coverage beginning in 2015. This would leave 22 states in which many low-income childless adults and some low-income parents could not qualify for Medicaid:

- Alabama
- Alaska
- Florida
- Georgia
- Idaho
- Indiana
- Kansas
- Louisiana
- Maine
- Mississippi
- Missouri
- Montana
- Nebraska
- North Carolina
- Oklahoma
- South Carolina
- South Dakota
- Tennessee
- Texas
- Utah
- Virginia
- Wyoming

Many of these people are caught in a so-called coverage gap because they do not qualify for Medicaid, but they also do not qualify for government subsidies provided to people who purchase certain medical insurance plans through the government-operated health exchanges. According to Rachel Garfield et al., in "The Coverage Gap: Uninsured Poor Adults in States That Do Not Expand Medicaid—An Update" (November 12, 2014, http://kff.org/health-reform/issue-brief/the-coverage-gap-uninsured-poor-adults-in-states-that-do-not-expand-medicaid-

FIGURE 8.4

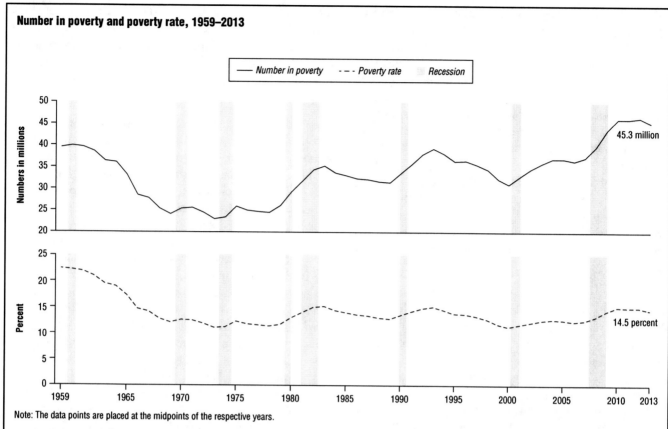

Number in poverty and poverty rate, 1959–2013

Note: The data points are placed at the midpoints of the respective years.

SOURCE: Carmen DeNavas-Walt and Bernadette D. Proctor, "Figure 4. Number in Poverty and Poverty Rate: 1959 to 2013," in *Income and Poverty in the United States: 2013*, U.S. Department of Commerce, U.S. Census Bureau, September 2014, http://www.census.gov/content/dam/Census/library/publications/2014/demo/p60-249.pdf (accessed September 17, 2014)

an-update), the subsidies are available only to applicants with an income between 100% and 400% of the federal poverty levels. The Centers for Medicare and Medicaid Services notes in "Incomes That Qualify for Lower Costs" (https://www.healthcare.gov/qualifying-for-lower-costs-chart) that for 2014 the federal poverty level for a family of four in almost all states was $23,850 per year; 138% of this value was $32,913 per year. When the ACA was written, it was expected that people with an income below 138% of the federal poverty levels would be covered nationwide by expanded Medicaid programs. Garfield et al. estimate that nearly 4 million Americans fell into the coverage gap as of November 2014. The vast majority (86%) were in the South, with the largest percentages living in Texas (25%), Florida (17%), North Carolina (9%), and Georgia (7%).

U.S. INCOME AND WEALTH INEQUALITY: CAUSES AND CONSEQUENCES

The data are clear that income and net worth inequality exist and that the gap between the higher and lower percentiles has widened in recent decades. However, there are varying opinions on the causes for this phenomena and whether it is a wrong that should be righted or an inevitable result of a growing economy.

In *Changes in the Distribution of Income among Tax Filers between 1996 and 2006: The Role of Labor Income, Capital Income, and Tax Policy* (December 29, 2011, http://www.fas.org/sgp/crs/misc/R42131.pdf), Thomas L. Hungerford of the Congressional Research Service lists causes that are commonly cited by analysts:

- High-paying union jobs have declined in availability
- The minimum wage has not risen fast enough to keep up with inflation
- The U.S. business sector increasingly favors highly educated workers and those skilled in particular fields, especially technological fields, over lower educated workers and workers lacking technological skills
- "Winner-take-all" markets have grown more common in which many people compete, but only a small few reap extremely large rewards
- Politicians have instituted changes, such as tax policies, that have primarily benefited wealthier Americans

Thomas Piketty is a professor at the Paris School of Economics and a well-known researcher in the field of income and wealth inequality. In his book *Capital in the Twenty-First Century* (2014), Piketty attributes much of the growing income inequality in the United States to a rise in "supersalaries"—extremely high salaries that are paid mostly to corporate executives. In addition, he discusses at length the role of ownership of capital in wealth building. Examples include corporate profits, rents, royalties, dividends, interest, and profits gained from the sale of property or stock. Piketty claims that the rate of return (i.e., profit on investment) for capital has been improving dramatically compared with the rate of growth of the U.S. economy as a whole. This change has allowed capital owners—who also tend to be high-income earners—to surge far ahead of the rest of the population in terms of wealth.

Even so, analysts disagree on what exactly has caused wealth inequality, and they also disagree over whether it is a problem that should be fixed or is simply a natural consequence of the growth that has occurred in the U.S. economy over time. The debate over wealth inequality has always been a politically partisan issue, with those on the left maintaining that widespread financial inequality causes many social and economic problems and those on the right arguing that the market should be allowed to adjust itself with regard to wages and income.

Wealth Inequality Seen as a Problem to Be Fixed

Paul Krugman, an economics professor at Princeton University, is an outspoken critic of wealth inequality in the United States. In "For Richer" (NYTimes.com, October 20, 2002), he asserts that "as the gap between the rich and the rest of the population grows, economic policy increasingly caters to the interests of the elite, while public services for the population at large—above all, public education—are starved of resources. As policy increasingly favors the interests of the rich and neglects the interests of the general population, income disparities grow even wider."

It is a common criticism that unequal wealth distribution in the United States concentrates too much political, social, and economic power at the top of the distribution. For example, G. William Domhoff of the University of California, Santa Cruz, espouses this viewpoint in "Wealth, Income, and Power" (February 2013, http://www2.ucsc.edu/whorulesamerica/power/wealth.html). He notes that wealth allows people to donate money to political parties, pay lobbyists, and take other actions to create and further "policies beneficial to the wealthy." In addition, Domhoff states, "Wealth also can be useful in shaping the general social environment to the benefit of the wealthy, whether through hiring public relations firms or donating money for universities, museums, music halls, and art galleries." The mechanisms of stock ownership are explained in Chapter 7. Individuals that own large portions of the stock in particular corporations can wield great influence over how those companies are operated. As Domhoff points out, this power has "a major impact on how the society functions."

FOCUS ON BIG BUSINESS. One factor contributing to income and net worth concentration at the top of the wealth distribution has been a surge in pay and benefits for high-level executives, such as the chief executive officers (CEOs) of companies. CEO pay is a hot topic for critics of wealth inequality. In "CEO Pay Continues to Rise as Typical Workers Are Paid Less" (June 12, 2014, http://www.epi.org/publication/ceo-pay-continues-to-rise), Lawrence Mishel and Alyssa Davis of the Economic Policy Institute, a nonprofit think tank devoted to advancing progressive causes, claim that the average compensation (including pay and stock options) of the CEOs of the "top" 350 U.S. companies in 2013 was $15.2 million. This data set excludes Facebook, Inc., whose CEO was compensated an estimated $24.8 million in 2013. According to Mishel and Davis, real CEO compensation in the United States grew 937% between 1978 and 2013. They say this compares to a 10.2% increase over the same time period for "a typical worker's compensation."

Income and wealth inequality sparked public protests in 2011 that were dubbed the Occupy Wall Street Movement or, more broadly, the Occupy Movement. The protests were also driven by the idea that greed and reckless behavior by Wall Street's large financial corporations caused the housing bust and subsequent Great Recession (which lasted from December 2007 to June 2009) that brought economic hardship to so many Americans. As noted in Chapter 4, dozens of struggling companies in the financial industry received government bailouts during the economic downturn. However, even as the companies were accepting the bailouts, many continued to pay out millions of dollars in bonuses to their employees. The net result was a groundswell of popular resentment against "big business." Occupy Movement activists adopted the slogan "We are the 99%" (i.e., the 99% of Americans at the bottom of the wealth distribution). As of December 2014, activists continued to hold or inspire protests in which people denounced income and wealth inequality and related social and political concerns.

FIXING WEALTH INEQUALITY. In general, people who consider wealth inequality to be a social and economic wrong believe the government should right the wrong. Progressives support government actions that alleviate wealth inequality, just as the government outlawed racial segregation in schools during the 1950s as a step toward eliminating racial inequality in society at large. Progressives typically advocate "bottom-up" measures, such as raising the minimum wage or improving public education, to "raise the floor" for Americans in the lower- and middle-income classes. Progressives also champion "top-down" measures that shrink the wealth of the higher-income classes to pay for the bottom-up measures. This concept is known as the redistribution of wealth.

Taxation is the primary tool by which the government can and does "redistribute" wealth among Americans. As is explained in Chapter 9, the U.S. income tax system is called progressive because it taxes individuals with higher incomes at higher rates than individuals with lower incomes. The tax rates only apply to what the IRS defines as "taxable income." The U.S. tax code allows hundreds of deductions and credits that lessen taxable income amounts. The most well-known example is the deduction for mortgage interest. However, critics complain that wealthy individuals and businesses benefit the most from income tax deductions and credits that have been built into the tax code over the years.

As noted earlier, people at the higher end of the income distribution own the most capital. Some analysts argue that tax rates should be increased on high incomes and capital gains. Others counter that high-income individuals and capital owners help grow the economy by creating jobs for people lower down in the distribution. In *The Price of Inequality: How Today's Divided Society Endangers Our Future* (2012), Joseph E. Stiglitz of Columbia University disputes the idea that taxes should be lowered on the wealthy because they are the people who create jobs and opportunities for others. Stiglitz notes, "The same old 'myth' that we should celebrate the wealth of those at the top because we all benefit from it has been used to justify the maintenance of low taxes on capital gains. But most capital gains accrue not from job creation but from one form of speculation or another. Some of this speculation is destabilizing and played a role in the economic crisis that has cost so many jobs."

Many progressives support raising the tax rates and reducing the tax deductions and credits for people in the higher tax brackets. In addition, they advocate eliminating tax deductions, tax credits, and subsidies (government-provided funds) for wealthy private corporations, particularly those in the oil and gas industry. They also want an increase in the capital gains tax. President Barack Obama (1961–) supports many progressive causes. For example, the White House states in "Taxes" (2014, http://www.whitehouse.gov/issues/taxes) that the president endorses a tax policy that "means asking those at the top to do their fair share and putting an end to special privileges and loopholes that benefit those who need them the least." Obama's tax policy is discussed at length in Chapter 9.

Wealth Inequality Seen as an Inevitable Result of Economic Growth

On the other side of this issue are those who believe that wealth inequality is an inevitable result of economic growth and does not pose a problem to the U.S. economy and society. This viewpoint is generally associated with conservative political thinking, which holds that market forces must be allowed to adjust themselves without government interference. One idea that is regularly expressed by politicians in this camp is that "what's good for the rich is good for the rest of us." This philosophy was espoused by President Ronald Reagan (1911–2004), who championed a trickle-down economic policy—economic actions (such as tax cuts beneficial to the wealthy) that encourage greater investment in business growth and thereby increase employment, wages, and other benefits for those in the middle and lower classes. Thus, conservatives argue that high-income and high-wealth individuals are job creators who provide opportunities and spur economic growth.

In "Rich Man, Poor Man: How to Think about Income Inequality" (NationalReview.com, June 16, 2003), Kevin A. Hassett states his belief that the unease over wealth inequality is driven by social views on the "basic justice" of society. However, he argues that there are similarly compelling arguments against taking from the rich to benefit the poor. Hassett notes that taking resources from the rich limits their ability and incentive to start and grow businesses, which will ultimately hurt American workers even more.

This argument is commonly cited by people who oppose wealth redistribution. They also point out that wealth redistribution is an improper term because wealth was not originally "distributed" to its owners, but earned by them.

Michael Tanner argues in "The Income-Inequality Myth" (NationalReview.com, January 10, 2012) that any income inequality that does exist has been exacerbated not by tax cuts or other preferential treatment for the rich, but by the large shift in American industry over the past half century from manufacturing to information and technology. This shift has benefited the more educated over the less educated. Tanner disputes the idea that income inequality is necessarily a bad thing. He states, "In what way does someone else's success harm me? Such a viewpoint stems from the misguided notion that the economy is a pie of fixed size. If one person gets a bigger portion of the pie, others of necessity get smaller pieces, and the role of government is to divide up the slices of that pie. In reality, though, the size of the pie is infinite."

Tanner believes that risk takers and innovators who expand the economic pie deserve the rewards that come to them for doing so, and notes that "such rewards inevitably lead to greater inequality."

In "Defending the One Percent" (*Journal of Economic Perspectives*, vol. 27, no. 3., Summer 2013), N. Gregory Mankiw of Harvard University also points to evolving technologies as one of the drivers behind growing income inequality in the United States. He suggests, "It seems that changes in technology have allowed a small number of highly educated and exceptionally talented individuals to command superstar incomes in ways that were not possible a generation ago." He quotes Erik Brynjolfsson and Andrew McAfee in their book *Race against the Machine* (2011) as saying, "Aided by digital technologies, entrepreneurs, CEOs, entertainment stars, and financial executives have been able to leverage their talents across global markets and capture reward that would have been unimaginable in earlier times."

PUBLIC OPINION ON WEALTH DISTRIBUTION

The Gallup Organization regularly polls Americans about their opinions on wealth distribution and the fairness of the tax system. Andrew Dugan of the Gallup Organization summarizes in *As Taxes Rise, Half in U.S. Say Middle-Income Pay Too Much* (April 14, 2014, http://www.gallup.com/poll/168521/taxes-rise-half-say-middle-income-pay.aspx) the poll results related to the tax burden on people of different income sectors. In April 2014 nearly two-thirds (61%) of those asked said "upper-income" people pay "too little" in federal taxes. Around a quarter (24%) said this sector pays a "fair share" and 13% said it pays "too much." By comparison, 49% of respondents said "middle-income" Americans pay "too much" in federal taxes, while 42% said this sector pays a "fair share." Another 7% believed this sector pays "too little." In regards to "lower-income" Americans, the respondents were more evenly split, with 41% saying this sector pays "too much," 33% saying it pays a "fair share," and 23% saying it pays "too little" in federal taxes.

According to Dugan, some of the opinions regarding tax fairness varied significantly based on the self-identified political party affiliations of the respondents. For example, more than twice as many Democrats (52%) said lower-income people are taxed "too much," compared with 25% of Republicans. Likewise, 76% of Democrats said upper-income people are taxed "too little," compared with 45% of Republicans.

During the same poll, respondents were asked about the fairness of the federal tax burden on corporations. Only one-fifth (20%) of poll participants said corporations pay a "fair share" in taxes. Eight percent thought corporations pay "too much" and 66% said corporations pay "too little" in taxes.

In *In U.S., 67% Dissatisfied with Income, Wealth Distribution* (January 20, 2014, http://www.gallup.com/poll/166904/dissatisfied-income-wealth-distribution.aspx), Rebecca Riffkin of the Gallup Organization describes the results of polling done in January 2014 concerning satisfaction with "the way income and wealth are currently distributed in the U.S." Overall, only 7% of respondents were very satisfied and 25% were somewhat satisfied. Larger percentages were somewhat dissatisfied (28%) or very dissatisfied (39%). There were stark differences between poll participants of different political parties, with 75% of Democrats and 54% of Republicans indicating they were somewhat or very dissatisfied with income and wealth distribution in the United States. Self-described Independents fell between these two values, with 70% of them expressing some level of dissatisfaction.

During the January 2014 poll the respondents were not asked about possible remedies for unequal income and wealth distribution; however, such questions were posed during a poll conducted in April 2013. According to Frank Newport of the Gallup Organization, in *Majority in U.S. Want Wealth More Evenly Distributed* (April 17, 2013, http://www.gallup.com/poll/161927/majority-wealth-evenly-distributed.aspx), 59% of those asked agreed that "money and wealth in this country should be more evenly distributed among a larger percentage of the people." Democrats (83%) were much more strongly in favor of redistribution than Independents (60%) or Republicans (28%). During the same poll the participants were asked whether the U.S. government should redistribute wealth by imposing "heavy taxes on the rich." Just over half (52%) of respondents agreed with this idea, while 45% did not agree.

WORLDWIDE COMPARISONS

Some analysts of income and wealth inequality compare the distribution in the United States to that of other countries, particularly those with highly developed economies, such as in western Europe. Anthony Shorrocks and Jim Davies estimate in *Global Wealth Databook 2014* (October 2014, https://publications.credit-suisse.com/tasks/render/file/?fileID=5521F296-D460-2B88-081889DB12817E02) the wealth holdings of people in various countries. The researchers define personal wealth as "individual net worth" and note that it includes "the marketable value of financial assets plus non-financial assets (principally housing and land) less debts." They began with national data reported by some countries in household balance sheets. Survey data—such as that obtained through the SCF in the United States—were used to estimate wealth distributions. Shorrocks and Davies note that they also collected data on high-wealth individuals from media sources, such as *Forbes* magazine. Comprehensive data were mainly available only for developed countries, such as the United States, Canada, and west European nations.

According to Shorrocks and Davies, global household wealth totaled $263 trillion in mid-2014. This

TABLE 8.9

Income inequality based on polling results, by country, aggregated data for 2006–12

	Percentage of poor holding 20% of total income	Percentage of rich holding 20% of total income
World (131 countries)	54%	3%
Sub-Saharan Africa	63%	2%
East Asia	59%	2%
Latin America	56%	3%
U.S. & Canada	54%	4%
Southeast Asia	53%	3%
Middle East & North Africa	53%	3%
Balkans	52%	3%
South Asia	51%	3%
Australia & New Zealand	51%	5%
Europe	48%	5%
Commonwealth of Independent States	46%	6%

SOURCE: Glenn Phelps and Steve Crabtree, "Income Inequality Highest in Sub-Saharan Africa," in *Worldwide, Richest 3% Hold One-Fifth of Collective Income*, The Gallup Organization, January 3, 2014, http://www.gallup.com/poll/166721/worldwide-richest-hold-one-fifth-collective-income.aspx (accessed August 25, 2014). Copyright © 2014 Gallup, Inc. All rights reserved. The content is used with permission; however, Gallup retains all rights of republication.

---

wealth was distributed very unevenly with the top decile (90th to 100th percentile) holding 87.4% of the total. The top decile in the United States held 74.6% of total wealth, giving the nation one of the most unequal distributions in the world. Top deciles holding high percentages of wealth were also found in Russia (84.8% of total), Turkey (77.7% of total), Hong Kong (77.5% of total), Indonesia (77.2% of total), the Philippines (76% of total), and Thailand (75% of total). In China and India, two countries with rapidly growing economies, the top deciles held 64% and 74%, respectively, of total wealth.

The United States is more often compared with other democratic nations and those with economies that highly favor free markets over government control. The top deciles in a dozen of these countries held the following percentages of total wealth in mid-2014:

- Switzerland—71.9%
- Sweden—68.6%
- Israel—67.3%
- Norway—65.8%
- Germany—61.7%
- Canada—57%
- Finland—54.5%
- United Kingdom—54.1%
- France—53.1%
- Australia—51.5%
- Italy—51.5%
- Japan—48.5%

Thus, the United States had a more unequal wealth distribution than any of these nations as of mid-2014.

There are a variety of reasons put forth to explain why the United States has greater wealth inequality than nations with similar political and economic systems. Some analysts focus on income factors, such as the differences between countries concerning CEO salaries, minimum wages, or poverty rates. These analyses are complicated due to a lack of comparable data. For example, in "Do CEOs Make Much More in the U.S. Than Elsewhere?" (Forbes.com, March 13, 2013), Pedro Matos suggests that "U.S. firms operate differently from their international counterparts, particularly in terms of corporate governance." In addition, he points out that U.S. firms rely more on nonsalary compensation, such as stock options and bonuses, for CEO compensation than corporations in other countries. Cost-of-living variations between nations make it difficult to meaningfully compare their minimum wages or poverty rates, which are defined differently from country to country.

Government transfers are the primary means by which nations redistribute wealth. The United States conducts much less redistribution than its west European counterparts, which tend to have higher tax rates and more extensive social welfare systems. Kimberly J. Morgan explains in "America's Misguided Approach to Social Welfare" (ForeignAffairs.com, vol. 92, no. 1, January–February 2013) that the United States relies heavily on "private social benefits" provided by employers. Examples include employer-subsidized health insurance, pensions, and paid sick leave. However, such benefits are more common for higher-paying jobs than for lower-paying jobs. By contrast, Morgan notes that in many other developed economies, "low-income families receive a much more generous and comprehensive array of tax subsidies and benefits, including family allowances, tax breaks for children, and subsidized child care."

In *Worldwide, Richest 3% Hold One-Fifth of Collective Income* (January 3, 2014, http://www.gallup.com/poll/166721/worldwide-richest-hold-one-fifth-collective-income.aspx), Glenn Phelps and Steve Crabtree of the Gallup Organization note that the results are based on surveys conducted in 131 countries and regions between 2006 and 2012. The researchers estimate that the world's top 3% of residents in terms of income accounted for 20% of the total income, whereas more than half (54%) of the lowest-income residents accounted for 20% of total income. (See Table 8.9.) Income inequality was highest in sub-Saharan Africa, where 63% of the lowest-income residents accounted for 20% of the total income for the region.

CHAPTER 9
THE ROLE OF THE GOVERNMENT

In general, the art of government consists in taking as much money as possible from one party of the citizens to give to the other.

—Voltaire, *Dictionnaire Philosophique* (1764)

The government has many roles in the U.S. economy. Like other businesses, the government spends, borrows, and takes in money, consumes goods and services, and employs people. Federal, state, and local governments raise funds directly through taxes and fees. Governments also disburse money via contracts with businesses or through social programs that benefit the public.

The federal government is a manipulator of the U.S. economy. It influences macroeconomic factors, such as inflation and unemployment, through fiscal and monetary policies. Fiscal policy revolves around spending and taxation. Monetary policy is concerned with the amount of money in circulation and the operation of the nation's central banking system. The federal government also engages in economic forecasting by predicting the course of certain key economic indicators into the future.

FUNDING GOVERNMENT SERVICES

Governments are responsible for providing services that individuals cannot effectively provide for themselves, such as military defense, roads, education, social services, and environmental protection. Some government entities also provide public utilities, such as water, sewage treatment, or electricity. To generate the revenue that is necessary to provide services, governments collect taxes and fees and charge for many services they provide to the public. If these revenues are not sufficient to fund desired programs, governments borrow money.

Taxation

The most common taxes levied by federal, state, and local governments are:

- Income taxes—charged on wages, salaries, and tips
- Payroll taxes—Social Security insurance, Medicare (a federal health insurance program for people aged 65 years and older and people with disabilities), and unemployment compensation; employee portions are withdrawn from payroll checks
- Property taxes—levied on the value of property owned, usually real estate
- Capital gains taxes—charged on the profit from the sale of an asset such as stock or real estate
- Corporate taxes—levied on the profits of a corporation
- Estate taxes—charged against the assets of a deceased person
- Excise taxes—collected at the time something is sold or when a good is imported
- Wealth taxes—levied on the value of assets rather than on the income they produce

Taxes are broadly defined as being either direct or indirect. Direct taxes (such as income taxes) are paid by the entity on whom the tax is being levied. Indirect taxes are passed on from the responsible party to someone else. Examples of indirect taxes include business property taxes, gasoline taxes, and sales taxes, which are levied on businesses but passed on to consumers via increased prices.

When individuals with higher incomes pay a higher percentage of a tax, it is called a progressive tax; when those with lower incomes pay a larger percentage of their income, a tax is considered regressive. The federal income tax is an example of a progressive tax because individuals with higher incomes are subject to higher tax rates. Sales and excise taxes are regressive because the same tax applies to all consumers regardless of income, so less prosperous individuals pay a higher percentage of their income.

Borrowing against the Future

Government entities sometimes spend more than they make. When cash revenues from taxes, fees, and other sources are not sufficient to cover spending, money must be borrowed. One method used by government to borrow money is the selling of securities, such as bonds, to the public. A bond is basically an IOU (an abbreviation for "I Owe yoU") that a government body writes to a buyer. The buyer pays money up front in exchange for the IOU, which is redeemable at some point in the future (the maturity date) for the amount of the original loan plus interest. In addition, the federal government has the ability to write itself IOUs—to spend money in the present that it expects to make in the future.

U.S. government bodies borrow money because they are optimistic that future revenues will cover the IOUs they have written. This optimism is based in part on the power that governments have to tax their citizens and to control the cost of provided government services. Although tax increases and cuts in services can be enacted to raise money, these actions have political and economic repercussions. Politicians who wish to remain in office are reluctant to displease their constituents. Furthermore, the more citizens pay in taxes, the less money they will have to spend in the marketplace or invest in private business, thereby hurting the overall economy. As a result, governments must weigh their need to borrow against the future likely consequences of paying back the loan.

LOCAL GOVERNMENTS

The U.S. Census Bureau performs a comprehensive Census of Government every five years. As of December 2014, the most recent data were from the 2012 census (http://www.census.gov/govs/cog). In between the censuses, annual surveys are conducted to collect certain data on government finances and employment.

According to the Census Bureau (http://factfinder2.census.gov/faces/tableservices/jsf/pages/productview.xhtml?src=bkmk), 90,056 local government units were in operation in 2012. These units consisted of counties, municipalities, townships, school districts, and special districts. Special district governments usually perform a single function, such as flood control or water supply. For example, Florida is divided into five water management districts, each of which is responsible for managing and protecting water resources and balancing the water needs of other government units within its jurisdiction.

Local Revenues

As of December 2014, the most recent comprehensive data for local government revenues were from 2011. In *State and Local Government Finances Summary: 2011* (July 2013, http://www2.census.gov/govs/local/summary_report.pdf), Jeffrey L. Barnett and Phillip M. Vidal of the Census Bureau indicate that local governments took in nearly $1.7 trillion in 2011. Most of the money came from three sources: intergovernmental revenue ($554.1 billion), property taxes ($429.1 billion), and current charges for services ($247.8 billion). Intergovernmental revenue consists of funds that are transferred to the local government from the federal and state governments. State funds accounted for the vast majority of intergovernmental transfers in 2011.

Taxes, particularly property taxes, are an important source of revenue for many local governments. Consumption taxes are also collected on sales and gross receipts. This includes selective taxes that are levied against particular goods, such as motor fuels, alcoholic beverages, and tobacco products. Miscellaneous taxes include individual and corporate income taxes, motor vehicle license taxes, and a wide variety of other taxes.

The revenue included in current charges for services comes from many local sources, including hospitals, sewage treatment facilities, solid waste management, parks and recreation areas, airports, and educational facilities.

Other revenue sources for local governments are miscellaneous general revenue, public utilities, and insurance trusts. Miscellaneous general revenue comes from a variety of sources, including interest payments and the sale of public property. Public utilities primarily supply water, electricity, natural gas, and public transportation (such as buses and trains). Insurance trusts are monies collected from the paychecks of local government employees to pay for worker programs and benefits, such as pensions.

Local Expenditures

According to Barnett and Vidal, local governments spent nearly $1.7 trillion on annual expenses in 2011. Education was the largest single component, accounting for $599.3 billion of the total. Lesser amounts were spent on public health, welfare, and hospitals; utilities; environmental and housing concerns; and public safety (e.g., police and fire).

Lisa Jessie and Mary Tarleton of the Census Bureau estimate in *2012 Census of Governments: Employment Summary Report* (March 6, 2014, http://www2.census.gov/govs/apes/2012_summary_report.pdf) that in 2012 local governments employed nearly 14 million full-time and part-time employees.

STATE GOVERNMENTS

State Revenues

Barnett and Vidal indicate that state governments had revenues of nearly $2.3 trillion in 2011. Intergovernmental revenue from the federal government accounted for

$573.4 billion of the total. Self-generated state revenue totaled almost $1.1 trillion and consisted largely of individual (personal) income taxes ($259.3 billion), general sales taxes ($235.9 billion), and revenues from institutions of higher learning ($100.8 billion).

The Federation of Tax Administrators (FTA) is a nonprofit organization that provides research services for the tax administrators of all 50 states and the District of Columbia. In "State Individual Income Taxes" (January 1, 2014, http://www.taxadmin.org/fta/rate/ind_inc.pdf), the FTA provides information on state income tax rates. As of January 2014, 43 states and the District of Columbia imposed income taxes. The states that did not tax income were Alaska, Florida, Nevada, South Dakota, Texas, Washington, and Wyoming. Two additional states, New Hampshire and Tennessee, taxed only personal income derived from dividends and interest.

According to the FTA, in "State Sales Tax Rates and Food and Drug Exemptions" (January 1, 2014, http://www.taxadmin.org/fta/rate/sales.pdf), as of January 2014, 45 states and the District of Columbia assessed sales taxes. The five states without a sales tax were Alaska, Delaware, Montana, New Hampshire, and Oregon.

State Expenditures

Barnett and Vidal note that state governments had expenditures of $2 trillion in 2011. Approximately $496.8 billion of this total was in the form of transfers to other governments, such as local governments within the state. The remainder was devoted to spending priorities at the state level. Public welfare composed the single largest expense ($439.3 billion), followed by education ($261.9 billion). State spending on education is primarily for higher education, such as colleges and universities.

Jessie and Tarleton estimate that state governments employed nearly 5.3 million full-time and part-time employees in 2012.

FEDERAL GOVERNMENT

For accounting purposes, the federal government operates on a fiscal year (FY) that begins in October and runs through the end of September. Thus, FY 2015 covers the period of October 1, 2014, to September 30, 2015. Each year by the first Monday in February the U.S. president must present a proposed budget to the U.S. House of Representatives. This is the amount of money that the president estimates will be required to operate the federal government during the next fiscal year.

It can take several months for the House to debate, negotiate, and approve a final budget. The budget must also be approved by the U.S. Senate. This entire process can take many months, and sometimes longer than a year. This means that the federal government can be well into, or beyond, a fiscal year before knowing the exact amount of its budget for that year. If a new fiscal year begins and the formal budget for that year has still not passed, then Congress passes temporary spending bills called continuing resolutions that fund federal government operations for a short time, typically a few weeks or a month or two, while negotiations continue on the formal budget. Continuing resolutions have expiration dates, which if reached cause the federal government to cease all nonessential operations. Such a government "shutdown" last occurred in October 2013.

Budget enactment delays are much more likely when Congress is divided by political partisanship (lack of cooperation between parties). The midterm elections of November 2010 gave Republicans control of the House and Democrats control of the Senate. Many of the newly elected Republican congressional members had been supported by the tea party, a political movement that strongly favors less taxation and smaller government. Consequently, Republican lawmakers as a whole took hardline positions against economic policies that were favored by President Barack Obama (1961–), a Democrat. This partisanship resulted in fierce disagreements about government spending that delayed passage of appropriations bills. These delays necessitated the passage of several continuing resolutions to keep the federal government operating. For example, in September 2014 a continuing resolution was passed by the House and the Senate and signed by the president to keep the federal government operating until December 11, 2014. The midterm elections of November 2014 saw Republicans win control of both the House and the Senate. It remains to be seen if this development will help ease the bipartisan bickering over budgets and appropriations that has plagued the federal government in recent years.

Detailed data on the finances of the federal government are maintained by the Office of Management and Budget (OMB), an executive office of the U.S. president. The OMB assists the president in preparing the federal budget and supervises budget administration. Budget information is available at http://www.whitehouse.gov/omb/budget/Overview. Budget documents include historical tables that provide annual data on federal government receipts, outlays, debt, and employment dating back to 1940 or earlier. The FY 2015 budget was proposed in February 2014. It includes final values for FY 2013 and estimates for FYs 2014 to 2019.

Federal Revenues

According to the OMB, in *Fiscal Year 2015 Historical Tables: Budget of the U.S. Government* (February 2014, http://www.whitehouse.gov/sites/default/files/omb/budget/fy2015/assets/hist.pdf), the federal government had revenues (receipts) of nearly $2.8 trillion in FY 2013. The two

FIGURE 9.1

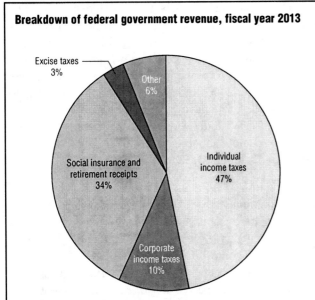

Breakdown of federal government revenue, fiscal year 2013

SOURCE: Adapted from "Table 2.1—Receipts by Source: 1934–2019," in *Fiscal Year 2015 Historical Tables, Budget of the U.S. Government*, Office of Management and Budget, 2014, http://www.whitehouse.gov/sites/default/files/omb/budget/fy2015/assets/hist.pdf (accessed August 27, 2014)

largest sources of revenue were individual income taxes (47%) and social insurance and retirement receipts (34%). (See Figure 9.1.)

TAXES ON INCOME. In the United States tax rates are approved by Congress and signed into law by the president; the Internal Revenue Service (IRS), a bureau of the U.S. Department of the Treasury, enforces the tax codes and collects tax payments, which are due each year on Tax Day, which is typically April 15.

The percentage of an individual's income that he or she pays in federal tax is based on his or her income level, which determines the individual's tax bracket. Tax brackets change as Congress modifies the tax codes, but individuals with higher incomes are always taxed at a higher rate than individuals with lower incomes. Through a variety of tax credits and deductions, individuals can lower the amount of income on which taxes are calculated, thereby lowering the amount of tax they pay.

As shown in Figure 9.1, corporate income taxes accounted for 10% of total federal government revenues in FY 2013. The federal tax on corporate income has been in effect since 1909. Because corporations are owned, and individuals derive income from them, the potential exists for "double taxation," that is, the same income is taxed twice, once as corporate income and, when the profits have been distributed to shareholders, again as individual income. To reduce the effects of double taxation, various credits and deductions have been enacted over the decades to allow income to pass through a corporation without being taxed until it reaches the individual. Credits and depreciation schedules reduce the amount of revenue that is subject to tax.

SOCIAL INSURANCE AND RETIREMENT RECEIPTS. Social insurance and retirement receipts are collected to fund specific programs for people who are retired, disabled, unemployed, or poor. The primary programs are Social Security and Medicare. Social Security provides funds to most workers who retire or become disabled. It also pays money to the survivors of workers who die. Medicare is a federal health insurance program for people aged 65 years and older and people with disabilities.

In "Update 2014" (January 2014, http://www.ssa.gov/pubs/EN-05-10003.pdf), the Social Security Administration explains that as of 2014 the federal government collected money to pay for these two programs as follows:

- Social Security—a tax of 12.4% on earned annual income up to $117,000. In other words, people who earn more than the annual limit pay the tax on $117,000, regardless of how much they earn.

- Medicare—a tax of 2.9% on all earned annual income. In addition, certain high-income individuals pay an additional amount called a surtax that was authorized by the Patient Protection and Affordable Care Act of 2012. (This law is commonly called the Affordable Care Act [ACA] or Obamacare.) The IRS notes in "Questions and Answers for the Additional Medicare Tax" (November 6, 2014, http://www.irs.gov/Businesses/Small-Businesses-&-Self-Employed/Questions-and-Answers-for-the-Additional-Medicare-Tax) that the 0.9% surtax went into effect in 2013. There are different income thresholds that trigger the surtax; for example, as of November 2014, it applied to married couples filing joint tax returns who made $250,000 or more in annual income.

Social Security and Medicare taxes are known as payroll taxes because they are assessed based on the amounts that businesses pay their workers. In 2014 employers paid 6.2% of the Social Security tax and 1.45% of the base Medicare tax for each employee. The other half was paid by wage earners via deductions from their paychecks. Likewise, high-income earners had the Medicare surtax deducted from their paychecks. Self-employed people had to pay their entire Social Security and Medicare tax bill but could deduct roughly half of the amount as a business expense when they filed their income tax returns.

EXCISE TAXES AND OTHER RECEIPTS. As shown in Figure 9.1, excise taxes accounted for only 3% of federal government revenues in FY 2013. Excise taxes are considered taxes on consumption. Although there is no federal sales tax, the federal government does levy excise taxes on items such as airplane tickets, gasoline, alcoholic beverages, firearms, and tobacco products.

The "other" category in Figure 9.1 accounted for 6% of federal government revenues in FY 2013. This category includes estate and gift taxes and customs duties. Estate and gift taxes are taxes on wealth. Estate taxes are levied against a person's estate after that person dies, whereas gift taxes are levied against the giver while the giver is alive. Estate and gift taxes only apply to amounts over specified limits. Customs duties are taxes charged on goods that are imported into the United States. The taxes vary by product and by exporting nation.

Federal Spending

According to the OMB, the federal government spent nearly $3.5 trillion in FY 2013. (See Table 9.1.) Overall, human resources were the largest expenditure for the federal government, accounting for $2.4 trillion in FY 2013. The largest single component of federal spending was for the Social Security program, which accounted for $813.6 billion of expenditures.

Jessie and Tarleton estimate that the federal government employed nearly 2.8 million civilian (nonmilitary) employees in 2012.

Federal Surpluses and Deficits

If the government spends less money than it takes in during a fiscal year, the difference is known as a budget surplus. Likewise, if spending is higher than revenues, the difference is called a budget deficit. A balanced budget occurs when spending and revenue are the same.

Figure 9.2 shows the annual budget surplus or deficit from 1901 to 2019 as reported in the OMB's FY 2015 budget (note that the data for 2014 to 2019 are estimated). In general, the federal government had a balanced budget for more than half of the 20th century, excluding slight deficits that occurred during World War I (1914–1918) and World War II (1939–1945). Beginning in 1970 the United States had an annual budget deficit for nearly three decades. The years 1998 to 2001 saw budget surpluses. In 2000 the surplus reached a record $236.2 billion. Budget deficits returned in 2002 and reached record lows over the following years. The budget deficit for 2009 was $1.5 trillion, which was a new record low. The OMB predicts that the 2014 deficit will be around $649 billion. Budget deficits of less than $1 trillion per year are expected through 2019.

National Debt

Whenever the federal government has a budget deficit, the Treasury Department must borrow money to cover the difference. The total amount of money that the Treasury Department has borrowed over the years is known as the federal debt, or more commonly the national debt. Budget surpluses cause the debt to go down, whereas deficits increase the debt.

The national debt was low until the early 1940s, when it jogged upward in response to government spending during World War II. (See Figure 9.3.) Over the next three decades the debt increased at a slow pace. During the late 1970s the national debt began a steep climb that continued into the second decade of the 21st century. The budget surpluses from 1998 to 2001 had a slight dampening effect on the growth of the debt but did not actually decrease the amount of debt. The budget deficits of the

TABLE 9.1

Breakdown of federal government spending, fiscal year 2013

Superfunction and function	2013
	In millions of dollars
National defense	633,385
Human resources	2,417,949
Education, training, employment, and social services	72,808
Health	358,315
Medicare	497,826
Income security	536,511
Social Security	813,551
Veterans benefits and services	138,938
Physical resources	89,997
Energy	11,042
Natural resources and environment	38,145
Commerce and housing credit	−83,199
Transportation	91,673
Community and regional development	32,336
Net interest	220,885
Other functions	185,174
International affairs	46,418
General science, space, and technology	28,908
Agriculture	29,492
Administration of justice	52,601
General government	27,755
Allowances	
Undistributed offsetting receipts	−92,785
Total, federal outlays	**3,454,605**
	As percentages of outlays
National defense	18.3
Human resources	70.0
Physical resources	2.6
Net interest	6.4
Other functions	5.4
Undistributed offsetting receipts	−2.7
Total, federal outlays	**100.0**
	As percentages of GDP
National defense	3.8
Human resources	14.5
Physical resources	0.5
Net interest	1.3
Other functions	1.1
Undistributed offsetting receipts	−0.6
Total, federal outlays	**20.8**

Note: GDP = gross domestic product.

SOURCE: Adapted from "Table 3.1—Outlays by Superfunction and Function: 1940–2019," in *Fiscal Year 2015 Historical Tables, Budget of the U.S. Government*, Office of Management and Budget, 2014, http://www.whitehouse.gov/sites/default/files/omb/budget/fy2015/assets/hist.pdf (accessed August 27, 2014)

FIGURE 9.2

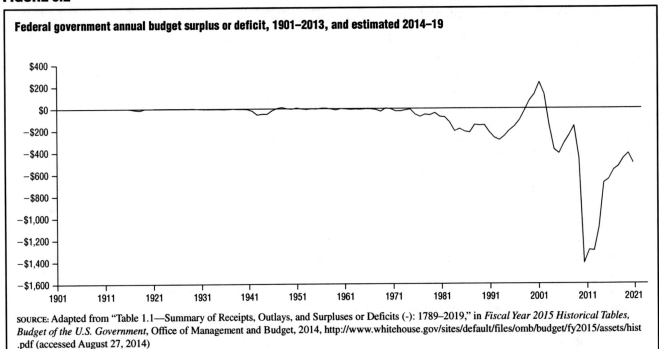

Federal government annual budget surplus or deficit, 1901–2013, and estimated 2014–19

SOURCE: Adapted from "Table 1.1—Summary of Receipts, Outlays, and Surpluses or Deficits (-): 1789–2019," in *Fiscal Year 2015 Historical Tables, Budget of the U.S. Government*, Office of Management and Budget, 2014, http://www.whitehouse.gov/sites/default/files/omb/budget/fy2015/assets/hist.pdf (accessed August 27, 2014)

FIGURE 9.3

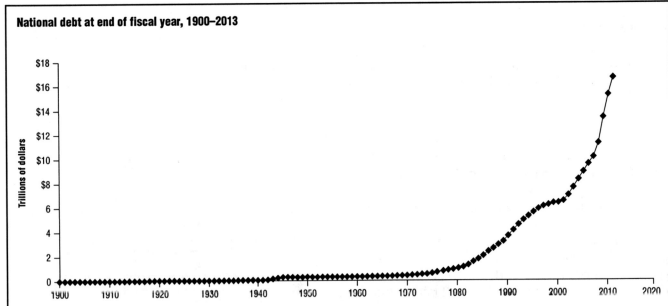

National debt at end of fiscal year, 1900–2013

SOURCE: Adapted from "Historical Debt Outstanding—Annual: 1900–1949," "Historical Debt Outstanding—Annual: 1950–1999," and "Historical Debt Outstanding—Annual: 2000–2010," in *Public Debt Reports*, U.S. Department of the Treasury, Bureau of the Public Debt, April 29, 2014, http://www.treasurydirect.gov/govt/reports/pd/histdebt/histdebt.htm (accessed August 28, 2014); and "Table I. Summary of Treasury Securities Outstanding, September 30, 2013," in *Monthly Statement of the Public Debt of the United States, September 30, 2013*, U.S. Treasury Department, September 30, 2013, http://www.treasurydirect.gov/govt/reports/pd/mspd/2013/opds092013.pdf (accessed August 28, 2014)

years that followed sent the debt into another rapid incline. By the end of FY 2013 the national debt stood at $16.7 trillion.

BORROWED MONEY AND IOUS. The national debt has two components: money that the federal government has borrowed from the public and money that the federal government has lent itself. The public lends money to the federal government by buying federal bonds and other securities. The government borrows the money with a promise to pay it back with interest after a set term. Some of the most common federal securities sold to the public are Treasury bills, Treasury notes, Treasury bonds, and savings bonds. These vary in value, interest paid, and set terms. Public investors include individuals and businesses (both domestic and foreign) and state and local governments.

The federal government also borrows from itself. This is debt owed by one Treasury account to another. Most of the so-called internal debt involves federal trust funds. For example, if a trust fund takes in more revenue in a year than is paid out, it lends the extra money to another federal account. In exchange, the lending trust fund receives an interest-bearing security (basically an IOU) that is redeemable in the future from the Treasury Department. The Congressional Budget Office (CBO) states in "Federal Debt and the Commitments of Federal Trust Funds" (May 6, 2003, http://www.cbo.gov/sites/default/files/cbofiles/ftpdocs/39xx/doc3948/10-25-longrangebrief4.pdf) that "what is in the trust funds is simply the government's promise to pay itself back at some time in the future."

As of December 5, 2014, the Treasury Department (http://www.treasurydirect.gov/NP/debt/current) reported that the national debt was nearly $18 trillion, broken down as follows:

- Owed to the public—$12.9 trillion (72% of the total)
- Intragovernmental—$5 trillion (28% of the total)

AS A PERCENTAGE OF THE GROSS DOMESTIC PRODUCT. Economists often discuss the national debt in terms of its percentage of the gross domestic product (GDP; the total market value of final goods and services that are produced within an economy in a given period) because a debt amount alone does not provide a complete picture of the effect of that debt on the one who owes it. According to the Central Intelligence Agency (CIA), in *World Factbook: United States* (June 20, 2014, https://www.cia.gov/library/publications/the-world-factbook/geos/us.html), the U.S. national debt owed to the public made up an estimated 71.8% of the nation's GDP in 2013.

The CIA ranks nations of the world in descending order of national debt owed to the public as a percentage of their GDP in 2013. The United States ranked 36th on this list. The nation with the largest percentage was Japan (226.1%).

THE BURDEN ON THE ECONOMY. The national debt represents a twofold burden on the U.S. economy. The debt owed to the public imposes a current burden. The federal government pays out interest to investors, and these interest payments are funded by current taxpayers. The debt that the federal government owes to itself is a future burden. At some point in the future the securities that were issued for intragovernmental debt must be redeemed for cash. The government will have to raise these funds by raising taxes, reducing spending, and/or borrowing more money from the public.

THE DEBT CEILING AND SEQUESTRATION. The debt ceiling is a limit imposed by Congress on the amount of the national debt. The U.S. Government Accountability Office (GAO) explains in *Debt Limit: Delays Create Debt Management Challenges and Increase Uncertainty in the Treasury Market* (February 2011, http://www.gao.gov/new.items/d11203.pdf) that a debt ceiling was first enacted by Congress during World War I. It has been raised numerous times since then. The GAO notes that "the debt limit does not control or limit the ability of the federal government to run deficits or incur obligations. Rather, it is a limit on the ability to pay obligations already incurred." In other words, the debt limit forbids the Treasury Department from borrowing funds to pay for already implemented government programs and activities. This includes paying interest due on bonds that have been sold.

In 2011 the process of raising the debt ceiling turned into a political tussle. Republican lawmakers demanded deficit reduction measures, including deep spending cuts from the Obama administration, as a condition for raising the debt ceiling. The battle raged for months and was "settled" in August 2011, just hours before the United States was going to hit the debt ceiling. The Budget Control Act of 2011 (BCA) allows the debt ceiling to be raised, but calls for trillions of dollars in spending cuts over coming years.

In *Approaches for Scaling Back the Defense Department's Budget Plans* (March 2013, http://www.cbo.gov/sites/default/files/cbofiles/attachments/43997_Defense_Budget.pdf), the CBO states that "the BCA initially created a set of caps that limited funding for discretionary programs and activities for each year over the 2012–2021 period. That act also established procedures that led to automatic spending reductions, which lowered those initial caps for the years 2014 to 2021 and cut funding for 2013 through a process known as sequestration."

Richard Kogan of the Center on Budget and Policy Priorities notes in "How the Across-the-Board Cuts in the Budget Control Act Will Work" (April 27, 2012, http://www.cbpp.org/cms/?fa=view&id=3635) that the law reduces funding by more than $1 trillion through 2021 for some government programs. It also calls for the establishment of a special committee to propose legislation to reduce deficits by another $1.2 trillion over the same period. In November 2011 the so-called supercommittee announced that it had failed to reach an agreement. Thus, the sequestration spending cuts went into effect in January 2013. Congress passed the Bipartisan Budget Act of 2013, which raised the spending caps for FYs 2014 and 2015, but extended sequestration through FY 2023.

GOVERNMENT SPENDING AND THE GDP

As noted in Chapter 2, one major component of the nation's GDP is government consumption expenditures and gross investment, which together represent the amount spent by the government (local, state, and federal) on final goods and services. The total does not include transfer payments made by the government to the public (such as Social Security and unemployment compensation) because these expenditures do not purchase goods or services. Table 9.2 shows government consumption expenditures and gross investment between 2010 and 2013 and annualized for 2014 based on quarterly data. The values are "real," meaning they have been adjusted for inflation. Overall, government agencies spent nearly $2.9 trillion on final goods and services in 2013. State and local governments accounted for $1.7 trillion, or 60% of the total. The federal government accounted for the remaining $1.1 trillion. Much of this money ($717.7 billion) was devoted to national defense.

As shown in Table 9.3, government consumption expenditures and gross investment declined 2% in 2013, echoing a declining spending trend that began in 2011. By contrast, during the late 1990s and the first decade of the 21st century the expenditures and investment totals increased each year primarily due to huge growth in spending on national defense.

PUBLIC INVESTMENT AND TAXES

To fund itself, the federal government uses the money of its constituents. The buying of federal securities, such as bonds, represents a voluntary investment in the government by the public. Federal securities are considered to be a safe low-risk investment because they are backed by an entity that has been in existence for more than 200 years and has a proven track record of fiscal soundness. However, money invested in government securities is not available for private investment. In general, private investments are seen as more stimulating for the economy because they provide direct funds for growth, such as the building of new factories and the hiring of new workers. Public (government) investment may or may not have a stimulating effect on the economy, depending on how the funds are spent.

Taxes represent an involuntary investment by the public in government. The effect of taxes on the economy is a source of never-ending debate in U.S. politics. Taxing personal income decreases the spending power of the public because people have less money to invest in private enterprise or to use to consume goods and services. Limited taxation is favored by those who believe that workers and companies with more available money to spend will participate to a greater extent in the economy. This, they say, will lead to economic growth. Others observe that cutting taxes without severely reducing government spending leads to large budget deficits and undermines the government programs that provide a social safety net to the disadvantaged.

Tax Breaks

Historically, the federal government has allowed tax breaks (such as deductions) on income or expenses that are related to specific activities. The prime example is mortgage interest. Most taxpayers can deduct the interest they pay each year on their mortgages. The government uses tax breaks to encourage activities it considers good for society in general, such as being a homeowner or donating to charities. However, tax breaks represent substantial

TABLE 9.2

Government consumption expenditures and gross investment, 2010–13 and first and second quarters 2014

	2010	2011	2012	2013	Seasonally adjusted at annual rates 2014	
					I	II
Government consumption expenditures and gross investment	3,091.4	2,997.4	2,953.9	2,894.5	2,868.5	2,880.0
Federal	1,270.7	1,236.4	1,214.4	1,145.3	1,117.4	1,115.3
National defense	813.5	795.0	768.7	717.7	693.9	695.7
Consumption expenditures	636.0	626.2	608.6	571.8	562.3	559.8
Gross investment	177.5	168.7	160.0	145.7	131.1	135.6
Nondefense	457.1	441.4	445.7	427.5	423.4	419.4
Consumption expenditures	339.2	323.5	330.3	318.2	317.7	315.0
Gross investment	117.9	118.0	115.3	109.3	105.5	104.3
State and local	1,820.8	1,761.0	1,739.5	1,748.4	1,750.2	1,763.7
Consumption expenditures	1,469.7	1,430.4	1,427.0	1,444.7	1,454.1	1,458.8
Gross investment	351.0	330.3	311.8	302.8	295.2	304.0

Note: Users are cautioned that particularly for components that exhibit rapid change in prices relative to other prices in the economy, the chained-dollar estimates should not be used to measure the component's relative importance or its contribution to the growth rate of more aggregate series. Values are in billions of chained (2009) dollars. Quarterly data are seasonally adjusted at annual rates.

SOURCE: Adapted from "Table 3B. Real Gross Domestic Product and Related Measures," in *National Income and Product Accounts—Gross Domestic Product: Second Quarter 2014 (Advance Estimate) Annual Revision: 1999 through First Quarter 2014*, U.S. Department of Commerce, Bureau of Economic Analysis, July 30, 2014, http://www.bea.gov/newsreleases/national/gdp/2014/pdf/gdp2q14_adv.pdf (accessed August 7, 2014)

TABLE 9.3

Percentage change in government consumption expenditures and gross investment, 1994–2013

	1994	1995	1996	1997	1998	1999	2000	2001	2002	2003	2004	2005	2006	2007	2008	2009	2010	2011	2012	2013
Government consumption expenditures and gross investment	0.1	0.5	1.0	1.9	2.1	3.4	1.9	3.8	4.4	2.2	1.6	0.6	1.5	1.6	2.8	3.2	0.1	−3.0	−1.4	−2.0
Federal	−3.5	−2.6	−1.2	−0.8	−0.9	2.0	0.3	3.9	7.2	6.8	4.5	1.7	2.5	1.7	6.8	5.7	4.4	−2.7	−1.8	−5.7
National defense	−4.9	−4.0	−1.6	−2.7	−2.1	1.5	−0.9	3.5	7.0	8.5	6.0	2.0	2.0	2.5	7.5	5.4	3.2	−2.3	−3.3	−6.6
Nondefense	−0.8	0.0	−0.5	2.8	1.3	2.7	2.3	4.7	7.4	4.1	2.0	1.3	3.5	0.3	5.5	6.2	6.4	−3.4	1.0	−4.1
State and local	2.8	2.7	2.4	3.6	3.8	4.2	2.8	3.7	2.9	−0.4	−0.1	0.0	0.9	1.5	0.3	1.6	−2.7	−3.3	−1.2	0.5

SOURCE: Adapted from "Table 7. Real Gross Domestic Product: Percent Change from Preceding Year," in *National Income and Product Accounts—Gross Domestic Product: Second Quarter 2014 (Advance Estimate) Annual Revision: 1999 through First Quarter 2014*, U.S. Department of Commerce, Bureau of Economic Analysis, July 30, 2014, http://www.bea.gov/newsreleases/national/gdp/2014/pdf/gdp2q14_adv.pdf (accessed August 7, 2014)

amounts of lost revenue to the federal government, perhaps as much as $1 trillion per year. In fact, many individuals and businesses—nearly half by one estimate—end up paying no federal income tax each year. Some analysts call for eliminating or reducing tax breaks so the federal government can collect more revenue to fund programs or reduce the federal deficit. Nevertheless, tax deductions are extremely popular with taxpayers, especially those who are able or potentially able to take advantage of them. Thus, eliminating or curtailing tax breaks would be politically unpopular.

Tax Cuts

In general, the Republican Party advocates smaller government and lower taxes. Although its preference for lower taxes includes Americans at all income levels, the party is especially concerned about Americans at the highest income levels. The argument is that high-income earners create jobs and invest money, actions that help all Americans and improve the overall economy. By contrast, the Democratic Party typically favors higher taxes on high-income earners.

The Economic Growth and Tax Relief Reconciliation Act of 2001 and the Jobs and Growth Tax Relief Reconciliation Act of 2003 were initiated by the first administration of President George W. Bush (1946–), a Republican, and are collectively referred to as the "Bush tax cuts." The laws instituted a series of tax rate reductions and incentive measures that were phased in over several years. The laws were designed to expire on January 1, 2011, at which time the higher tax rates in effect before the laws were implemented would return.

When President Obama took office in January 2009, the economy was in the midst of the Great Recession (which lasted from December 2007 to June 2009). Republican lawmakers strongly favored extending the Bush tax cuts, arguing that government should not increase taxes during the economic slowdown. Obama and many fellow Democratic lawmakers favored extending the tax cuts, but only for wage earners with an annual income of less than $200,000 for individuals or $250,000 for married couples. This idea received stiff resistance from Republicans. In a compromise, the Obama administration supported temporary extensions of the tax cuts through the Tax Relief, Unemployment Insurance Reauthorization, and Job Creation Act of 2010. It extended the Bush tax cuts through year-end 2012. In January 2013 the American Taxpayer Relief Act of 2012 was passed by Congress and signed by Obama. It permanently extended the tax cuts only for individuals making less than $400,000 to $450,000 per year, depending on their tax filing status.

THE FUTURE OF SOCIAL SECURITY AND MEDICARE

Social Security and Medicare are two of the most expensive programs that are operated by the federal government. (See Table 9.1.) Together, they accounted for more than $1.3 trillion of spending in FY 2013, or 38% of total expenditures.

As of 2014, people born in 1929 or later qualify for retirement benefits once they have worked for 10 years. Benefit amounts are based on wage history; thus, higher-paid workers will have higher retirement benefits than lower-paid workers. The Social Security Administration explains in "Retirement Planner: Benefits by Year of Birth" (2014, http://www.socialsecurity.gov/retire2/agereduction.htm) the age at which full benefits can be paid:

- People born in or before 1937—aged 65
- People born between 1938 and 1959—sliding age scale ranging from 65 years and 2 months to 66 years and 10 months
- People born in 1960 or after—aged 67

People who have worked for at least 10 years are eligible for permanently reduced retirement benefits starting at age 62. Benefits for widows, widowers, and family members have varying age requirements and other conditions that must be met. Medicare coverage begins at age 65 for everyone except for certain disabled people who can qualify earlier.

Since their inception, the Social Security and Medicare programs have been a source of partisan contention and debate. Much of the debate has centered on how the programs should be funded and the role of government in social welfare. During the 1990s concerns began to arise about how the nation can afford these programs in the future as the population ages and as the number of earners contributing to the plans decreases. A huge increase in the aged population is expected to take place because of the baby boom that followed World War II. However, as these workers retire, there will be fewer workers contributing to Social Security because succeeding generations have been smaller due to declining birth rates. At the same time, life expectancies have been increasing, meaning that elderly people are living longer past retirement age and collecting benefits for more years.

Funding Social Security

In *Status of the Social Security and Medicare Programs: A Summary of the 2014 Annual Reports* (July 2014, http://www.ssa.gov/oact/TRSUM/tr14summary.pdf), an annual report on the status of the Social Security trust funds—Old-Age and Survivors Insurance (OASI), Disability Insurance, and Health Insurance—the Social Security Board of Trustees shows historical trust fund ratios and estimates future ratios. (See Figure 9.4.) The trustees note that assets as a percentage of annual expenditures peaked for the trust funds during the first decade of the 21st century. The OASI trust fund peaked above 400% in 2011. Table 9.4 shows the first

FIGURE 9.4

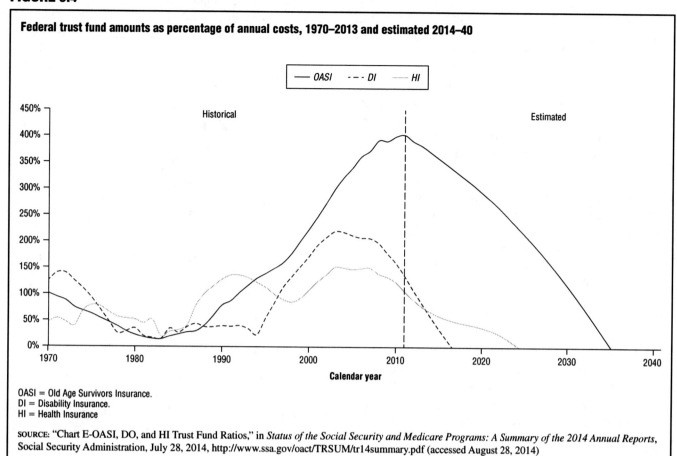

Federal trust fund amounts as percentage of annual costs, 1970–2013 and estimated 2014–40

OASI = Old Age Survivors Insurance.
DI = Disability Insurance.
HI = Health Insurance.

SOURCE: "Chart E-OASI, DO, and HI Trust Fund Ratios," in *Status of the Social Security and Medicare Programs: A Summary of the 2014 Annual Reports*, Social Security Administration, July 28, 2014, http://www.ssa.gov/oact/TRSUM/tr14summary.pdf (accessed August 28, 2014)

TABLE 9.4

Key dates for the Social Security and Medicare trust funds as of July 2014

	OASI	DI	OASDI[a]	HI
Year of peak trust fund ratio[b]	2011	2003	2008	2003
First year cost exceeds income excluding interest[c]	2010	2005	2010	2021
First year cost exceeds income including interest[c]	2022	2009	2020	2023
Year trust fund reserves are depleted	2034	2016	2033	2030

[a]Column entries represent key dates for the theoretical combined OASI and DI funds.
[b]Dates pertain to the post-2000 period.
[c]Dates indicate the first year that a condition is projected to occur and to persist annually thereafter through 2088.
OASI = Old Age Survivors Insurance.
DI = Disability Insurance.
OASDI = Old Age Survivors Insurance and Disability Insurance.
HI = Health Insurance.

SOURCE: "Key Dates for the Trust Funds," in *Status of the Social Security and Medicare Programs: A Summary of the 2014 Annual Reports*, Social Security Administration, July 28, 2014, http://www.ssa.gov/oact/TRSUM/tr14summary.pdf (accessed August 28, 2014)

years in which the outgoing amounts from the trust funds will exceed the income (including interest) into the trust funds. It also shows the years in which the trust funds will be exhausted. All the trust funds face exhaustion between 2016 and 2034. If that should happen, the programs will then rely solely on incoming, rather than saved, funds. The trustees estimate that payroll tax income would cover around 77% of the benefits scheduled to be paid out in 2033.

FIXING THE PROBLEM. The issue of preparing for future shortfalls in Social Security has become a fierce debate. Republicans generally favor allowing some privatization of the Social Security system, meaning that workers would be able to partially opt out of the Social Security plan and establish their own retirement savings accounts. However, President Obama is against privatization because he believes that such a plan would tie benefits to the stability of the stock market, which has a history of experiencing major fluctuations. Other possible options include increasing the retirement age and/or decreasing benefits, particularly for wealthier retirees. However, neither of these options would be politically popular with the American public.

FEDERAL GOVERNMENT MANIPULATION OF MACROECONOMICS

The federal government plays a role in the national economy as a tax collector, spender, and employer. Federal policy makers also engage in purposeful manipulation of the U.S. economy at the macroeconomic level—for example, by influencing supply and demand factors. This was not always the case. Before the 1930s the government mostly maintained a hands-off approach to macroeconomic affairs—a tradition that dated back to the founding of the nation. However, the ravages of the Great Depression (1929–1939) brought a level of desperation that encouraged leaders to attempt to influence macroeconomic factors. Although these efforts were largely futile at soothing deep economic depression, they accustomed Americans to the idea of government interference in economic affairs.

When massive federal spending during World War II helped end the Great Depression, policy makers believed they had discovered a new solution, a government solution, for economic downturns. Government efforts to manage macroeconomic factors became a routine matter over the following decades. These manipulations are commonly divided into two categories: fiscal policy and monetary policy.

Fiscal Policy

The word *fiscal* is derived from the Latin term *fiscalis*, meaning "treasury." It is believed that a fiscalis was originally a woven basket in which money was kept. In modern English, the word *fiscal* has become synonymous with the word *financial*. The federal government's fiscal policy is concerned with the collection and spending of public money so as to influence macroeconomic affairs. Examples of fiscal policy include:

- Increasing government spending to spur businesses to produce more and hire more; this can lower the unemployment rate
- Increasing taxes to pull money out of the hands of consumers; this can lower excessive demand that is driving high inflation rates
- Decreasing taxes to put more money in the hands of consumers; this can increase demand and consequently increase supply (production)

These examples illustrate optimistic outcomes. In reality, the actions of fiscal policy can have complicated and unforeseen effects on the U.S. economy. The situation described in the first example can backfire if production does not grow fast enough to satisfy consumer demand. The result will be rising prices and high inflation rates. Likewise, tax increases and decreases can have unexpected and undesirable consequences. The relationships between the major macroeconomic factors—unemployment, inflation, and supply and demand—are complex and difficult to keep in balance.

Monetary Policy

Monetary policy is concerned with influencing the supply of money and credit and the demand for them to achieve specific economic goals. The actions of monetary policy are not as direct and obvious as the tax and spend activities that are associated with fiscal policy. Monetary changes are achieved indirectly through the nation's banking system. The following are some of the results of monetary policy changes:

- Increasing the amount of money that banks lend to the public—this leads to greater borrowing, which puts more money into the hands of consumers, increasing the demand for goods and services
- Decreasing the amount of money that banks lend to the public—this leads to less borrowing, which slows the growth of the money supply and dampens demand, which can reduce high inflation rates
- Lowering interest rates on loans—this encourages borrowing, which increases the money supply and consumer demand
- Having higher interest rates on loans—this discourages people from borrowing more money, which slows the growth of the money supply and can reduce high inflation rates

Just as in fiscal policy, it is difficult to achieve the desired results. An oversupply of money and credit will aggravate price inflation if production cannot meet increased consumer demand. Likewise, an undersupply can lower consumer demand too much and stifle economic growth. The challenge for the federal government is deciding when, and by how much, money supply and credit availability should be changed to maintain a healthy economy. These decisions and manipulations are made by the Federal Reserve System, the nation's central bank.

THE FEDERAL RESERVE SYSTEM. In 1913 Congress passed the Federal Reserve Act to form the nation's central bank. The Federal Reserve System was granted power to manipulate the money supply—the total amount of coins and paper currency in circulation, along with all holdings at banks, credit unions, and other financial institutions.

The Federal Reserve includes a seven-member board of governors headquartered in Washington, D.C., and 12 Reserve Banks located in major cities around the country:

- Boston, Massachusetts
- New York City, New York
- Philadelphia, Pennsylvania
- Cleveland, Ohio
- Richmond, Virginia
- Atlanta, Georgia
- Chicago, Illinois

- St. Louis, Missouri
- Minneapolis, Minnesota
- Kansas City, Missouri
- Dallas, Texas
- San Francisco, California

In *The Federal Reserve System: Purposes and Functions* (June 2005, http://www.federalreserve.gov/pf/pdf/pf_complete.pdf), the Federal Reserve explains that it uses three techniques to indirectly achieve "maximum employment, stable prices, and moderate long-term interest rates":

- Open market operations—the Federal Reserve buys and sells government securities on the financial markets. The resulting money transfers ultimately lower or raise the amount of money that banks have available to lend to the public and the associated interest rates.

- Discount rate adjustments—the Federal Reserve raises or lowers the discount rate. This is the rate that it charges banks for short-term loans. In response, the banks adjust the federal funds rate, the rate they charge each other for loans. Then the banks adjust the prime rate, the interest rate they charge their best customers (typically large corporations). In the end, these adjustments affect the interest rates that are paid by the general public on mortgages, car loans, credit cards, and so on.

- Reserve requirement adjustments—the Federal Reserve raises or lowers the reserve requirement, the amount of readily available money that banks must have to operate. Each bank's reserve requirement is based on a percentage of the total amount of money that customers have deposited at that bank. Money above the reserve requirement can be lent by the banks. Changes in the reserve requirement influence bank decisions about loans to the public.

During and after the Great Recession the Federal Reserve relied heavily on open market operations and discount rate adjustments to boost the lagging economy. Chapter 4 describes the very low interest rates for loans, including mortgages, that resulted from the national bank's manipulation of the discount rate. The Federal Reserve also created "new" money that it essentially transferred to banks by buying investment assets, such as bonds, from them. Terry Burnham indicates in "So You Thought Quantitative Easing Was Over? Think Again" (PBS.org, November 24, 2014) that as of November 2014 the Federal Reserve had purchased $4.5 trillion worth of bonds and securities through what it calls "quantitative easing." Burnham notes that the central bank's actions "created more new money in the past six years than in the entire history of the Republic."

ECONOMIC FORECASTING

The vast complexities of the macroeconomy make it difficult for forecasters to make predictions about future business cycles. Nonetheless, government and private sources do conduct economic forecasting, primarily with regards to individual economic indicators, such as the GDP or unemployment rates.

The CBO explains in "Overview" (2014, http://www.cbo.gov/about/overview) that since 1975 it has "produced independent analyses of budgetary and economic issues to support the Congressional budget process." The CBO publishes annual budget and economic outlooks that are updated as needed to reflect changing economic conditions. As of December 2014, the most recent update, *An Update to the Budget and Economic Outlook: 2014 to 2024* (http://cbo.gov/sites/default/files/cbofiles/attachments/45653-OutlookUpdate_2014_Aug.pdf), was published in August 2014.

As shown in Figure 9.5, the CBO expects the nation's GDP to grow at nearly 4% per year in 2015 and 2016. Thereafter, the GDP is projected to increase by around 2% annually through 2024. Unemployment rate projections are depicted in Figure 9.6. The unemployment rate is expected to remain relatively flat at just below 6% through 2024. Both Figure 9.5 and Figure 9.6

FIGURE 9.5

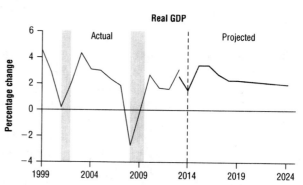

Percentage annual change in real gross domestic product, 1999–2013 and predicted for 2014–24

Notes: Real gross domestic product (GDP) is the output of the economy adjusted to remove the effects of inflation. The unemployment rate is a measure of the number of jobless people who are available for work and are actively seeking jobs, expressed as a percentage of the labor force. The overall inflation rate is based on the price index for personal consumption expenditures; the core rate excludes prices for food and energy. Data are annual. For real GDP and inflation, values from 1999 through 2013 (the thin lines) reflect revisions to the national income and product accounts that the Bureau of Economic Analysis made on July 30, 2014. Values from 2013 through 2024 (the thick lines) reflect the data available and projections made before July 30. Percentage changes are measured from the fourth quarter of one calendar year to the fourth quarter of the next year. Shaded bars indicate U.S. recessions.

SOURCE: "Summary Figure 2. Actual Values and CBO's Projections of Key Economic Indicators: Real GDP," in *An Update to the Budget and Economic Outlook: 2014 to 2024*, Congressional Budget Office, August 2014, http://cbo.gov/sites/default/files/cbofiles/attachments/45653-OutlookUpdate_2014_Aug.pdf (accessed August 28, 2014)

FIGURE 9.6

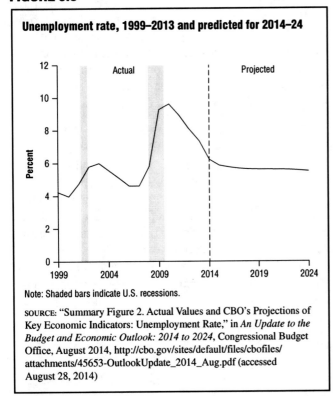

Unemployment rate, 1999–2013 and predicted for 2014–24

Note: Shaded bars indicate U.S. recessions.

SOURCE: "Summary Figure 2. Actual Values and CBO's Projections of Key Economic Indicators: Unemployment Rate," in *An Update to the Budget and Economic Outlook: 2014 to 2024*, Congressional Budget Office, August 2014, http://cbo.gov/sites/default/files/cbofiles/attachments/45653-OutlookUpdate_2014_Aug.pdf (accessed August 28, 2014)

FIGURE 9.7

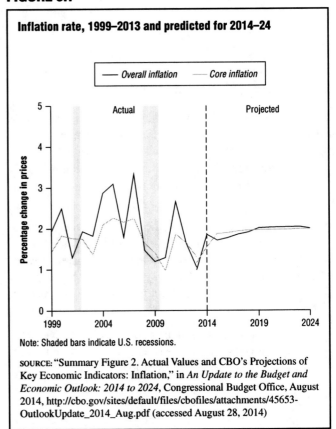

Inflation rate, 1999–2013 and predicted for 2014–24

Note: Shaded bars indicate U.S. recessions.

SOURCE: "Summary Figure 2. Actual Values and CBO's Projections of Key Economic Indicators: Inflation," in *An Update to the Budget and Economic Outlook: 2014 to 2024*, Congressional Budget Office, August 2014, http://cbo.gov/sites/default/files/cbofiles/attachments/45653-OutlookUpdate_2014_Aug.pdf (accessed August 28, 2014)

illustrate greatly improved conditions compared with the low GDP growth and high unemployment rate associated with the Great Recession. The CBO projects that the nation's inflation rate will rise only marginally through 2024. (See Figure 9.7.) Prices are expected to rise at around 2% per year. As explained in Chapter 3, inflation rates of up to 3% per year are considered appropriate for a healthy growing economy.

Figure 9.8 shows CBO projections for federal government outlays on major budget categories through 2024. The outlays are expressed as a percentage of the nation's expected GDP. For example, net interest payments—which are mostly due to the accumulated national debt—are expected to increase from just over 1% of GDP in 2013 to more than 3% of GDP in 2024. Social Security outlays are projected to rise from around 4.9% of GDP to around 5.6% of GDP. Spending on major health care programs is also expected to grow from approximately 4.6% of GDP in 2013 to approximately 5.9% of GDP in 2024. These programs include Medicare, Medicaid (a state and federal health insurance program for low-income people), the Children's Health Insurance Program (CHIP), and subsidies paid to qualified people who purchase medical insurance through the new health insurance exchanges authorized through the ACA. As described in Chapter 8, as of 2014 roughly half of the states had expanded their Medicaid programs in accordance with an optional provision of the ACA. Figure 9.9 indicates the CBO expects enrollment in Medicaid and CHIP to increase by 13 million people through 2024. In addition, 25 million people are projected to purchase health insurance through the exchanges; many of them will qualify for the federal subsidies. According to the CBO, in *An Update to the Budget and Economic Outlook: 2014 to 2024*, the subsidies were expected to total around $17 billion in 2014.

FIGURE 9.8

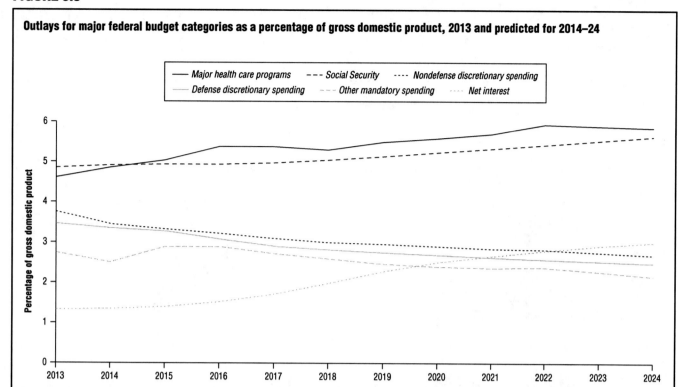

Outlays for major federal budget categories as a percentage of gross domestic product, 2013 and predicted for 2014–24

Notes: The category "Major health care programs" consists of Medicare, Medicaid, the Children's Health Insurance Program, and subsidies for health insurance purchased through exchanges and related spending. (Medicare spending is net of premiums paid by beneficiaries and other offsetting receipts.) Spending for Social Security and the federal government's major health care programs is projected to rise from 9.5 percent of GDP in 2013 to 11.5 percent in 2024. "Other mandatory spending" is all mandatory spending other than that for major health care programs, Social Security, and net interest.

SOURCE: "Figure 1-4. Projected Outlays for Major Budget Categories," in *An Update to the Budget and Economic Outlook: 2014 to 2024*, Congressional Budget Office, August 2014, http://cbo.gov/sites/default/files/cbofiles/attachments/45653-OutlookUpdate_2014_Aug.pdf (accessed August 28, 2014)

FIGURE 9.9

Predicted health insurance coverage of the nonelderly population in 2024 with and without the Affordable Care Act in effect

[Millions of nonelderly people]

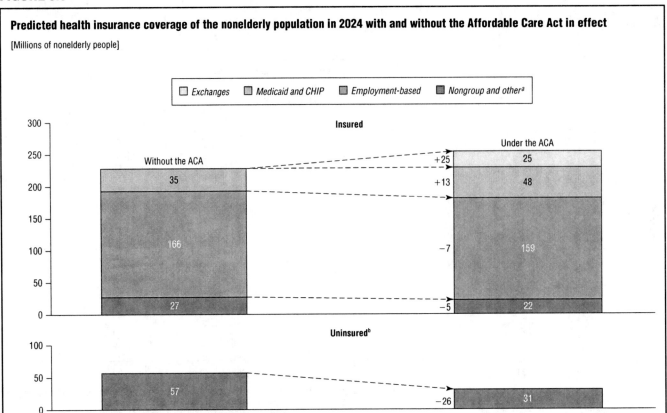

ACA = Affordable Care Act; CHIP = Children's Health Insurance Program.
[a]"Other" includes Medicare; the changes under the ACA are almost entirely for nongroup coverage.
[b]The uninsured population includes people who will be unauthorized immigrants and thus ineligible either for exchange subsidies or for most Medicaid benefits; people who will be ineligible for Medicaid because they live in a state that has chosen not to expand coverage; people who will be eligible for Medicaid but will choose not to enroll; and people who will not purchase insurance to which they have access through an employer, an exchange, or directly from an insurer.
Notes: The nonelderly population consists of residents of the 50 states and the District of Columbia who are younger than 65.

SOURCE: "Figure 1. Effects of the Affordable Care Act on Health Insurance Coverage, 2024," in *Updated Estimates of the Effects of the Insurance Coverage Provisions of the Affordable Care Act, April 2014*, Congressional Budget Office, April 2014, https://www.cbo.gov/sites/default/files/cbofiles/attachments/45231-ACA_Estimates.pdf (accessed August 14, 2014)

CHAPTER 10
INTERNATIONAL TRADE AND THE UNITED STATES' PLACE IN THE GLOBAL ECONOMY

Those who have money go abroad in the world.

—Chinese proverb

Technology has made it easier to go abroad in the world. U.S. companies can sell their goods and services on a global market. Likewise, U.S. consumers can purchase merchandise made around the world—and they do so in large numbers. Global trade is driven by the same forces that control the U.S. market: supply and demand. However, there is the added complication of many very different national governments trying to exert influence over trade and market factors in their favor. The U.S. economy is very large based on national production. However, the United States buys far more from other countries than it sells to them. Economists disagree about whether this trade imbalance is good or bad for the U.S. economy.

CURRENCY COMPARISONS

One of the difficulties inherent in comparing national economies around the world is that there are different currencies in use. Examples include the U.S. dollar, the euro, the Canadian dollar, the British pound, the peso (Mexico), the yen (Japan), and the yuan (China). According to the European Central Bank (https://www.ecb.eur opa.eu/euro/intro/html/map.en.html), as of 2014, the euro was the official currency of 18 European countries, including France and Germany.

An exchange rate is the amount of one currency that can be exchanged for a like value in another currency. A U.S. dollar can be exchanged for equivalent amounts of any other foreign currency. The exchange rate for any given foreign currency at any given time depends on many complex economic factors, and exchange rates can vary widely over time.

Exchange rates are important to tourists. A U.S. visitor to France would exchange his or her U.S. dollars for euros, the currency in local use in France. Exchange rates are also key to international trade because currency conversions are typically required when goods and services are traded between countries with different currencies. Like all commodities, the value of a particular currency is subject to supply and demand factors. Most countries exert some influence over their exchange rates. Some national governments interfere little in the market forces that influence the value of their currency internationally. This is particularly true for Western developed nations such as the United States and Canada. They are said to have floating or fluctuating exchange rates. Other countries such as China, Saudi Arabia, and Venezuela wield more control over how their currencies are valued internationally. Their exchange rates may be described as fixed or pegged because the rates are less flexible than those of other nations.

China has a relatively fixed exchange rate in comparison to other nations. Critics claim that China has long depressed the value of the yuan to keep the country's labor costs low and to lower the prices of Chinese goods in the international marketplace. This makes it more difficult for other countries to compete against China in international trade. In "China Allows Wider Currency Fluctuation" (WSJ.com, March 15, 2014), Lingling Wei indicates that in 2014 the Chinese government took actions to allow the yuan to float a little more freely. Wei notes that the move "is seen as an important step toward establishing a market-based exchange-rate system, whereby the yuan would move up and down just like any other major currency."

ECONOMY SIZE COMPARISONS

International economies are often compared in terms of gross domestic product (GDP; the total market value of final goods and services that are produced within an economy in a given period). Chapter 2 describes in detail the derivation of U.S. GDP values. Comparing the GDPs of different nations is complicated by various factors including currency differences. One method for comparison is based on market exchange rates. For example, a Chinese GDP expressed in yuan would simply be converted to U.S. dollars using the prevailing exchange rate for the two currencies. However, this method does not take into account important differences (such as prices) between economies. An alternative method is based on purchasing power parity (PPP). The World Bank explains in "Price Level Ratio of PPP Conversion Factor (GDP) to Market Exchange Rate" (2014, http://data.worldbank.org/indicator/PA.NUS.PPPC.RF) that a PPP conversion factor is "the number of units of a country's currency required to buy the same amount of goods and services in the domestic market as a U.S. dollar would buy in the United States." Each method has its supporters and opponents, and neither method completely equalizes the economic differences between nations.

The Central Intelligence Agency uses the PPP method to rank national GDP values in *The World Factbook* (2014, https://www.cia.gov/library/publications/the-world-factbook). In 2013 the United States ($16.7 trillion) had the largest GDP of any single nation, followed by China ($13.4 trillion), India ($5 trillion), Japan ($4.7 trillion), and Germany ($3.2 trillion). (See Table 10.1.)

TABLE 10.1

Gross domestic product (purchasing power parity) for the world's 25 largest economies, 2013

Rank	Economy	GDP in trillions of dollars
1	United States	16.72
2	European Union	15.85
3	China	13.39
4	India	4.99
5	Japan	4.73
6	Germany	3.23
7	Russia	2.55
8	Brazil	2.42
9	United Kingdom	2.39
10	France	2.28
11	Mexico	1.85
12	Italy	1.81
13	Korea, South	1.67
14	Canada	1.52
15	Spain	1.39
16	Indonesia	1.29
17	Turkey	1.17
18	Australia	1.00
19	Iran	0.99
20	Saudi Arabia	0.93
21	Taiwan	0.93
22	Poland	0.81
23	Argentina	0.77
24	Netherlands	0.70
25	Thailand	0.67

GDP = gross domestic product.

SOURCE: Adapted from "Country Comparison: GDP (Purchasing Power Parity)," in *The World Factbook*, Central Intelligence Agency, 2014, https://www.cia.gov/library/publications/the-world-factbook/rankorder/2001rank.html?countryname=China&countrycode=ch&regionCode=eas&rank=3#ch (accessed August 29, 2014)

The combined nations of the European Union (EU) had a GDP of $15.9 trillion, putting the EU in a position just below the United States in terms of economic strength.

The World Bank also uses the PPP method to compare global GDPs, but it reports results in "international dollars." In "Gross Domestic Product 2013, PPP" (September 22, 2014, http://databank.worldbank.org/data/download/GDP_PPP.pdf), the World Bank ranks the United States ($16.8 trillion) first, followed by China ($16.2 trillion), India ($6.8 trillion), Japan ($4.6 trillion), and Germany ($3.5 trillion). By this reckoning, the Chinese economy was a very close second to the U.S. economy in 2013 in terms of GDP. China is a developing nation and its economy has been growing at a fast rate. As such, its GDP is widely expected to surpass that of the United States in coming years. The article "China Set to Overtake U.S. as Biggest Economy in PPP Measure" (Bloomberg.com, April 30, 2014) provides estimates from various analysts that China will take the top position sometime between 2014 and 2024 regardless of whether the market exchange rate or the PPP method is used.

GLOBAL AND U.S. TRADE

The World Trade Organization (WTO) indicates in *International Trade Statistics 2014* (2014, http://www.wto.org/english/res_e/statis_e/its2014_e/its2014_e.pdf) that world trade totaled $22.9 trillion in 2013. The value of merchandise trade was $18.3 trillion and trade in commercial services was $4.6 trillion.

According to the WTO, China was the world's leading exporter of merchandise in 2013 with $2.2 trillion in value, or 11.7% of the total. The United States ($1.6 trillion) and Germany ($1.5 trillion) ranked second and third, respectively. The United States was the leading importer of merchandise in 2013 with $2.3 trillion in value or 12.3% of the total. It was followed by China ($2 trillion) and Germany ($1.2 trillion).

The WTO notes that the United States was the top trader in commercial services with $662 billion in exports (or 14.3% of the total) and $432 billion in imports (or 9.8% of the total) in 2013. Other top exporters included the United Kingdom ($293 billion) and Germany ($286 billion). China ($329 billion) and Germany ($317 billion) ranked second and third, respectively, in terms of commercial service imports in 2013.

U.S. Trade in Goods and Services

Table 10.2 lists the values of U.S. exports and imports of goods and services as measured by the U.S. Department of Commerce's Bureau of Economic Analysis (BEA) between 2010 and 2013 and annualized for 2014 based on quarterly data. The values are in real (inflation-adjusted) dollars. In 2013 the United States exported $2 trillion in goods and services. The vast

TABLE 10.2

Breakdown of net exports of goods and services, 2010–13 and first and second quarters 2014

	2010	2011	2012	2013	Seasonally adjusted at annual rates 2014	
					I	II
Net exports of goods and services	−458.8	−459.4	−452.5	−420.4	−447.2	−470.3
Exports	1,776.6	1,898.3	1,960.1	2,019.8	2,026.9	2,073.4
Goods	1,218.3	1,297.6	1,344.9	1,382.9	1,388.1	1,430.7
Services	558.0	600.6	614.7	636.6	638.4	642.0
Imports	2,235.4	2,357.7	2,412.6	2,440.3	2,474.1	2,543.7
Goods	1,826.7	1,932.1	1,973.1	1,991.5	2,017.7	2,081.7
Services	407.8	424.2	438.7	448.4	456.3	461.0

Note: Users are cautioned that particularly for components that exhibit rapid change in prices relative to other prices in the economy, the chained-dollar estimates should not be used to measure the component's relative importance or its contribution to the growth rate of more aggregate series. Values are in billions of chained (2009) dollars. Quarterly values are seasonally adjusted at annual rates.

SOURCE: Adapted from "Table 3B. Real Gross Domestic Product and Related Measures," in *National Income and Product Accounts—Gross Domestic Product: Second Quarter 2014 (Advance Estimate) Annual Revision: 1999 through First Quarter 2014*, U.S. Department of Commerce, Bureau of Economic Analysis, July 30, 2014, http://www.bea.gov/newsreleases/national/gdp/2014/pdf/gdp2q14_adv.pdf (accessed August 7, 2014)

majority ($1.4 trillion, or 68%) of the total was in goods. The remaining $636.6 billion (32% of the total) was in services. Overall, U.S. exports increased each year between 2010 and 2013. U.S. imports totaled $2.4 trillion in 2013. Again, the largest component ($2 trillion, or 82%) of the total was in goods. The United States imported $448.4 billion (18% of the total) in services. Overall, the value of U.S. imports grew annually between 2010 and 2013.

Table 10.3 shows the percentage change in U.S. export and import values between 1994 and 2013. In 2010 the value of U.S. exports increased 11.9% over the year before. This was the highest annual growth rate since 1997. By contrast, in 2009 the value of U.S. exports fell 8.8% over the year before. That year witnessed the tail end of the Great Recession, a time of greatly reduced economic output for the nation. The value of U.S. imports increased 13.5% in 1997 and declined 13.7% in 2009. The latter percentage reflects decreased spending by U.S. consumers during the Great Recession.

U.S. Trading Partners

In "Top Trading Partners—December 2013" (February 6, 2014, http://www.census.gov/foreign-trade/statistics/highlights/top/top1312yr.html), the U.S. Census Bureau reports that the United States' top-10 goods trading partners in 2013 were:

- Canada—$632.4 billion
- China—$562.4 billion
- Mexico—$506.6 billion
- Japan—$203.7 billion
- Germany—$162.1 billion
- South Korea—$103.8 billion
- United Kingdom—$100 billion
- France—$77.3 billion
- Brazil—$71.7 billion
- Saudi Arabia—$70.8 billion

In 2013 Canada was the leading export market for U.S. goods ($300.3 billion), followed by Mexico ($226.2 billion) and China ($122 billion). The United States imported the highest value of goods from China ($440.4 billion), Canada ($332.1 billion), and Mexico ($280.5 billion).

U.S. TRADE BALANCE

Table 10.2 shows U.S. net exports (imports minus exports) of goods and services between 2010 and 2013 and annualized for 2014. These values are calculated as part of the nation's GDP, as explained in Chapter 2. Net exports in 2013 totaled −$420.4 billion. Nonzero values for net exports indicate a difference between import and export values. A positive balance of trade is called a surplus. This is a situation in which the value of exports is greater than the value of imports. A negative balance of trade is called a deficit. This occurs when the value of imports exceeds the value of exports. Between 2010 and 2013 U.S. imports were valued higher than exports. The annualized quarterly data for 2014 also predict a trade deficit for the year as a whole. Figure 10.1 shows the trade balance for goods and services on a monthly basis between June 2012 and July 2014. During this period the value of U.S. imports was approximately $40 billion to $45 billion greater than U.S. exports each month.

Historical data from the BEA (September 17, 2014, http://www.bea.gov/iTable/iTable.cfm?ReqID=62&step=1#reqid=62&step=1&isuri=1&6210=1&6200=1) indicate the United States had a trade surplus almost every

TABLE 10.3

Percentage change in net exports of goods and services, 1994–2013

	1994	1995	1996	1997	1998	1999	2000	2001	2002	2003	2004	2005	2006	2007	2008	2009	2010	2011	2012	2013
Net exports of goods and services																				
Exports	8.8	10.3	8.2	11.9	2.3	2.6	8.6	−5.8	−1.7	1.8	9.8	6.3	9.0	9.3	5.7	−8.8	11.9	6.9	3.3	3.0
Goods	9.6	11.6	8.9	14.5	2.2	4.2	10.1	−6.2	−3.4	1.9	8.6	7.3	9.4	7.5	6.1	−12.1	14.4	6.5	3.7	2.8
Services	7.0	6.8	6.3	5.3	2.8	−1.4	4.7	−5.0	2.7	1.5	12.7	3.8	8.1	13.7	4.8	−1.1	6.8	7.6	2.4	3.6
Imports	11.9	8.0	8.7	13.5	11.7	10.1	13.0	−2.8	3.7	4.5	11.4	6.3	6.3	2.5	−2.6	−13.7	12.7	5.5	2.3	1.1
Goods	13.4	9.0	9.4	14.4	11.8	12.8	13.1	−3.2	3.7	4.9	11.2	6.7	5.9	1.8	−3.7	−15.8	14.9	5.8	2.1	0.9
Services	5.3	3.0	5.2	8.7	10.9	−3.0	12.6	−0.6	3.3	2.1	12.7	4.5	8.6	6.2	3.7	−3.8	3.8	4.0	3.4	2.2

SOURCE: Adapted from "Table 7. Real Gross Domestic Product: Percent Change from Preceding Year," in *National Income and Product Accounts—Gross Domestic Product: Second Quarter 2014 (Advance Estimate) Annual Revision: 1999 through First Quarter 2014*, U.S. Department of Commerce, Bureau of Economic Analysis, July 30, 2014, http://www.bea.gov/newsreleases/national/gdp/2014/pdf/gdp2q14_adv.pdf (accessed August 7, 2014)

FIGURE 10.1

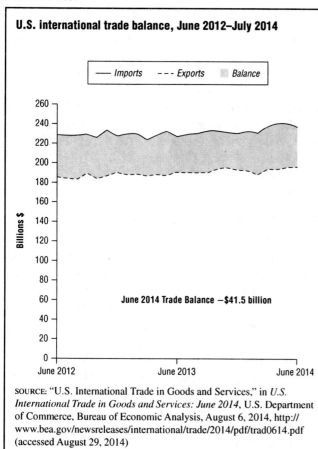

U.S. international trade balance, June 2012–July 2014

SOURCE: "U.S. International Trade in Goods and Services," in *U.S. International Trade in Goods and Services: June 2014*, U.S. Department of Commerce, Bureau of Economic Analysis, August 6, 2014, http://www.bea.gov/newsreleases/international/trade/2014/pdf/trad0614.pdf (accessed August 29, 2014)

year from 1960 to the mid-1970s. After that time trade deficits became much more common. From 1982 onward the United States had a trade deficit nearly every year through 2013.

The Trade Deficit and the U.S. Dollar

The trade deficit is directly linked to the value of the U.S. dollar on foreign exchange markets. When the U.S. dollar weakens compared with a foreign currency, it means that each dollar buys less of the foreign currency than it did before. Consequently, each dollar buys fewer goods from that nation. By contrast, each unit of the foreign currency is now worth more in U.S. dollars and has more purchasing power of U.S. goods. For example, when the U.S. dollar weakens compared with the Japanese yen, Japanese goods cost more for Americans, but U.S. goods become cheaper for Japanese consumers. As a result, imports from Japan to the United States are likely to decrease, whereas exports from the United States to Japan are likely to increase.

Likewise, when the U.S. dollar strengthens, it buys more foreign currency and more foreign goods than it did before. Thus, a stronger dollar is associated with higher imports into the United States and fewer exports to other countries.

Is the Trade Deficit Good or Bad?

The United States' enormous trade deficit is a subject of great debate among economists and politicians. Some believe the deficit is bad for the U.S. economy, particularly the country's manufacturing sector, and that steps should be taken by the government to correct the imbalance. Others contend the deficit is a natural consequence of a strong U.S. economy and should not be an issue of concern.

In *The U.S. Trade Deficit: Causes, Consequences, and Cures* (October 11, 2007, http://www.au.af.mil/au/awc/awcgate/crs/rl31032.pdf), Craig K. Elwell of the Congressional Research Service (CRS) outlines the perceived good and bad effects of a large trade deficit. There is a direct link between the trade deficit and the flow of capital. As Americans buy more foreign goods, foreigners have more money to invest in U.S. financial instruments (e.g., stocks, bonds, and Treasury notes). On the plus side, these investments indicate strong foreign confidence in the security and future growth of the U.S. economy. Also, much of the money flowing into the country is invested in productive capital—that is, invested in growing U.S. industry. However, in many cases these purchases represent debt obligations that the United States will have to pay in the future. In essence, the United States is becoming indebted to foreign nations. Elwell states that "borrowing from abroad allows the United States to live better today, but the payback must mean some decrement to the rate of advance of U.S. living standards in the future."

In a broader sense, large capital inflows demonstrate that Americans prefer spending their money on foreign imports rather than investing it in domestic financial instruments. Put simply, Americans prefer spending to saving. Elwell explains that "so long as domestic saving in the United States falls short of domestic investment and an inflow of foreign saving is available to fill all or part of the gap, the United States will run a trade deficit."

Many critics of the trade deficit claim it hurts the U.S. economy overall, particularly by raising unemployment. Elwell disputes this claim, explaining that the dramatic growth of the trade deficit during the 1990s and the first decade of the 21st century coincided with a generally healthy U.S. economy and relatively low unemployment rates. However, Elwell acknowledges that extensive foreign imports have hurt some U.S. manufacturing industries, particularly textiles, apparel, and steel.

FREE TRADE AGREEMENTS

The U.S. government has long been part of free trade agreements with other individual countries (known as bilateral agreements) and with groups of countries (known as trading blocs). In this context, free trade means the ability to buy and sell goods across international borders with a minimum of tariffs or other interferences. Tariffs (import taxes) are fees charged by a country to import goods into that country.

One of the most historically notable trade agreements is the General Agreement on Tariffs and Trade (GATT), which was first signed by the United States and 22 other countries in 1947. This agreement dealt primarily with industrial products and marked a trend toward the increasing globalization of the world economy. The agreement reduced tariffs, removed other obstacles to international trade, and clarified rules surrounding barriers to free trade. A series of GATT negotiations that concluded in 1994 created the WTO, which replaced GATT. The WTO now functions as the principal international body for administering trade rules among member countries. As of June 2014, the WTO (http://www.wto.org/english/theWTO_e/whatis_e/tif_e/org6_e.htm) consisted of 160 member countries, including the United States.

U.S. priorities regarding trade policy have shifted over the decades according to the state of the economy. During the recession of the late 1970s, U.S. producers called for the government to institute measures—such as high tariffs—to protect them from international competition. During the growth period of the 1980s, however, the focus of companies turned to their own international expansion, and by the 1990s a push for free trade had gained increased momentum in the United States.

In 1994 the United States, Canada, and Mexico implemented the North American Free Trade Agreement (NAFTA) with the ultimate goal of completely eliminating barriers to trade among the three countries. By 2008 all tariffs had been phased out. Before NAFTA was signed, the United States insisted on assurances from Canada and Mexico that they would enforce labor and environmental laws before ratifying the agreement. NAFTA was very controversial in the United States when it was first implemented. Critics claimed the agreement would encourage U.S. companies to move their manufacturing facilities to Mexico to take advantage of lower labor costs and less government regulation. Although this did happen to some degree, the U.S. economy experienced strong growth overall during the late 1990s and into the early years of the first decade of the 21st century.

European Union

In 1957 six European countries signed the Treaty of Rome, which established the European Economic Community. In 1992 the Maastricht Treaty was signed, which officially established the EU. The leaders of European countries hoped that by engaging in commerce they could create long-term stability and enforce the rule of law in cooperative democratic societies. The EU, which is one of the most important trading partners of the United States, expanded in 2004 from 15 nations to 25, creating the largest trading bloc in history.

According to the EU, in "The Member Countries of the European Union" (http://europa.eu/about-eu/countries/member-countries/index_en.htm), in 2014 the EU consisted of 28 member countries: Austria, Belgium, Bulgaria, Croatia, Cyprus, the Czech Republic, Denmark, Estonia, Finland, France, Germany, Greece, Hungary, Ireland, Italy, Latvia, Lithuania, Luxembourg, Malta, the Netherlands, Poland, Portugal, Romania, Slovakia, Slovenia, Spain, Sweden, and the United Kingdom.

INTERNATIONAL MONETARY FUND

The International Monetary Fund (IMF), a global financial system, was established in July 1944, during the United Nations (UN) Monetary and Financial Conference (more commonly known as the Bretton Woods Conference because it took place in Bretton Woods, New Hampshire). The IMF extends short-term loans to members experiencing economic instability. As a condition of receiving its credit assistance, the IMF requires the debtor country to enact significant reform of its economic structure, and often of its political structure as well. The conditions for being granted a loan can include drastic cuts in government spending; privatizing government-owned enterprises, such as railroads and utilities; establishing higher interest rates; increasing taxes; and eliminating subsidies on necessities such as food and fuel.

Supporters of the IMF indicate that these reforms oftentimes eliminate corruption and help establish effective institutions such as courts. By contrast, critics argue that for some countries, these reforms can have devastating social consequences, including severe unemployment, crippling price increases in the cost of basic goods, and political instability resulting from widespread dissatisfaction. Despite this criticism, the IMF membership has grown considerably since its founding. In 1944 it had 45 members. By 2014 the IMF (http://www.imf.org/external/about.htm) consisted of 188 member countries, including the United States.

WORLD BANK

At the same conference that created the IMF in July 1944, the International Bank for Reconstruction and Development (IBRD) was established. In 1960 the International Development Association (IDA) was created. The IBRD and the IDA are commonly known as the World Bank. The World Bank is not a bank in the traditional sense of the word but an agency of the UN. The World Bank works to combat world poverty by providing low-interest loans, interest-free credit, and grants to developing countries. According to the World Bank (http://go.worldbank.org/Y33OQYNE90), in 2014 the

IBRD and the IDA consisted of 188 and 172 member countries, respectively.

In its early days, the World Bank often participated in large projects such as dam building. During the first decade of the 21st century it supported the efforts of governments in developing countries to build schools and health centers, provide water and electricity, fight disease, and protect the environment. It is one of the world's largest sources of development assistance.

GLOBALIZATION AND THE ANTIGLOBALIZATION MOVEMENT

The move toward global free trade, or globalization, has generated intense controversy. Proponents maintain that globalization can improve living standards throughout the world. Their arguments include the following:

- Countries and regions will become more productive by concentrating on industries in which they have a natural advantage and trading with other nations for goods in which they do not have an advantage.

- Multinational corporations will be able to realize economies of scale—that is, operate more economically because they are buying in bulk, selling to a much larger market, and utilizing a much larger labor pool. This will increase productivity and lead to greater prosperity.

- Free trade will lead to faster growth in developing countries.

- Increased incomes and the development of job-related skills among the citizens of poorer nations will foster the spread of information, education, and, ultimately, democracy.

Critics of globalization point out the negative effects that multinational corporations have on people in the developing world. They argue that:

- Most of the profits from free trade flow to the United States and to other industrialized countries.

- Local industries can be destroyed by competition from wealthier nations, causing widespread unemployment and social disruption.

- Centuries of cultural tradition can be quickly obliterated by the influence of international companies.

- Multinational corporations often impinge on national sovereignty to protect their profits.

Critics also note that the free trade policies are often applied unfairly, as the United States insists that other countries open their markets to U.S. goods while it protects its own producers from competition. For example, the U.S. government has established many tariffs and regulations that raise the prices of imported food products, denying poor farmers in the developing world access to the lucrative U.S. market. In addition, opponents of globalization point out that the spread of multinational corporations can be detrimental to workers in industrialized nations by exporting high-paying jobs to countries with lower labor costs and that international competition in the labor market could actually lead to lower living standards in the industrialized world.

The antiglobalization movement is not an organized group but an umbrella term for many independent organizations that oppose the pursuit of corporate profits at the expense of social justice in the developing world. These groups often protest the actions of organizations such as the WTO, the IMF, and the World Bank for their perceived bias toward corporations and wealthy nations. For example, in 1999 a WTO conference in Seattle, Washington, drew more than 40,000 protestors in a massive demonstration that generated intense media attention and completely overshadowed the meeting itself.

ECONOMIC AND TRADE SANCTIONS

The United States uses economic and trade sanctions (stopping some or all forms of financial transactions and trade with a country) as a political tool against countries that are thought to violate human rights, tolerate drug trafficking, support terrorism, and produce or store weapons of mass destruction. Sanctions are enforced by the U.S. Department of the Treasury's Office of Foreign Assets Control (OFAC). As of December 2014, the OFAC (http://www.treasury.gov/resource-center/sanctions/Programs/Pages/Programs.aspx) listed economic and/or trade sanctions against dozens of countries, including Iran, North Korea, and Syria.

Overall, the sanctions have little effect on U.S. consumers because most of the nations involved are not regular trading partners of the United States. One sanctioned nation that is engaged in trade with the United States is Russia; however, the sanctions against it are very narrow. The U.S. and Russian governments have been at odds since Ukraine experienced a revolution in February 2014, and pro-Russian militants (allegedly aided by the Russian military) seized control of the Crimean peninsula in Ukraine. Russia subsequently "annexed" the area, meaning that it absorbed the territory as part of its own territory. This move prompted trade sanctions from the United States and the EU against particular individuals in Russia and Ukraine. In addition, the U.S. government has imposed sanctions on specific Russian companies and banks. Russia is a member of the WTO, having joined in 2012. It has complained to the organization about the sanctions. The article "Russia's Putin Says Sanctions Violate Principles of WTO" (Reuters.com, September 18, 2014) notes that in September 2014 Vladimir Putin (1952–), the president of Russia, asserted that the "Western sanctions against Russia violated the principles of the World Trade Organization."

INTELLECTUAL PROPERTY

Technological advancements have posed new challenges to world trade. As private-sector investment in information technology continues, world economies are becoming even more interconnected. Proponents of free trade, including the United States, have pushed for more protection of intellectual property rights, abuse of which poses a major barrier to world trade. As defined by the UN in the Convention Establishing the World Intellectual Property Organization (2014, http://www.wipo.int/treaties/en/convention/trtdocs_wo029.html), which was signed on July 14, 1967, and amended on September 28, 1979, intellectual property includes:

- Literary, artistic, and scientific works
- Performances of performing artists, phonograms, and broadcasts
- Inventions in all fields of human endeavor
- Scientific discoveries
- Industrial designs
- Trademarks, service marks, and commercial names and designations
- Protection against unfair competition and all other rights resulting from intellectual activity in the industrial, scientific, literary, or artistic fields

Challenges for the international community include establishing minimum standards for protecting intellectual property rights and procedures for enforcement and dispute resolution. These challenges are not new. The World Intellectual Property Organization (WIPO) explains in "Major Events 1883 to 2002" (2014, http://www.wipo.int/treaties/en/general) that as early as 1883, with the founding of the 14-member Paris Convention for the Protection of Industrial Property, countries recognized the special nature of creative works, including inventions, trademarks, and industrial designs. In 1886 the Berne Convention for the Protection of Literary and Artistic Works extended the model of international protection to copyrighted works such as novels, poems, plays, songs, operas, musicals, drawings, paintings, sculptures, and architectural works.

In 1893 the Paris Convention and the Berne Convention combined to form the United International Bureaus for the Protection of Intellectual Property, which maintained its headquarters in Berne, Switzerland. This organization eventually evolved into the WIPO, located in Geneva, Switzerland, which carries out a program that is designed to:

- Harmonize national intellectual property legislation and procedures
- Provide services for international applications for industrial property rights
- Exchange intellectual property information
- Provide legal and technical assistance to developing and other countries
- Facilitate the resolution of private intellectual property disputes
- Marshal information technology as a tool for storing, accessing, and using valuable intellectual property information

As of December 2014, the WIPO (http://www.wipo.int/members/en) consisted of 188 member nations, including the United States.

INTERNATIONAL INVESTMENT

International investment involves the buying and selling of foreign investments (such as stocks, bonds, and other financial instruments) and the investment of cash directly in foreign companies. Also included are investment assets such as foreign-owned gold and foreign currencies.

International investment data are collected by the BEA, the Treasury Department, and the Federal Reserve System (the nation's central bank). The two primary types of international investments that they track are financial investments and direct investments. In *Direct Investment Positions for 2011: Country and Industry Detail* (July 2012, http://www.bea.gov/scb/pdf/2012/07%20July/0712_dip.pdf), Kevin B. Barefoot and Marilyn Ibarra-Caton of the BEA explain that a direct investment is an investment in which a resident of one country "obtains a lasting interest in, and a degree of influence over the management of a business enterprise in another country." Legally, a resident can be a person or other entity, such as a corporation or government body. The U.S. government defines a direct investment as ownership or control of at least 10% of a foreign business enterprise.

U.S. Investment in Foreign Assets

The BEA reports in the press release "U.S. Net International Investment Position: End of the Fourth Quarter and Year 2013" (March 26, 2014, http://www.bea.gov/newsreleases/international/intinv/2014/pdf/intinv413.pdf) that U.S.-owned assets abroad totaled $22 trillion at year-end 2013. This value was up from $21.6 trillion at year-end 2012.

Foreign Investment in U.S. Assets

The United States has historically allowed and often encouraged foreign investment in U.S. assets. The U.S. Government Accountability Office (GAO) states in *Sovereign*

Wealth Funds: Laws Limiting Foreign Investment Affect Certain U.S. Assets and Agencies Have Various Enforcement Processes (May 2009, http://www.gao.gov/new.items/d09608.pdf) that "the United States has an overall policy of openness to foreign investment through policy statements and treaties and international agreements addressing investment." The GAO explains, however, that there are federal laws that limit or otherwise restrict the amount of foreign ownership or control in certain industries—specifically the transportation, energy, natural resources, banking, agriculture, and national defense industries.

Foreign-Owned Assets in the United States

In "U.S. Net International Investment Position: End of the Fourth Quarter and Year 2013," the BEA notes that foreign-owned assets in the United States totaled $26.5 trillion at year-end 2013. This value was up from $25.5 trillion at year-end 2012.

It should be noted that the purchase of U.S. assets by foreign entities is funded in large part by U.S. imports. When Americans buy more foreign goods and services, more U.S. dollars flow into foreign countries. This provides greater opportunities for foreigners to invest in U.S. assets.

FOREIGN HOLDINGS OF U.S. FINANCIAL ASSETS. Every five years the U.S. government conducts a comprehensive survey called a full benchmark survey to measure foreign holdings of U.S. securities. As of December 2014, the most recent data available were published in *Foreign Portfolio Holdings of U.S. Securities as of June 30, 2013* (April 2014, http://www.treasury.gov/ticdata/Publish/shla2013r.pdf) by the Treasury Department. It should be noted that securities include stocks (also known as equities), bonds, and other financial instruments (such as U.S. Treasury bills and notes) and securities sold by U.S. agencies.

As of June 2013, foreign holdings of U.S. securities totaled $14.4 trillion. (See Table 10.4.) The vast majority of the holdings ($13.5 trillion) consisted of long-term securities (securities with maturity time more than one year). A much smaller value ($878 billion) of short-term debt was in foreign hands.

As shown in Table 10.4, foreign holdings of U.S. securities have increased dramatically since 2006, when they totaled $7.8 trillion. This trend is of major concern to some analysts and politicians, who fear that the United States has become too dependent on foreign money. The danger to the U.S. economy as a whole lies in the possibility that foreigners might suddenly decide to pull out of U.S. financial assets. This could destabilize the financial market and harm U.S. economic growth.

The Treasury Department estimates ownership by country of foreign-owned U.S. securities. However, it warns that these estimates should be viewed as "rough indicators" because the underlying data are imperfect and are muddied by the complexities of the modern financial system. Certain countries, such as Luxembourg and Switzerland, contain major financial industries that manage or hold securities for residents of other countries. These securities are reported to the U.S. government as owned by Luxembourg and Switzerland, when in fact they are owned by other foreigners. In addition, certain types of

TABLE 10.4

Foreign holdings of U.S. securities, by type of security, as of June 2006–June 2013

[Billions of dollars]

	2006	2007	2008	2009	2010	2011	2012	2013
Long-term securities	7,162	9,136	9,463	8,492	9,736	11,561	12,451	13,532
Equities[a]	2,430	3,130	2,969	2,252	2,814	3,830	4,237	5,070
Debt	4,733	6,007	6,494	6,240	6,921	7,731	8,213	8,462
U.S. Treasury	1,727	1,965	2,211	2,604	3,343	4,049	4,673	4,916
U.S. agency[b]	984	1,304	1,464	1,196	1,086	1,031	991	874
Corporate[c]	2,021	2,738	2,820	2,440	2,493	2,651	2,549	2,672
Short-term debt	615	635	858	1,150	956	878	811	878
U.S. Treasury	253	229	379	862	743	658	637	679
U.S. agency	147	109	174	90	61	43	29	25
Corporate	215	297	306	197	152	177	145	173
Total long-term and short-term	7,778	9,772	10,322	9,642	10,691	12,440	13,261	14,410

[a]Equities include common and preferred stock, all types of investment company shares, such as open-end funds, closed-end funds, money market mutual funds, and hedge funds, as well as interests in limited partnerships and other equity interests that may not involve stocks or shares.
[b]Agencies include U.S. Government agencies and corporations as well as Federally Sponsored Enterprises, such as the Federal National Mortgage Association.
[c]Corporate debt includes all other non-Treasury and non-Agency debt, such as certificates of deposit with a maturity of over one year, and U.S. municipal debt securities.
Note: Components may not sum to totals due to rounding.

SOURCE: "Table 1. Foreign Holdings of U.S. Securities, by Type of Security, as of End-June, Selected Survey Dates," in *Foreign Portfolio Holdings of U.S. Securities as of June 30, 2013*, U.S. Department of the Treasury, Federal Reserve Bank of New York, and Board of Governors of the Federal Reserve System, April 2014, http://www.treasury.gov/ticdata/Publish/shla2013r.pdf (accessed August 29, 2014)

TABLE 10.5

Foreign holdings of U.S. securities, by holder, as of June 2013

[Billions of dollars]

Country	Total	Equity	Treas. LT debt	Agency LT debt ABS[a]	Agency LT debt Other	Corp. LT debt ABS[a]	Corp. LT debt Other	ST debt
Japan	1,766	316	1,023	152	44	19	149	63
China[b]	1,735	261	1,272	153	21	6	17	5
Cayman Islands	1,168	628	66	27	6	109	237	96
United Kingdom	1,116	584	131	6	3	30	334	30
Luxembourg	990	378	107	15	6	42	362	80
Canada	814	617	47	1	2	19	105	24
Switzerland	581	266	157	9	8	14	96	32
Ireland	575	141	91	32	19	41	139	114
Middle East oil exporters[c]	545	275	151	10	3	5	19	82
Belgium	490	29	163	3	5	20	257	13
Country unknown	51	[d]	0	[d]	0	[d]	51	[d]
Rest of world	4,578	1,576	1,708	251	101	99	503	341
Total	**14,410**	**5,070**	**4,916**	**657**	**217**	**403**	**2,268**	**878**
Of which: holdings of foreign official institutions	5,406	799	3,648	318	133	19	108	381

[a]Asset-backed securities. Agency ABS are backed primarily by home mortgages; corporate ABS are backed by a wide variety of assets, such as car loans, credit card receivables, home and commercial mortgages, and student loans.
[b]Excludes Hong Kong and Macau, which are reported separately.
[c]Bahrain, Iran, Iraq, Kuwait, Oman, Qatar, Saudi Arabia, and the United Arab Emirates.
[d]Greater than zero but less than $500 million.
Note: LT = long term. ST = short term.

SOURCE: "Table 6. Value of Foreign Holdings of U.S. Securities, by Major Investing Country and Type of Security, as of June 30, 2013," in *Foreign Portfolio Holdings of U.S. Securities as of June 30, 2013*, U.S. Department of the Treasury, Federal Reserve Bank of New York, and Board of Governors of the Federal Reserve System, April 2014, http://www.treasury.gov/ticdata/Publish/shla2013r.pdf (accessed August 29, 2014)

---

securities are allowed to be unregistered as to country of ownership. With these caveats, the Treasury Department provides estimates of foreign holdings of U.S. securities by country in Table 10.5. Japan had the largest amount ($1.8 trillion), followed by China ($1.7 trillion) and the Cayman Islands ($1.2 trillion).

China's large holdings of U.S. securities are of particular concern to some analysts. As noted earlier, a country or countries holding large amounts of U.S. securities could, in theory, severely disrupt the U.S. economy by suddenly selling those securities. The probability and consequences of this possible event are examined regularly by the CRS. James K. Jackson of the CRS explains in *Foreign Ownership of U.S. Financial Assets: Implications of a Withdrawal* (April 8, 2013, http://digitalcommons.ilr.cornell.edu/cgi/viewcontent.cgi?article=2062&context=key_workplace) that a sudden sell-off of foreign-owned U.S. securities could, in theory, cause a dramatic spike in interest rates in the United States. However, Jackson explains that this scenario is highly unlikely because the financial industry responds quickly to market fluctuations. As such, any attempted sell-off would lead to a sudden increase in supply, which would drive down the price of the securities and cause the seller or sellers to lose huge amounts of money. In addition, Jackson notes that the Federal Reserve would not "sit by idly" in such a circumstance, but would act quickly in cooperation with other national banks to stabilize credit markets.

In *China's Holdings of U.S. Securities: Implications for the U.S. Economy* (August 19, 2013, http://www.fas.org/sgp/crs/row/RL34314.pdf), Wayne M. Morrison and Marc Labonte of the CRS echo these assurances and note that the severe problems experienced by the U.S. financial markets during the Great Recession did not scare off foreign investors. They state, "If these events failed to cause a sudden flight from U.S. assets ... by China or other countries, it is hard to imagine what would." However, Morrison and Labonte acknowledge that the broader problem for the U.S. economy is the nation's lack of saving. They note that "the United States must boost its level of savings in the long run in order to reduce its vulnerability to a potential shift away from U.S. assets by foreign investors."

IMPORTANT NAMES AND ADDRESSES

Agency for Healthcare Research and Quality
540 Gaither Rd., Ste. 2000
Rockville, MD 20850
(301) 427-1104
URL: http://www.ahrq.gov/

American Bankruptcy Institute
66 Canal Center Plaza, Ste. 600
Alexandria, VA 22314
(703) 739-0800
FAX: (703) 739-1060
URL: http://www.abiworld.org/

Bureau of Economic Analysis
1441 L St. NW
Washington, DC 20230
(202) 606-9900
E-mail: CustomerService@bea.gov
URL: http://www.bea.gov/

Congressional Budget Office
Ford House Office Bldg., Fourth Floor
Washington, DC 20515-6925
(202) 226-2602
E-mail: communications@cbo.gov
URL: http://www.cbo.gov/

Consumer Federation of America
1620 I St. NW, Ste. 200
Washington, DC 20006
(202) 387-6121
E-mail: cfa@consumerfed.org
URL: http://www.consumerfed.org/

Economic Policy Institute
1333 H St. NW, Ste. 300, East Tower
Washington, DC 20005
(202) 775-8810
FAX: (202) 775-0819
E-mail: epi@epi.org
URL: http://www.epi.org/

Federal Communications Commission
445 12th St. SW
Washington, DC 20554
1-888-225-5322
FAX: 1-866-418-0232
URL: http://www.fcc.gov/

Federal Home Loan Mortgage Corporation
8200 Jones Branch Dr.
McLean, VA 22102
(703) 903-2000
URL: http://www.freddiemac.com/

Federal Housing Finance Agency
400 Seventh St. SW
Washington, DC 20024
(202) 649-3800
FAX: (202) 649-1071
URL: http://www.fhfa.gov/

Federal National Mortgage Association
3900 Wisconsin Ave. NW
Washington, DC 20016-2892
(202) 752-7000
1-800-732-6643
URL: http://www.fanniemae.com/

Federal Reserve System, Board of Governors
20th St. and Constitution Ave. NW
Washington, DC 20551
1-888-851-1920
URL: http://www.federalreserve.gov/

Federal Trade Commission
600 Pennsylvania Ave. NW
Washington, DC 20580
(202) 326-2222
URL: http://www.ftc.gov/

Federation of Tax Administrators
444 N. Capitol St. NW, Ste. 348
Washington, DC 20001
(202) 624-5890
FAX: (202) 624-7888
URL: http://www.taxadmin.org/

Internal Revenue Service
1111 Constitution Ave. NW
Washington, DC 20224
(202) 622-5000
1-800-829-1040
URL: http://www.irs.gov/

National Bureau of Economic Research
1050 Massachusetts Ave.
Cambridge, MA 02138
(617) 868-3900
E-mail: info@nber.org
URL: http://www.nber.org/

Office of Management and Budget
725 17th St. NW
Washington, DC 20503
(202) 395-3080
FAX: (202) 395-3888
URL: http://www.whitehouse.gov/omb/

Organisation for Economic Co-operation and Development
2 rue André Pascal, 75775
Paris, Cedex 16 France
(011-33) 1-45-24-82-00
FAX: (011-33) 1-45-24-85-00
URL: http://www.oecd.org/

Social Security Administration
1100 West High Rise
6401 Security Blvd.
Baltimore, MD 21235
1-800-772-1213
URL: http://www.ssa.gov/

U.S. Bureau of Labor Statistics
Postal Square Bldg.
2 Massachusetts Ave. NE
Washington, DC 20212-0001
(202) 691-5200
URL: http://www.bls.gov/

U.S. Census Bureau
4600 Silver Hill Rd.
Washington, DC 20233

(301) 763-4636
1-800-923-8282
URL: http://www.census.gov/

U.S. Commodities Futures Trading Commission
Three Lafayette Centre
1155 21st St. NW
Washington, DC 20581
(202) 418-5000
FAX: (202) 418-5521
E-mail: Questions@cftc.gov
URL: http://www.cftc.gov/

U.S. Consumer Product Safety Commission
4330 East West Hwy.
Bethesda, MD 20814
(301) 504-7923
1-800-638-2772
FAX: (301) 504-0124
URL: http://www.cpsc.gov/

U.S. Department of Commerce
1401 Constitution Ave. NW
Washington, DC 20230
(202) 482-2000
E-mail: TheSec@doc.gov
URL: http://www.commerce.gov/

U.S. Department of Health and Human Services
200 Independence Ave. SW
Washington, DC 20201
1-877-696-6775
URL: http://www.hhs.gov/

U.S. Department of Housing and Urban Development
451 Seventh St. SW
Washington, DC 20410
(202) 708-1112
URL: http://www.hud.gov/

U.S. Department of Labor
Frances Perkins Bldg.
200 Constitution Ave. NW
Washington, DC 20210
1-866-487-2365
URL: http://www.dol.gov/

U.S. Department of the Treasury
1500 Pennsylvania Ave. NW
Washington, DC 20220
(202) 622-2000
FAX: (202) 622-6415
URL: http://www.treasury.gov/

U.S. Equal Opportunity Employment Commission
131 M St. NE
Washington, DC 20507
(202) 663-4900
E-mail: info@eeoc.gov
URL: http://www.eeoc.gov/

U.S. Government Accountability Office
441 G St. NW
Washington, DC 20548
(202) 512-3000
E-mail: contact@gao.gov
URL: http://www.gao.gov/

U.S. International Trade Commission
500 E St. SW
Washington, DC 20436
(202) 205-2000
URL: http://www.usitc.gov/

U.S. Securities and Exchange Commission
100 F St. NE
Washington, DC 20549
(202) 942-8088
E-mail: https://tts.sec.gov/oiea/
QuestionsAndComments.html
URL: http://www.sec.gov/

U.S. Small Business Administration
409 Third St. SW
Washington, DC 20416
1-800-827-5722
E-mail: answerdesk@sba.gov
URL: http://www.sba.gov/

World Bank
1818 H St. NW
Washington, DC 20433
(202) 473-1000
FAX: (202) 477-6391
URL: http://www.worldbank.org/

World Intellectual Property Organization
34, chemin des Colombettes, CH-1211
Geneva 20 Switzerland
(011-41-22) 338-9111
FAX: (011-41-22) 733-5428
URL: http://www.wipo.int/

World Trade Organization
Centre William Rappard
Rue de Lausanne 154, CH-1211
Geneva 21, Switzerland
(011-41-22) 739-5111
FAX: (011-41-22) 731-4206
E-mail: enquiries@wto.org
URL: http://www.wto.org/

RESOURCES

Several government agencies provided invaluable economic data and information for this book: the U.S. Department of Commerce's Bureau of Economic Analysis (BEA), the U.S. Census Bureau, the U.S. Department of Labor's Bureau of Labor Statistics (BLS), and the Federal Reserve System.

The BEA compiles the National Income and Product Accounts, which include detailed financial information on gross domestic product, personal income and outlays, saving, corporate profits, and international trade and balance of payments.

The BLS publishes statistical data on wages, benefits, and income; inflation and economic indexes; employment and unemployment; industries and occupations; employment demographics; and worker health and safety standards. In addition, the BLS posts many of its publications online, including *Employment Situation*, *Occupational Outlook Handbook*, *Monthly Labor Review*, and *Occupational Outlook Quarterly*.

The Census Bureau provides comprehensive economic and demographic data. Particularly useful for the study of the U.S. economy are the census publications *Historical Statistics of the United States, Colonial Times to 1970, Bicentennial Edition, Part 1* (September 1975), *Income and Poverty in the United States: 2013* (Carmen DeNavas-Walt and Bernadette D. Proctor, September 2014), and *Statistical Abstract of the United States: 2012* (2012).

The Federal Reserve System publishes economic data and papers on a variety of economic subjects, including housing, consumer spending, interest rates, consumer credit, net worth, wealth distribution, and debt. Especially useful was the series *Federal Reserve Statistical Release*.

Other government agencies and offices consulted during the compilation of this book include the U.S. Government Accountability Office, the Congressional Research Service, the Social Security Administration, and the White House. The latter provided budgetary information and the annual *Economic Report of the President*.

Important information was also obtained from the Federal Trade Commission; the U.S. Departments of Agriculture, Education, Energy, Health and Human Services, Housing and Urban Development, and the Treasury; the U.S. Department of Homeland Security's Office of Immigration Statistics; the Internal Revenue Service; the Congressional Budget Office; the Office of Management and Budget; the Federal Housing Finance Agency; the Central Intelligence Agency; and the U.S. Small Business Administration.

International organizations that provided input include the World Trade Organization and the World Intellectual Property Organization.

A number of independent, nonpartisan think tanks and private organizations were consulted to obtain various points of view on socioeconomic issues. Finally, the Gallup Organization was the source for numerous public opinion polls that were conducted to gauge American attitudes on economic topics.

INDEX

Page references in italics refer to photographs. References with the letter t following them indicate the presence of a table. The letter f indicates a figure. If more than one table or figure appears on a particular page, the exact item number for the table or figure being referenced is provided.

A

ACA. *See* Affordable Care Act
Accounting scandals, 123–124
Adjustable rate mortgages (ARMs), 61–62, 63f, 66–67, 102
Affordable Care Act (ACA)
    health care spending and prices, 50–53
    insurance coverage effects, 15–16, 16t, 20–21, 21t, 22t, 162f
    labor costs, 107–108
    Medicaid eligibility, 139–141
    provisions, 19–20, 20t
African Americans, 7–8
Age
    family net worth, 140t
    financial concerns survey, 38f
    health insurance coverage, 16(t1.2)
    poverty, 135–136
    unemployment, 80
    wealth distribution, 135
Aging population, 156
Agriculture, 7, 108, 110t
AIG (American International Group), 102
Airline Deregulation Act, 111
American International Group (AIG), 102
American Recovery and Reinvestment Act, 14
Annualization, 26–27
Antiglobalization movement, 169
Antitrust law, 111
Appropriations bills, 149
Arizona, 93
ARMs (adjustable rate mortgages), 61–62, 63f, 66–67, 102
Arms race, 8–9
Assets
    financial assets, 117–122, 118t, 131–132, 134f
    international investment, 170–172, 171t, 172t
    mortgage applications, 60–61
    net worth, 133t
    nonfinancial assets, 132, 135f, 136t
AT&T, 111
Auto loans, 72–73
Automotive industry, 102, 111

B

Bacanovic, Peter, 123
Bad debt, 57
Bailouts, 14, 102, 113
Bankruptcy, 74–76, 75f
Bankruptcy Abuse Prevention and Consumer Protection Act, 73
Banks and the banking industry
    credit cards, 73
    federal manipulation of macroeconomics, 158–159
    financial crisis, 67–69, 101–103
    panics, 2–3
    prime loan rate, 12f
    public confidence in, 112t, 113, 113f
    transaction accounts and certificates of deposit, 117
Berne Convention, 170
Big business
    international comparisons of CEO salaries, 145
    monopolies, 111
    public confidence, 97, 112–113, 112t, 113f
    wealth inequality, 142
    world's biggest firms, 100–101
BLS. *See* Bureau of Labor Statistics
Bonds, 119, 120, 148, 152–153
Border security, 93
Bretton Woods Conference, 168
Budget, federal government, 11–12, 13f, 149–153, 152(f9.3)
Budget Control Act of 2011, 153
Bureau of Economic Analysis
    Affordable Care Act, 52
    business sector data, 23
    corporate profits, 104, 106
    employee compensation, 127
    gross domestic product estimates, 25–26
    international investment, 171
    Principal Federal Economic Indicators, 24
    trade data, 164–165, 167
Bureau of Labor Statistics (BLS)
    Consumer Expenditure Survey, 42
    consumer price index, 44
    employee benefits, 89–90
    employment data, 77–78
    unemployment measures, 81–83
    wages and salaries, 85–89, 127–128
Bush, George H. W., 12
Bush, George W.
    Great Recession, 14
    tax cuts, 156
    Troubled Asset Relief Program, 69
    unemployment measures, 82–83
Business
    agriculture, 110t
    competition and monopolies, 111
    corporate behavior, 112–113
    corporate bonds, 119
    corporate income taxes, 150
    corporate profits, 104, 106, 106t, 107f
    deregulation, 111
    economic indicators, 23–24
    employment cost index, 109f
    executive pay, 142, 144
    federal regulation, 108–109, 111
    Great Recession bailouts, 14

gross private domestic investment, 103–104, 104*t*, 105*t*
international comparisons of CEO salaries, 145
labor costs, 106–108, 107*t*, 108*t*, 109*f*
legal structures, 97–98
public confidence, 112*t*, 113*f*, 114*f*
public opinion on corporate taxes, 144
statistics, 98–100, 99*t*
wealth inequality, 142
world's biggest firms, 100–101
Business cycles, 31–33, 31*t*
"Buying on margin," 4

C

Canada, 168
Capital gains, 11, 143
Capitalism, 1
Car loans, 72–73
Carter, Jimmy, 10, 11
Cash value life insurance, 121
Census Bureau, 128, 137–139
Centers for Disease Control and Prevention, 75–76
Centers for Medicare and Medicaid Services (CMS), 106–107
Central Intelligence Agency, 164
Certificates of deposit, 118–119
Checking accounts, 117
Chief executive officers, 142, 144, 145
Child labor, 3
Children's Health Insurance Program, 160
China, 163, 164, 172
Clinton, Bill, 12
CMS (Centers for Medicare and Medicaid Services), 106–107
Coincident indicators, 32
Cold War, 8–9, 11–12
Collections, debt, 57, 72
College financing, 38*f*
Commodities, 121
Commonwealth Fund, 76
Compensation
earnings, by gender and race/ethnicity, 127–128
employee benefits, 89–90
labor costs, 106–108, 108*t*
types, 84–85
wages and salaries, 85–89, 86(*t*5.8), 87*t*
wealth inequality, 141–144, 145
Competition, business, 11, 111
Conference Board, 32, 34
Congress, U.S. and the budget process, 149, 153
Congressional Budget Office, 88, 102, 130–131, 159–160
Conservatorship, 68
Consumer confidence, 34, 39
Consumer credit, 11, 70–74, 72*t*, 74*t*

Consumer Expenditure Survey, 42, 45*t*–46*t*
Consumer goods, 139
Consumer price index, 47*f*
health care, 51*t*, 52*t*
health care prices, 49–51
inflation, 44–49
real earnings calculation, 87–88
Consumer Product Safety Commission, 108–109
Consumer spending
Consumer Expenditure Survey, 45*t*–46*t*
consumer price index for health care, 51*t*
effects, 39
gross domestic product calculation, 27
health care spending and prices, 16–17, 17*f*, 19*f*, 49–53, 49*t*, 50*t*
inflation rates, 48–49
local governments, 148
personal consumption expenditures, 41–42, 42*t*, 43*t*, 44*f*
personal income and outlays, 40*t*
purchasing power and real earnings, 88
Consumption expenditures, government, 28, 153–154, 154*t*
Convention Establishing the World Intellectual Property Organization, UN, 170
Corporate bonds, 119
Corporate income taxes, 150
Corporate profits, 104, 106, 106*t*, 107*f*
Corporations, 98
Cost-of-living adjustments, 48
Costs, labor. *See* Labor costs
Council of Economic Advisers, 50
Court cases
Standard Oil Co. of New Jersey v. United States, 111
United States v. Philip Morris Inc., 112
Credit, consumer. *See* Consumer credit; Debt, personal
Credit CARD Act, 73
Credit cards, 73–74, 74*t*
Credit default swaps, 101–102
Credit market debt, 58
Credit reports and scores, 56, 68
Credit unions, 117
Currency comparisons, global, 163, 164, 167
Current Population Survey, 127–128
Customs duties, 151

D

Data. *See* Economic indicators; Statistical information
Debt, national, 151–153, 152(*f*9.3)
Debt, personal
bad debt and collections, 57
bankruptcies, 74–76, 75*f*
categories of debt, 55–56
consumer credit, 70–74, 72*t*, 73*f*

credit cards, 74*t*
credit reports and scores, 56–57
data, 57–59
family-level data, 59
household debt, 58*t*, 59*t*, 60*f*
household liabilities, 132
interest rates and payments, 56, 56*t*
See also Mortgages
Debt ceiling, 153
Debt ratio, 61
Deductions, tax, 154, 156
Defaults, loan
creditor actions, 57
mortgages, 63, 67
student loans, 72
Defense industry. *See* National defense and the defense industry
Deferred action immigration programs, 93
Deficits, federal, 11–12, 13*f*, 151, 152(*f*9.2)
Deflation, 47
Delinquency, loan
Great Recession, 59
mortgages, 63, 67, 70
student loans, 72
Demand for housing, 65, 66
Democrats, 144, 149, 156
Demographics
earnings, 127–128
financial assets, holdings of, 118*t*
financial concerns survey, 38*f*
net worth, 140*t*
poverty, 135–137
unemployment, 7–8, 9(*f*1.7), 80, 83(*t*5.4), 127
wealth distribution, 135
Depressions. *See* Great Recession; Recessions and depressions
Deregulation, 11, 111
Developing countries, 93, 168–169
Discouraged workers, 80–81, 84(*t*5.5)
Discrimination, employment, 95, 109
Disposable personal income, 39–40, 40*f*, 115, 116*t*
Distribution of wealth. *See* Wealth and wealth distribution
Documentation of income and assets, 60–61
Domhoff, G. William, 142
Double taxation, 150
Dow Jones Industrial Average, 12
Down payments, mortgage, 61
DREAMers, 93

E

Economic confidence index, 34, 35(*f*2.7)
Economic forecasting
employment, 84, 86(*t*5.7)
federal budget outlays, 161*f*
gross domestic product, 159–160, 159*f*
health insurance coverage, 22*t*, 162*f*

inflation, 160(f9.7)
Social Security and Medicare, 156–157, 157f, 157t
unemployment rate, 160(f9.6)
Economic growth, 12–13, 143–144, 159–160
Economic Growth and Tax Relief Reconciliation Act, 156
Economic indicators
  business cycles, 31–33
  consumer confidence index, 34
  corporate profits, 104, 106
  data sets, revisions, and adjustments, 24–27
  employment cost index, 107
  gross domestic product, 27–28, 28t, 29f, 30–31
  gross private domestic investment, 103–104
  household debt service ratio, 58, 60f
  housing starts, 69, 70f
  Principal Federal Economic Indicators, 24, 25t–26t
  by sector, 23–24
  World War II, 7
Economic policies, federal. *See* Federal government
Economic Recovery Tax Act of 1981, 11
Economic sanctions, 169–170
Economic well-being survey, 137–139
Economics, Keynesian, 6
Economy types, 1–2
Education, financing, 38f, 71–72
Educational attainment
  family net worth, 140t
  unemployment, 80, 83(t5.4)
  wealth distribution, 135
Eisenhower, Dwight D., 7
Elwell, Craig K., 167
Emergency Economic Stabilization Act, 14, 69
Employee benefits, 89–90, 106–107
Employer firms, 98, 99(t6.2)
Employer-subsidized health insurance plans, 90
Employment
  Bureau of Labor Statistics data, 77
  business sizes by employee numbers, 98–100
  employer-sponsored health insurance, 91t, 92t
  employment cost index, 109f
  family net worth by employment status, 140t
  federal regulation, 109
  financial concerns survey, 38f
  immigrants, 91–93
  individual role in the economy, 33
  by industry, 78–79, 79f, 81t
  labor costs, 108t
  labor force participation rate decline, 77–78
  labor laws, 95
  labor unions, 93–95, 94f
  nontraditional work arrangements, 90–91
  by occupations, 79
  offshoring and outsourcing, 93
  private *vs.* government employees, 80f
  projections, 84, 86(t5.7)
  public opinion, 84, 85(f5.6)
  trends, 78t
Employment Standards Administration, 109
Energy crisis, 10–11
Energy industry, 111
Enron, 123–124
Environmental Protection Agency, 109
Equal Employment Opportunity Commission, 95, 109
Equifax, 56
Estate taxes, 151
European Union, 168
Exchange rates, 163, 164, 167
Excise taxes, 150–151
Executives, compensation of, 145
Expenditures. *See* Consumer spending; Spending, business; Spending, government
Experian, 56
Exports. *See* Trade

F
Fair Isaac Company, 57
Fair Labor Standards Act, 88, 95
Families
  debt data, 59
  financial assets, 118t
  income, 138t–139t
  net worth, 140t
  nonfinancial assets, 136t
  poverty, 137
  wealth distribution, 135
Family Smoking Prevention and Tobacco Control Act, 112
Fannie Mae, 64, 68, 69
Fatalities, work-related, 95
FDA (Food and Drug Administration), 109, 112
FDIC (Federal Deposit Insurance Corporation), 101, 102–103, 117
Federal Communications Commission, 111
Federal Deposit Insurance Corporation (FDIC), 101, 102–103, 117
Federal Energy Regulatory Commission, 111
Federal government
  bonds, 119, 148, 152–153
  border security, 93
  budget, 11–12, 13f, 149–153, 152(f9.3), 153
  business regulation, 108–109, 111
  debt ceiling and sequestration, 153
  deficits and surpluses, 11–12, 13f, 151, 152(f9.2)
  financial crisis, 68, 69
  financial industry regulation, 102–103
  Great Recession policies, 14
  Keynesian economics, 6
  labor laws, 95
  macroeconomics, manipulation of, 157–159
  minimum wage, 88–89
  mortgages, role in, 64
  national debt, 151–153, 152(f9.3)
  national defense spending, 9–10, 9(f1.8)
  New Deal, 5
  outlays as a percentage of gross domestic product, 161f
  public investment and taxes, 154, 156
  revenues, 149–151, 150f
  revenues and expenditures, 5f, 8t
  securities and taxes, 154, 156
  Social Security and Medicare, future of, 156–157, 157f, 157t
  spending breakdown, 151t
  stock market regulation, 122–123
  undocumented workers policies, 92–93
  wealth inequality and economic policies, 142–144
  World War I debt, 3
Federal Home Loan Mortgage Corporation. *See* Freddie Mac
Federal Housing Administration, 64
Federal Housing Finance Agency, 68
Federal Maritime Commission, 111
Federal National Mortgage Association. *See* Fannie Mae
Federal Reserve
  establishment of, 3
  financial assets of households, 131–132
  Great Recession policies, 14
  interest rates, 11, 12f, 56
  international investment, 172
  macroeconomics, manipulation of, 158–159
  personal debt data, 57–59
  regional economies, 33
  Survey of Consumer Finances, 117, 129–130
  Survey of Household Economics and Decisionmaking, 35, 36, 49, 50, 124
  total consumer debt, 71
Federal Reserve Bank of New York, 58–59, 72
Federal Trade Commission, 109
Federation of Tax Administrators, 149
FICO, 57, 68
Finances, personal
  financial concerns about retirement, college, and employment, 38f

home equity, 137f
net worth, 137t, 140t
nonfinancial assets, 136t
survey of personal financial well-being, 34–36, 36f, 36t, 37f, 37t
See also Consumer spending; Debt, personal; Income; Savings, personal

Financial assets
family holdings, 118t
households and nonprofit organizations, 134f
international investment, 171–172
investments, 117–122
net worth, 131–132, 133t

Financial industry
federal manipulation of macroeconomics, 158–159
financial crisis, 67–69, 101–103
Great Recession, 97
investment fraud, 123–124
misconduct and the mortgage crisis, 68, 69
public confidence in, 113
stock market regulations, 122–123
transaction accounts and certificates of deposit, 117–119

Fiscal policy, 158
Fixed interest rates, 56
Fixed rate mortgages, 61, 62f
"Flipping houses," 66
Food and Drug Administration (FDA), 109, 112
Forbes magazine, 103
Ford, Gerald R., 10
Forecasting, economic. *See* Economic forecasting
Foreclosures, 57, 63, 69, 70
Fraud, 67, 68, 97, 123–124
Freddie Mac, 64, 68, 69
Free market economies, 1
Free trade agreements, 167–168
Funding, government. *See* Spending, government
Futures, 121

G

Gasoline prices, 27, 27f
GDI (gross domestic income), 30–31, 103, 104
GDP. *See* Gross domestic product
Gender, 127–128, 135
General Agreement on Tariffs and Trade, 168
Generic drugs, 51
Gift taxes, 151
Gini index, 128
Globalization, 93, 169
GNP (gross national product), 7, 30
Goods-producing industries employment, 78–79, 79f, 81t

Government Accountability Office, 153, 170–171
Government assistance, 10, 137–139, 145
Government bonds, 119, 148, 152–153
Government employees, 79, 80f, 94
Government sector economic indicators, 24
Government transfers, by income quintile, 130

Great Depression
events leading to, 3–4
federal government measures, 158
New Deal, 5
stock market and banking industry, 4–5, 123

Great Recession
consumer price index, 47
delinquent loans, 59
federal government measures, 159
financial industry, 97
goods-providing industries employment, 79
history, 14
net worth of households, 132
public confidence in big business, 113
public opinion on home values, 36–37
real earnings calculation, 88
tax cuts, 156
undocumented workers, 93
unemployment, 77, 81

Greenspan, Alan, 13
Gross domestic income (GDI), 30–31, 103, 104
Gross domestic product (GDP)
agriculture sector, 108
business cycles, 32
consumer spending, 39
economic forecasting, 159–160, 159f
economic indicators, 27–28, 30–31
federal budget outlays as percentage of, 161f
global comparisons, 163–164, 164t
government spending, 153–154
Great Recession, 14
national debt as percentage of, 153
national health spending as percentage of, 19f
real gross domestic product, 28t, 29f, 29t, 30f
recessions, 32f
revisions, 25–26

Gross national product (GNP), 7, 30
Gross private domestic investment, 103–104, 104t, 105t
Growth Tax Relief Reconciliation Act, 156
Guestworkers, 91

H

HARP (Home Affordable Refinance Program), 69
Hassett, Kevin A., 143

Health care and health insurance
Affordable Care Act provisions, 19–20, 20t
consumer price index, 51t, 52(t3.9)
employer-sponsored health insurance, 89–90, 89t, 91t, 92t
employment cost index, 109f
health care spending and prices, 50–53
insurance coverage, 15–16, 16t, 20–21, 21t, 22t, 162f
labor costs, 106–108
Medicaid eligibility, 139–141
medical debt and bankruptcies, 75–76
Medicare and Medicaid, establishment of, 10
national health spending, 16–17, 17f, 18f, 19f
personal consumption expenditures, 49–53, 49t
poverty, 139–141
price indexes, 50t
privately funded *vs.* publicly funded plans, 15
projections, 22t, 160, 162f
uninsured persons, 20f

Health Insurance Portability and Accountability Act, 15
Home Affordable Modification Program, 69, 102
Home Affordable Refinance Program (HARP), 69
Home equity, 132, 137f
Home equity loans, 63, 65
Home improvements, 63
Home offices, 90
Home price index, 69–70
Homeownership, 122, 122(t7.3)

Households
debt, 58–59, 58t, 59t, 60f
financial assets, 134f
income, 128–129, 129t
net worth, 131–135, 133t–134t, 137t
nonfinancial assets, 135f

Housing
home equity, 132, 137f
homeownership as an investment, 122
house price appreciation, 71t
house price index, 66f
market booms and busts, 64–70
median home values, 67f
net worth, 140t
public opinion on home values, 36–37, 37(t2.7)
sales of new single-family homes, 65f

Housing and Economic Recovery Act, 68, 69
Housing starts, 69, 70f

I

Illegal foreign workers, 92–93
Illegal insider trading, 123
Immigrant workers, 91–93
Imports. *See* Trade

Income
  disposable personal income, 39–40, 40f, 115, 116t
  distribution, 132t
  family income, 138t–139t
  health insurance coverage, 16(t1.3)
  household income dispersion, 129t
  income inequality, 128–129
  international comparisons, 145
  mortgage applications, 60–61
  net worth distribution by, 133–135
  personal income and outlays, 40t
  quintile shares, 130–131, 131t
  Survey of Consumer Finances, 130
  wealth inequality, 141–144
  See also Wages and salaries
Income taxes, 3, 150, 154, 156
Individual mandate, 19, 52
Individuals, economic effects of, 33
Industry
  employment, 78–79, 79f, 81t, 86(t5.7)
  foreign competition, 11
  unemployment, 80, 83(t5.3)
Inflation
  annual inflation rate, 4f
  consumer price index, 47f
  economic indicator adjustments, 27
  health care, 17, 18(f1.13), 49–51
  prices, 44–49
  projections, 160, 160(f9.7)
  stagflation, 10
  World War I, 3
  World War II, 6
Insider trading, 123
Insurance, mortgage, 64
Intellectual property, 170
Interest and interest rates
  auto loans, 72–73
  bank prime loan rate, 12f
  consumer credit, 70
  credit cards, 73
  Great Recession, 14
  housing market booms and busts, 65
  inflation, 11
  loan rate types, 56
  macroeconomics, 158–159
  mortgages, 61–62, 62f, 63f
  payments, effects on, 56t
Internal federal debt, 153
Internal Revenue Service (IRS), 117, 129–130
International agreements, 167–168, 170
International Bank for Reconstruction and Development, 168–169
International Development Association, 168–169
International issues
  currency comparisons, 163
  economic and trade sanctions, 169–170
  economy size comparisons, 163–164
  free trade agreements, 167–168
  globalization and the antiglobalization movement, 169
  gross domestic product by country, 164t
  intellectual property, 170
  International Monetary Fund, 168
  investment, international, 170–172, 171t, 172t
  national debt as percentage of gross domestic product, 153
  trade, 164–165, 165t, 166t, 167–168, 167f
  wealth inequality, 144–145, 145t
  World Bank, 168–169
International Monetary Fund, 168
Inventories, business, 104
Investment fraud, 123–124
Investments
  bonds, 119
  federal securities, 154
  financial assets, 117–122, 118t
  government investments, 154, 154t, 155t
  gross domestic product calculation, 27
  gross private domestic investment, 103–104, 104t, 105t
  home equity loans for home improvements, 63
  homeownership as an investment, 122
  international investment, 170–172, 171t, 172t
  liquidity, risk, and reward, 116–117
  pooled investment funds, 120
  stocks, 12, 119–120
IRS (Internal Revenue Service), 117, 129–130

J

Johnson, Lyndon B., 10
JPMorgan Chase, 69, 103

K

Kennedy, John F., 10
Kessler, Gladys, 112
Keynes, John Maynard, 6
Keynesian economics, 6–7
Krugman, Paul, 142

L

Labor costs, 106–108, 107t, 108t, 109f
Labor force. See Employment
Labor laws, 95
Labor unions, 93–95, 94f
Lagging indicators, 32
Laissez-faire economies, 2, 3, 5, 6, 14
Lawsuits. See Litigation
Leading indicators, 32–33
Legal issues
  business structure types, 97–98
  employment discrimination, 95
  undocumented workers, 93
  See also Litigation
Legislation
  Airline Deregulation Act, 111
  Bankruptcy Abuse Prevention and consumer Protection act, 73
  Budget Control Act of 2011, 153
  Credit CARD Act, 73
  Economic Growth and Tax Relief Reconciliation Act, 156
  Economic Recovery Tax Act of 1981, 11
  Emergency Economic Stabilization Act, 69
  Fair Labor Standards Act, 88, 95
  Family Smoking Prevention and Tobacco Control Act, 112
  Growth Tax Relief Reconciliation Act, 156
  Housing and Economic Recovery Act, 68, 69
  labor laws, 95
  Racketeer Influenced and Corrupt Organizations Act, 112
  Securities Exchange Act, 123
  Sherman Antitrust Act, 111
  Small Business Act, 98
  Tax Reform Act of 1986, 11
Lehman Brothers, 102
Liabilities, household, 132, 133t–134t
Life insurance, 121
Limited liability companies, 98
Liquidity of investments, 116–119
Litigation
  employment discrimination, 95
  federal private-label securities litigation cases, 68t
  financial industry, 68t, 69, 103
  tobacco industry, 112
Loans. See Debt, personal; Mortgages
Loan-to-value ratios, 61, 66, 68
Local economies, 33
Local government, 94, 148

M

Maastricht Treaty, 168
Macroeconomics, federal manipulation of, 157–159
Madoff, Bernard, 124
Major money flows, 2f
Making Home Affordable program, 69
Mankiw, Gregory, 144
Manufacturing. See Industry
Market economies, 1
Market income, 130–131, 131t
Market power, 111
Medicaid
  Affordable Care Act, 19, 52
  eligibility, 139–141
  establishment of, 10
  projections, 160
  provisions, 15
  Texas, 21

Medicare
   establishment of, 10
   expansion of, 15
   future of, 156, 157f, 157t
   household income, 130
   projections, 160
   taxes, 150
Mexico, 168
Microeconomics, 33
Microsoft, 111
Minimum wage, 88–89
Minorities, 7–8
Misconduct of financial institutions, 68, 69
Misery index, 10, 10t
Mixed economies, 2
Monetary policy, 158
Money flows, 2f
Money market accounts, 117
Money market mutual funds, 117–118
Monopolies, 111
Mortgage insurance, 64
Mortgage-backed securities, 64, 68t, 101–102
Mortgages
   debt outstanding, 122(t7.4)
   federal government role, 64
   financial crisis, 101–102
   foreclosures, defaults, and delinquencies, 63
   housing market booms and busts, 64–70
   interest rates, 61–62, 62f
   monthly payment comparisons, 63f
   refinancing, 62–63
   secondary market, 63–64
   underwriting standards, 60–61
Mutual funds, 117–118, 120

N

NASDAQ, 13–14
National Bureau of Economic Research, 7, 14, 31
National Compensation Survey, 89
National Conference of State Legislatures, 93
National Credit Union Administration, 117
National debt, 151–153, 152(f9.3)
National defense and the defense industry, 8–10, 9(f1.8), 11–12
National Health Interview Survey, 75–76
National Highway Traffic Safety Administration, 111
National Labor Relations Board, 94–95
Net worth
   aggregate net worth, 131–132
   distribution of wealth, 132–135
   families, 140t
   households, 133t–134t, 137t
   income, by source and percentile of net worth, 138t–139t

19th-century panics and depressions, 2–3
1920s, 3–4
1960s, 10
1970s, 10–11
1980s, 11–12
1990s, 12–13
Nixon, Richard M., 10
Nonemployer firms, 98
Nonfinancial assets, 132, 133t, 135f, 136t
Nonprofit organizations, 133t–134t, 134f, 135f
Nonrevolving credit and loans, 55, 70–71
Nontraditional work arrangements, 90–91
North American Free Trade Agreement, 168
North American Industry Classification System, 98
Northwestern University, 94–95

O

Obama, Barack
   Affordable Care Act, 107
   debt ceiling and sequestration, 153
   federal budget process, 149
   financial crisis intervention, 69
   Great Recession, 14
   minimum wage, 88
   Social Security, future of, 157
   stimulus package, 102
   tax policies, 143, 156
   undocumented workers policies, 93
Occupational Safety and Health Administration (OSHA), 95, 109
Occupations, employment by, 79
Occupy Movement, 142
Office of Management and Budget, 24, 149
Offshoring, 93
Oil industry, 10–11
Okun, Arthur, 10
Options on futures, 121
Outsourcing, 93
Overvaluing scandals, 123–124

P

Panics, 2–3
Paris Convention, 170
Partnerships, business, 97
Patient Protection and Affordable Care Act. *See* Affordable Care Act
Payroll taxes, 90, 100, 150
Perry, Rick, 21
Personal consumption expenditures, 27, 41–42, 42t, 43t, 48–49, 51
Personal economic well-being survey, 36f, 36t, 37f, 37t, 38f
Personal finances. *See* Finances, personal
Personal income. *See* Income
Personal savings. *See* Savings, personal
PFEIs (Principal Federal Economic Indicators), 24–27, 25t–26t

Philip Morris Inc., United States v., 112
Piketty, Thomas, 142
Planned economies, 1–2
Politics, 92–93, 101
Ponzi, Charles, 124
Ponzi schemes, 124
Pooled investment funds, 120
Post–World War II era, 7–8
Poverty, 141f
   demographics, 135–137
   economic well-being survey, 137–139
   War on Poverty, 10
   World Bank, 168
Prescription drugs, 51
Prices
   consumer price index, 47f
   consumer price index for health care, 51t, 52(t3.9)
   employer-subsidized health insurance plans, 90
   health care and health insurance, 49–53, 50t, 52t
   house prices, 64–70, 66f, 67f, 71t
   inflation, 44–49
   real earnings calculation, 87–88
   real *vs.* nominal prices, 27, 27f
   stock shares, 120
Principal Federal Economic Indicators (PFEIs), 24–27, 25t–26t
Private industry employment, 87t
Private *vs.* government employees, 79, 80f
Private-label mortgage-backed securities, 64, 68
Privatization of Social Security, 157
Production and nonsupervisory employee compensation, 86–88, 87t
Profits, corporate, 104, 106, 106t, 107f
Progressive Era, 3
Progressives, 142–143
Projections. *See* Economic forecasting
Protests, 142
Public opinion
   confidence in banks, 113f
   confidence in business, 112–113, 114f
   confidence in various U.S. institutions, 112t
   credit card habits, 73–74
   economic confidence index, 35(f2.7)
   employment situation, 84, 85(f5.6)
   financial concerns about retirement, college, and employment, 38f
   home values, 37(t2.7)
   labor unions, 95
   minimum wage, 88–89
   on the national economy, 34
   personal economic well-being, 34–36, 36f, 36t, 37f, 37t
   problems facing the country, 33–34, 34t, 35(f2.6)
   wealth distribution, 144

Public protests, 142
Purchasing power, 88, 164
Putin, Vladimir, 169–170

Q

Quintile measures of income, 128, 130–131, 131*t*, 132*t*

R

Race/ethnicity
  earnings by, 128
  family net worth, 140*t*
  poverty, 136
  unemployment, 7–8, 9(*f*1.7), 80, 83(*t*5.4), 127
  wealth distribution, 135
Racketeer Influenced and Corrupt Organizations Act, 112
RAND Corporation, 53
Ratio of income percentiles, 128–129
Reagan, Ronald, 11–12, 143
Reaganomics, 11–12
Real estate. *See* Housing
Real gross domestic product (GDP), 28, 28*t*, 29*f*, 30*f*, 32*f*
Recessions and depressions
  National Bureau of Economic Research, 31
  19th-century depressions, 2–3
  1980s, 11–12
  real gross domestic product, 32*f*
  See also Great Depression; Great Recession
Redistribution of wealth, 143, 145
Refinancing, mortgage, 62–63, 69
Regional economies, 33
Republicans, 144, 149, 153, 156
Retirement and retirement savings, 38*f*, 120–121, 125*t*, 156–157
Return on investments, 116–119
Revenues, government
  budget deficits, 13*f*
  federal government, 5*f*, 8*t*, 149–151, 150*f*
  federal securities and taxes, 154, 156
  local governments, 148
  state governments, 148–149
Revolving credit, 55, 70, 71
Risk, investment, 116–119
Robo-signing, 69
Roosevelt, Franklin D., 5, 122–123
Russia, 169–170

S

Safety issues, 95, 108–109
Salaries. *See* Wages and salaries
Sales of new single-family homes, 65*f*
S&P/Case-Shiller indexes, 65, 70

Savings, personal
  bonds, 119
  financial assets, family holdings of, 118*t*
  international economics, 172
  retirement savings, 120–121, 125*t*
  savings, for specific purposes, 124*t*
  transaction accounts and certificates of deposit, 117–119
  trends, 115–116, 116(*f*7.1)
Scandals, 113, 123–124
Secondary mortgage market, 63–64
Secondary stock market, 120
Section 125 cafeteria plans, 90
Sectors, economic, 1
Secured loans, 55–56
Securities
  federal regulation, 122–123
  federal securities, 154, 159
  international investment, 171–172, 171*t*, 172*t*
  investment considerations, 117
  mortgage-backed securities, 64, 68, 101–102
  Securities and Exchange Commission, 111
Securities and Exchange Commission, 111, 123–124
Securities Exchange Act, 123
Self-employment, 90, 100
Sequestration, 51, 153
Service-providing industries employment, 78–79, 79*f*, 81*t*
Settlements
  federal private-label securities cases, 68*t*
  financial industry, 103
  tobacco industry, 112
Sherman Antitrust Act, 111
Sixteenth Amendment, 3
Small Business Act, 98
Small Business Administration, 98–99
Small businesses, 98–100, 112, 114*f*
Smoking, 112
Social Security
  future of, 156–157, 157*f*, 157*t*
  government transfers, 130
  projections, 160
  retirement, 120–121
  taxes, 150
Social welfare programs. *See* Government assistance
Sole proprietorships, 97
Soviet Union, 8–9, 12
Spending, business, 106–108
Spending, consumer. *see* Consumer spending
Spending, government
  budget deficits, 13*f*
  consumption expenditures, 28, 153–154, 154*t*, 155*t*
  federal budget outlays, 161*f*

  federal revenues and expenditures, 5*f*, 8*t*
  federal spending breakdown, 151*t*
  government services, 147–148
  local governments, 148
  national defense spending, 8–10, 9(*f*1.8), 11–12
  national health spending, 16–17, 17*f*, 19*f*
  Social Security, 156–157
  state governments, 149
Stagflation, 10
Standard & Poor's home price indexes, 65
Standard Oil Co. of New Jersey v. United States, 111
States
  Affordable Care Act provisions, 19
  banking regulation, 103
  financial crisis lawsuits, 69
  government revenues and expenditures, 148–149
  health insurance coverage, 21
  house prices, 69–70, 71*t*
  Medicaid, 15, 139–141
  minimum wage, 89
  undocumented workers policies, 93
  uninsured persons, 21*t*
Statistical information
  annual inflation rate, 4*f*
  balance sheet of households and nonprofit organizations, 133*t*–134*t*
  bank prime loan rate, 12*f*
  bankruptcy filings, 75*f*
  business statistics, 99*t*
  consumer credit outstanding, 72*t*
  Consumer Expenditure Survey, 45*t*–46*t*
  consumer price index, 47*f*
  consumer price index for health care, 51*t*, 52(*t*3.9)
  corporate profits, 106*t*, 107*f*
  credit card debt, 74*t*
  discouraged workers, 84(*t*5.5)
  disposable personal income, 40*f*, 116*t*
  economic confidence index, 35(*f*2.7)
  employees, by industry category, 79*f*
  employer-sponsored health insurance, 91*t*, 92*t*
  employment, by industry sector, 81*t*
  employment cost index, 109*f*
  employment projections, 86(*t*5.7)
  employment status trends, 78*t*
  exports, 165*t*, 166*t*
  family income, by source and percentile of net worth, 138*t*–139*t*
  family net worth, 140*t*
  federal government revenue breakdown, 150*f*
  federal government revenues and expenditures, 1901–1929, 5*f*
  federal government revenues and expenditures, 1930–1950, 8*t*

federal private-label securities litigation cases, 68t
federal spending breakdown, 151t
federal surpluses and deficits, 152(f9.2)
federal trust funds, 157f, 157t
financial assets, family holdings of, 118t
financial assets of households and nonprofit organizations, 134f
financial concerns about retirement, college, and employment, 38f
fixed rate mortgage interest rates, 62f
foreign holdings of U.S. securities, 171t, 172t
gasoline prices, 27f
government consumption expenditures and gross investment, 154t, 155t
gross domestic product, by country, 164t
gross domestic product projections, 159f
gross private domestic investment, 104t, 105t
health insurance coverage, by age, 16(t1.2)
health insurance coverage, by income, 16(t1.3)
health insurance coverage projections, 22t, 162f
home equity, 137f
home mortgage debt outstanding, 122(t7.4)
homeownership, 122(t7.3)
house price index, 66f
household debt, 58t, 59t
household debt service ratio, 60f
household income dispersion, 129t
housing starts, 70f
income distribution, 132t
inflation projections, 160(f9.7)
interest payments, by interest rate and years of loan, 56t
labor costs, 107t, 108t
labor unions, 94f
median home values, 67f
national debt, 152(f9.3)
national defense spending, 9(f1.8)
national health spending, 17f, 19f
net worth of households and nonprofit organizations, 137t
nonfinancial assets, family holdings of, 136t
nonfinancial assets of households and nonprofit organizations, 135f
personal consumption expenditures, 42t, 43t, 44f
personal income and outlays, 40t
poverty and the poverty rate, 141f
price indexes for personal health care and health insurance expenditures, 50t
private vs. government employees, 80f
production and nonsupervisory employee earnings, 87t
public confidence in banks, 113f

public confidence in small business and big business, 114f
public confidence in various U.S. institutions, 112t
public opinion on personal economic well-being, 36f, 36t
public opinion on problems facing the country, 34t, 35(f2.6)
public opinion on the employment situation, 85(f5.6)
real gross domestic product, 28t, 29t, 30f
real gross domestic product and recessions, 32f
retirement savings, 125t
sales of new single-family homes, 65f
savings, for specific purposes, 124t
savings, personal, 116f
unemployment, by demographic characteristics, 83(t5.4)
unemployment, by industry and worker class, 83(t5.3)
unemployment, by race, 9(f1.7)
unemployment claims, 85(f5.5)
unemployment projections, 160(f9.6)
unemployment rate, 6f
unemployment rate measures, 84(t5.6)
unemployment trends, 82f
uninsured persons, 2008–14, 20f
uninsured persons, 2013–14, 21t
wages and salaries, 86(t5.8)
Stewart, Martha, 123
Stocks and the stock market
federal regulation, 122–123
Internet bubble burst, 13–14
investment basics, 119–120
investment considerations, 117
1990s, 12
Securities and Exchange Commission, 111
Student loans, 71–72
Subprime mortgages, 66–67, 102
Supersalaries, 142, 144
Supply-side economics, 11
Surpluses, federal, 13f, 151, 152(f9.2)
Survey of Consumer Finances, 117, 129–135
Survey of Household Economics and Decisionmaking, 35–36, 49–50, 124, 124t
Survey of Income and Program Participation, 137–139

T

Tanner, Michael, 143
TARP (Troubled Asset Relief Program), 14, 69, 102
Tax breaks, 154, 156
Tax cuts, 156
Tax Reform Act of 1986, 11

Taxes
disposable personal income, 40
employer-sponsored health insurance, 90
federal government, 131t, 154, 156
funding government services, 147–148
local governments, 148
public opinion, 144
Reagan era tax reductions, 11
redistribution of wealth, 143
self-employment, 100
state governments, 148–149
wealth inequality, 145
Technology, 144, 170
Telecommunications industry, 111
Temporary foreign workers, 91
Texas, 21
Third-party debt collection, 57
Tobacco industry, 112
Toxic mortgages, 101–102
Trade, 28, 164–165, 165t, 166t, 167–168, 167f
Trade sanctions, 169–170
Transaction accounts, 117–119
TransUnion, 56
Treasury bonds, 119
Treasury Department, 153, 171–172
Treaty of Rome, 168
TRICARE program, 15
Trickle-down economics, 143
Troubled Asset Relief Program (TARP), 14, 69, 102
Trust busting, 3
Trust funds, federal government, 153, 156–157, 157f, 157t
Tyco International, 123–124

U

Ukraine, 169
Uncollectible debt, 57
Underwater mortgages, 67–68, 69
Underwriting standards, mortgage, 60–61
Undocumented workers, 92–93
Unemployment
demographics, 83(t5.4), 127
discouraged workers, 80–81, 84(t5.5)
Great Depression, 5
Great Recession, 77
by industry and worker class, 83(t5.3)
measures, 81–83, 84(t5.6)
post–World War II era, 7–8, 9(f1.7)
projections, 160(f9.6)
rates, 6f, 78t, 79–80, 82f
Reagan era, 11
stagflation, 10
worker categories and demographics, 80
World War II, 6

Unemployment compensation, 83, 85(f5.5), 100
Uninsured persons
  Affordable Care Act, 15–16, 20–21
  medical debt and bankruptcies, 76
  numbers of, 20f, 21t
  projections, 22t
Union membership, 93–95, 94f
United Nations, 168–169, 170
United States, Standard Oil Co. of New Jersey v., 111
United States v. Philip Morris Inc., 112
Unsecured loans, 56
Upside-down mortgages, 67–68, 69
U.S. Department of Defense, 15
U.S. Department of Labor, 91
U.S. Department of Veterans Affairs, 15, 64

V

Variable interest rates, 56
  See also Adjustable rate mortgages
Veterans, 15, 64
Volcker, Paul A., 11

W

Wages and salaries
  Bureau of Labor Statistics data, 85–89
  earnings, by gender and race/ethnicity, 127–128
  employee compensation, by industry type, 86(t5.8)
  international comparisons, 145
  labor costs, 106
  production and nonsupervisory employees, 87t
  See also Income
War on Poverty, 10
Wealth and wealth distribution
  family net worth, 140t
  home equity, 137f
  income and wealth inequality, 141–145
  income distribution, 127–131, 132t
  income inequality, 128–129
  net worth distribution, 132–135
  net worth of households, 131–135, 137t
  nonfinancial assets, family holdings of, 136t
Welfare. *See* Government assistance
Whistle-blowers, 95
Workers' compensation, 100
World Bank, 164, 168–169
World Intellectual Property Organization, 170
World Trade Organization, 164, 168
World War I, 3
World War II, 5–6
WorldCom, 123–124